Morphological
Integration

Morphological Integration

By

EVERETT C. OLSON

and

ROBERT L. MILLER

With a new Afterword by Barry Chernoff
and Paul M. Magwene

 THE UNIVERSITY OF CHICAGO PRESS

THE UNIVERSITY OF CHICAGO PRESS,
CHICAGO 60637
THE UNIVERSITY OF CHICAGO PRESS, LTD.,
LONDON

© 1958 by The University of Chicago
Foreword and Afterword © 1999 by The University of Chicago
All rights reserved. Published 1958

Paperback edition published 1999
Printed in the United States of America

05 04 03 02 01 00 99   1 2 3 4 5
ISBN: 0-226-62905-8

Library of Congress Cataloging-in-Publication Data

Olson, Everett Claire, 1910–
    Morphological integration / by Everett C. Olson and
    Robert L. Miller ; with a new afterword by Barry Chernoff
    and Paul M. Magwene.
        p.   cm.
    Includes bibliographical references.
    ISBN 0-226-62905-8 (paperback : alk. paper)
    1. Morphology (Animals)   2. Evolution (Biology)
I. Miller, Robert Lee, 1920–   .   II. Title.
QL799.045   1999
571.3′1—dc21   99-24323
                                                    CIP

⊗ The paper used in this publication meets the minimum re-
quirements of the American National Standard for Information
Sciences—Permanence of Paper for Printed Library Materials,
ANSI Z39.48-1992.

Foreword, 1999
BARRY CHERNOFF AND PAUL M. MAGWENE

Olson and Miller's *Morphological Integration* was a landmark publication, so much so, that it merits reprinting forty years hence. Despite the many advances in genetics, development, anatomy, systematics, and morphometrics, the synthesis of ideas and research agenda put forth in their book remain remarkably fresh, timely and relevant. Though other works essential to the field of morphological evolution, such as Gould (1977) or Raff (1996), are predicated upon many of the ideas expressed by Olson and Miller, few contemporary authors credit Olson and Miller as they richly deserve and fewer have attempted so broad a synthesis including theory and hypotheses, novel methods, and data of both a paleontological and neontological nature.

Olson and Miller attempted to focus attention on the covariation and variability of traits. To them morphological integration was the covariation among morphological structures due to development and function. Although this spirit of covariation was well established in quantitative genetics, it was an important departure in paradigm for systematics, paleontology, or evolutionary studies (cf. Mayr 1969, 1970). Olson and Miller expressed a strong view of the importance of organismal—*whole organism*—evolution in contrast to the reductive tendencies emerging in the 1950's. Additionally they recognized that a new evolutionary paradigm was needed because the expression and covariation of morphological structures was not well explained by evolutionary theories; a fact still true today.

The book begins with a seemingly simple premise that we can gain a profound understanding of the process of evolution from an understanding of how anatomical features covary spatially over the organism and temporally over ontogenetic and evolutionary time. By adding to this the consideration of function, Olson and Miller constructed a hierarchical framework using factors of ontogeny, function, and phylogeny to account for patterns of covariation among morphological traits.

Olson and Miller were far ahead in their thinking compared to many architects of the Modern Synthesis because they recognized phylogenetic hypotheses as the necessary orienting propositions for interpreting macroevolutionary processes. Their analyses of morphological integration provided tests of theories about patterns that have underlying causal factors of phylogeny, development, and function with the implication of natural selection as the agent and adaptation as the process. Olson and Miller mapped the changing patterns of morphological integration onto their best phylogenetic estimates of several groups; their graphical analyses foreshadowed the field of comparative methods. As such they test one of their fundamental hypotheses that integration within a lineage should increase towards the Recent.

A large portion of this classic book is devoted to analyses of growth and the significance for evolutionary biology of the stages intervening embryos and adults. Like Hennig (1966), the authors regarded such stages as bearing critical signals (semaphorants *fide* Hennig). They analyzed the changing patterns of trait covariation to test another of their fundamental hypotheses that the degree of integration should increase during the course of an ontogeny.

This book contains another subtle yet important element for students of quantitative morphology. Olson and Miller detail the precise nature of their data and spend much time discussing the significance of the measurements, their error rates, and the details of the specimens. Though the authors were paleontologists, they obtained neontological data from frogs, rats, squirrels, monkeys, and pigeons in order to control for age and population structure and to demonstrate the efficacy of their statistical methods. Furthermore, they imposed high standards for statistical precision, and demonstrated clearly the importance of parameter estimations from population samples. Although their methods are now surpassed, the spirit of having exacting methods with built-in flexibility to account for the biology is refreshing. It offers a good mixture of both deduction and heurism.

Our recommendation is to read the book for its ideas, for its agenda, and for its synthesis. Pay less attention to the statistical analyses other than for their basic lesson in estimations of correlation. In the Afterword, we will bring matters up to date, suggest new methods, and attempt to leave you with the importance of pursuing a research program in morphological integration.

Foreword: Sources Cited

GOULD, S. J. 1977. Ontogeny and phylogeny. Cambridge, MA: The Belknap Press of Harvard University Press.

HENNIG, W. 1966. Phylogenetic systematics. Chicago, IL: University of Illinois Press.

MAYR, E. 1969. Principles of systematic zoology. New York, NY: McGraw-Hill Book Company.

———. 1970. Populations, species and evolution. Cambridge, MA: The Belknap Press of Harvard University Press.

RAFF, R. A. 1996. The shape of life. Chicago, IL: University of Chicago Press.

Errata

Apart from the new Foreword, Afterword, and biographical information for Olson and Miller, this 1999 edition has been photographically reproduced from the original 1958 printing. The following are corrections for typographical errors in the book.

Page 45: Under the diagram labeled "F" in Fig. 1, ρ-group should read ρF-group.

Page 50, beginning of line 4 and last line: Ta_1 should read Tl_1.

Page 61, line 4 (Class I): $x = \dfrac{a_1}{a_2}(y)$ should read $x = \dfrac{a_2}{a_1}(y)$.

Page 61, line 5 (Class II): should read

$$\text{Class II } (b_1 = k_1 = k_2), \text{ simple power function, } x = a_2 \left(\frac{y}{a_1}\right)^{b_2/k_1}.$$

Page 61, line 14: in the equation for $\dfrac{dx}{dy}$, a_2 in the denominator should read k_1.

Page 117, end of line 4: Ff_h should read $F'f_{hl}$.

Page 117, lines 9, 10, 22, and the second-to-last line on the page: all instances of Ff should read F'f (to indicate sample F groups).

Page 154, line 4: $n = 10$ should read $n = 5$.

Page 154, line 5: $(n^2 - n)/2 = 45$ should read $(n^2 - n)/2 = 10$.

Foreword

The idea of form is implicit in the term "morphological." Integration connotes the process of making whole or entire; as used in the phrase "morphological integration," it refers to the summation of the totality of characters which, in their interdependency of form, produce an organism. That an organism is or has been a living entity, however, extends the implications of the phrase beyond the realm of static form to include the dynamic attributes of life. Each component of the organism must be formed so that the part it plays in the existence and function of the whole is carried out properly with respect to all other parts. It would seem logical that the degree of interdependency of any two or more morphological components in development and function would bear a direct relationship to the extent of their particular morphological integration. This concept is the basic hypothesis that we accepted early in our studies and tested later and that we discuss in this book.

The concept itself and the means of concrete expression are readily confused, but a clear separation is necesary for proper interpretation of the research reported in these pages. For reasons that will be made apparent in the text, we have chosen to express morphological integration by the association of measures taken on organisms. The coefficient of correlation is the expression of association that has been employed. As a consequence, populations, represented by samples suitable for estimation of the coefficient of correlation of a population rather than single individuals, are basic to all analyses, and statistical inference is used throughout. It must remain clear, however, that the integration with which we are concerned is a real thing, that it exists as a moving force in evolution, and that the method of detection constitutes but a tool for recognition and interpretation.

The primary interest of the writers lies in the field of paleozoology, the field that Simpson has termed "four-dimensional biology." The materials of this field encompass the totality of life today and of the past, but direct researches tend to be guided by the problems of past life and are pertinent to the extent that they directly or indirectly enrich the knowledge and interpretation of organisms known from the fossil record. We feel that the techniques and principles set forth in this book should have application to problems in many areas of neozoology, but our primary interests and competences have maintained past life as the focal point of our efforts. Modern materials have been used in various experiments for the sake of better controls and more complete morphological information, but measurements have been confined to parts of animals that are subject to preservation in the fossil record. Samples of species of vertebrates have provided the major part of the materials studied. In part, this practice is the result of our primary interests. The complexities of the skeletons, the intimate relationships of hard and soft parts of the anatomy, and the highly functional aspects of many of the hard parts make vertebrates particularly attractive for our type of approach. One extensive study has been made on a series of samples of the blastoid, *Pentremites*, and there is no reason why many groups of invertebrates would not yield equally interesting results. Our materials have been selected with specific purposes in mind and are not intended to be illustrative of problems that are pertinent to one or another taxonomic group. We consider the conclusions and the methods more significant than the contributions to the knowledge of a particular group of animals.

The earliest study in the series was undertaken by the junior author on a series of skulls of the squirrel, *Sciurus niger*. It was prompted primarily by dissatisfaction with studies that considered only one or a few characters and by an interest in the various types of multivariate analysis that had been previously applied to organisms. A search for basic groups of measures in the skull was made by imposing a network of 235 measures on a sample of 45 skulls and lower jaws. The existence of such groups of measures and a biological basis for the grouping were reported by Miller (1950) and also in a short paper by Robert L. Miller and J. Marvin Weller of the University of Chicago (Miller and Weller, 1952). Subsequently the writers published the results of an analysis of samples of skulls of three species of captorhinomorph reptiles from the North American Permian (Olson and Miller, 1951*b*). In this paper a model that set forth the general concept of measures bonded by values of the correlation coefficient was presented. At the time this work was in progress, various types of multivariate analysis were considered as possible means for studying the association of measures. All were rejected as unsatisfactory for our purposes, generally because they tended to obscure the networks of measures which were the very things that we wished to examine.

The results of these early studies prompted further work which was generously supported by the National Science Foundation for a period of three years (NSF Grant 183). It is this aid that has made it possible for us to carry the studies to the state of maturity represented by this book. One paper that reports work done under this grant has been published (Olson, 1953). The decision to present the remainder of the work in book form was made when it became apparent that single papers were failing to make clear the general objectives of the investigation and were not reaching many readers who might be interested in the results. The somewhat difficult formal mathematical structure of the model which is basic to the work tended to obscure the more strictly biological aspects of the study, and the extensiveness of the data masked the essential meaning of the conclusions. The organization of this book is largely the result of the problems brought to our attention by the early papers.

The first two chapters present a verbal, non-technical account of the concepts basic to the studies and a summary statement of the position that morphological integration, as we conceive it, may come to occupy in the field of organic evolution. We have attempted to explain why the work was undertaken, the thinking essential to the methods that have been used, and the future that we feel such work might have. The next five chapters include formal and technical aspects of the theoretical framework, with explanations and discussions of specific studies pertinent to various aspects of the investigation. The final chapters take up the subject of morphological integration in evolution and include both the results of the rather limited studies in this area and theoretical extrapolations beyond the concrete evidence of the data from the measurements and calculations. The final chapter is followed by detailed appendixes. We feel that it is necessary that the data and critical technical information for each of the studies be made available. These could not be included in the text without cumbersome interruptions of the general theme of the various chapters. Problems of taxonomy, stratigraphy, and the nature of the samples are included in the appendixes. Measurements used in the experiments are presented in tabular form.

No study such as this can be carried out without extensive co-operation. It is a pleasure to express our sincere appreciation of the aid given by many individuals and by institutions that made materials available to us. The generous financial aid of the National Science Foundation for technical aid, travel, equipment, and calculation has been indispensable. Eban Matlis of the University of Chicago has acted as technical assistant for three years and has, in addition, contributed much as a statistical consultant. Members of the Committee on Statistics of the University of Chicago have given freely of their time whenever aid has been sought. In particular Professor W. Kruksal has generously contributed suggestions throughout most of the formative stages of this work. We should hasten to add that the writers assume full responsibility for all matters presented in the book, and the onus of criticism falls on them, not on those who have so kindly advised. Mathew Nitecki, Robert Demar, and Phillip Harrison, graduate students at the University of Chicago, have contributed importantly as laboratory assistants. Ralph G. Johnson, of the Department of Geology, University of Chicago, has been an active consultant throughout the whole course of study, particularly on various biological aspects of the work. As the study was being carried out, an active and stimulating group of graduate students in paleozoology and geology was studying at the University of Chicago. Among these, Robert Bader, James Richard Beerbower, Richard Konezeski, and Ernest Lundelius were constant critics and consultants. It is difficult to evaluate their contributions, for they were both detailed and casual, but the results are far richer for the aid which they have given.

The collections of the Chicago Natural History Museum, both geological and zoological, were made freely available for our use. The staff of the American Museum of Natural History in New York gave us access to the museum's very extensive collections of fossil vertebrates and aided us in many ways. We are particularly indebted to Bryan Patterson, George Gaylord Simpson, Bobb Schaeffer, and Alfred S. Romer, not only for their aid with the collections, but for their advice in various phases of the study.

To all of these persons and institutions and to others who have aided us in many ways, we offer our sincere appreciation, for without them the work could not have been carried out.

EVERETT C. OLSON
ROBERT L. MILLER

Table of Contents

List of Figures

List of Tables

CHAPTER I

Integration and Quantification

The initial study, reported by the authors in a short paper in 1951, was designed to demonstrate a method for detection, description, and interpretation of the intricate small morphological changes that occur during the evolution of species and subspecies. The characters treated were necessarily numerical dimensions, for the procedures required the formation of groups of measures—groups that were considered biologically significant. It seemed evident to the writers, as it has to others, that character changes occurring in evolution of species could not be considered to be independent of each other and that studies which did not consider this dependency ignored a significant aspect of change. Not only should the interrelationships of changing characters be a primary point of interest, but the nature and intensity of the relationships should remain evident at all stages of their study.

To accomplish these ends, the measures, which are the characters, were arrayed into groups by two different techniques. One grouping was based on a selected level of the correlation coefficient, ρ. It was assumed that such groups had biological significance. Groups so formed were called "ρ-groups." The other method of group formation was based on qualitative biological considerations. Measures were associated on the basis of a common biological property related to function and/or development. Such a group, for example, might consist of measures related to the function of mastication. Groups of measures formed in this way were called "F-groups." Intersections of these two groups, the ρ-group and the F-group, formed what were termed "ρF-groups." The model proposed to implement this procedure was given in the cited paper and is repeated in chapter iii of this book. At present it is necessary only to note that both ρ-groups and F-groups were essential to the study and that it was the ρF-group, formed by their intersection, that was the object of study. That is, the ρF-group contained those measures common to the independently derived ρ-group and F-group.

Although groups formed under this plan provided an avenue to interpretation of the relationships of measures, it was found as study continued that they introduced various limitations that restricted their general value. There were technical difficulties in the development of mathematical extensions because a ρF-group was both quantitative and qualitative. Also, F-groups, which could be defined clearly in some cases, were often difficult to derive where biological aspects were unclear, particularly in fossil materials. A strictly mathematical development of groups was the obvious way of overcoming these difficulties. Early studies had suggested that the existence of ρ-groups implied the existence of F-groups and that it might be possible to arrive at biological associations by the use of a model that did not involve qualitative biological considerations in its formal structure. Several studies on samples of animals were carried out to test the hypothesis that ρ-groups implied F-groups, and, as taken up in chapter vi, the hypothesis was strongly supported. In the course of these studies a modified model was developed, and logical extensions were undertaken. This model is presented in chapter iii, where it is compared in detail to the formal structure of the original ρF-model.

The new model and the experiments that demonstrate the relationships of mathematical groups provide the logical framework for the concept of *morphological integration*. All measurements used express form, or morphology. Relationships in the groups, in which measures are bound by correlations, are derived through the mathematical procedures of the model. These groups depict relationships that exist under the concept of F, broadly defined to include measures related by a wide variety of biological phenomena, such as growth, development, topographic proximity, and function. Morphology is thus explicitly related to these biological aspects in a definitive way that permits logical manipulation, by extensions of the model, in the search for broader meanings of association of measures within and between taxons in evolution. Interrelationships of the several groups of measures within a species, or other biological assemblage as may be appropriate, allows an approach to the total organism, with the limits imposed only by the measures taken. Measures, the groups they form, and the interrelationships of the groups may be viewed at any level of complexity without loss of the individuality of characters and without encountering the difficulties that arise when characters are viewed as if they behaved discretely in evolution, irrespective of other characters. When desirable, the totality of integration can be expressed as a single number.

The advantages that have accrued from the development of the modified model, with respect both to the original model and to other types of multivariate analysis, outweigh the difficulties introduced by the more complex structure, which was inevitable in the formulation. The remaining pages of this chapter and chapter ii are devoted to a general development of the idea of morphological integration and its relationships to various phases of biological research and theory.

THE UNITS OF STUDY

Thoday's attempt (in 1953) to arrive at the criteria that define a suitable unit for studies of evolution is of extreme interest to both the neo- and the paleozoologist. In a specific problem, however, it may be difficult to accept as a guide a theoretical ideal such as he proposes, for the problem itself tends to limit and define the scope and nature of the unit. A suitable unit may be any of a number of things: a chromosome, an enzyme, an organ or a system, an individual animal, a deme, a species population, or some larger taxonomic association. When faced with the solution of a problem by some specified means, the researcher must decide what unit or units are best adapted to his own ends.

It is simple, in our own case, to say that the unit most satisfactory for our purposes is an abstraction based on the interrelationships of a large number, theoretically the *total* number, of measures derived from individual measurements on members of a population. The model presented in chapter iii details the development of such an abstraction, which can be described as a network of numerical dimensions substituted part by part for anatomical dimensions. Perhaps this is enough; but, unless the lines of reasoning and the methods used to derive this abstraction are clear, the subsequent manipulations based upon it are sure to have little meaning. What is the population? What is a measure? Or who are the individuals? These are all legitimate questions. Why go to all this trouble to arrive at a complex abstraction several times removed from what can be seen in the materials as they lie on the laboratory table? We shall try to answer these and other pertinent questions.

It has been our conviction for a number of years that an understanding of the total animal from the standpoint of morphology would offer a profitable avenue of approach to a large suite of evolutionary problems. It should provide access to phenomena not evident from studies that consider single characters or suites of characters for which only the grossest relationships are evident. Somewhat the same idea is found in various other studies, for example, in some of the work on rates of evolution by Simpson (1944, 1953), who takes cognizance of total change by using generic assignment as an indication of degree of difference. We have approached the problem from a different direction, making an effort to know as much as possible about the animal through a systematic representation of its characters and their interrelationships. The expression "total animal" is, of course, shorthand and possibly confusing in spite of its convenience. As we have used the expression, two somewhat different things are meant. One is the individual organism in its morphological entirety, and the other is the animal as a representation of the population. There are thus at this point two units, one which is definite—the individual—and the other which is conceptually more flexible—the population.

The individual animal or some part of it we shall call the "morphological unit." This unit is the point of departure in any morphological research. The mind is quite incapable of grasping the totality of an animal in a way that brings biological meaning to most problems that involve morphology. The animal is more readily conceived of as an array of many characters, each in itself a morphological unit. An approach toward totality entails an initial particularization and subsequent synthesis of the discrete characters. Subjectivity is inevitable at various levels in this process, in the way that the animal is broken down, in the characters selected, in how they are expressed, and in how interrelationships are established. Theoretically, morphological totality could be preserved during these steps to reduce to some extent the role of subjectivity, but practically such a goal is far beyond reach or reason.

Single characters are used as the basis for syntheses throughout our studies. Characters are expressed as linear dimensions in all but a few cases. The morphological unit thus may be considered an abstraction that represents the animal, or some part of it, as a series of linear measurements. There is, of course, nothing sacred about linear measurements. They happen to be simple to take, easy to understand, and generally meaningful. Areal, angular, or volumetric measurements, among others, could be used without any change in basic philosophy or methodology.

Expression of an individual animal as an array of measurements is in one sense equivalent to a verbal description or a series of drawings. An unordered array of measurements has about as much meaning as sketches of parts of the body entered on paper without regard to a reasonable anatomical organization. Ordering by systems of either measurements or drawings provides a synthesis in the form of a series of groups that have some biological meaning. Measurements can be arrayed to form series of networks which are subsets of the morphological unit, the total network, and are, for some purposes, in themselves useful units of study. Single systems or subsets are used in studies of the dentitions of *Aotus*, the owl monkey, and *Hyopsodus*, an Eocene condylarth, later in this book. While the view of totality is maintained, single systems may be studied separately, preliminary to final collation of the data. This step in some instances is a necessary prelude to the development of total representation. Even in a single study, then, the unit may vary from one stage to another.

In our study, the individual as a morphological unit was measured to arrive at the expression of a second-order unit, "the population." The approximation of the population unit provides basic descriptive materials for studies of factors internal to population units and for comparisons between these units. The complete array of a measurement of some anatomical character on all individuals of a population is defined as a "measure." For simplicity, the same term is used when only a sample is involved, without the repetitive use of the expression "sample of a measure." Measures may be related in various ways to produce the unit, but, in considering the problems of morphological integration, we have found the coefficient of correlation to be the most meaningful expression of association.

Before the matter of units is dismissed, some comments on a less formalistic aspect of the populations that are pertinent to this work are necessary. It is more to such units that the comments of Thoday, noted earlier, were directed. The species seems to be the most suitable population for many studies of evolution, since it represents, under most circumstances, the highest taxonomic unit integrated by gene exchange, that is, it is the largest unit with direct hereditary continuity. It is by no means always the best or most efficient taxonomic unit, but, because it plays an important role in our studies as well as many others, special consideration is justified. This is hardly the place to depart into the controversial subject of the precise nature of a species; yet the question of the species population among fossil organisms is so much a part of later studies that some clarification of our position is required.

A community between living and extinct species exists by virtue of the fact that the great majority of species in both categories are defined morphologically. Even though definitive characters used in different supra-specific categories of modern animals and within the same higher categories of living and fossil species differ widely, the common basis of morphological recognition appears to provide a reasonable chance of consistency in categorization. The widest diversity occurs between species grouped under different high categorical levels such as classes or phyla. These are not of particular importance in most evolutionary studies, for rarely are such great gaps bridged by studies at the species level. Morphological definition takes cognizance only inferentially of the genetic basis for species, but errors introduced by this fact alone are probably more or less commensurate among moderns and ancients.

A more serious problem of the species in fossil organisms is introduced by the time factor. It is rarely possible to be sure of precise contemporaneity of members of a sample of fossils or to be sure of continuity of populations over gaps in the record. The whole problem of gene exchange becomes cloudy and subject to a variety of interpretations when the factor of geological time is introduced, and the purist in species definition probably can never be satisfied with the necessary compromise. This difficulty has been met in some instances by the assumption that samples that appear to be temporally discontinuous because of a zone of non-occurrence or an unrecorded interval are specifically distinct, even though morphological distinctions cannot be made. The same distinction has been made where geographic isolation seems probable. When fossils are involved, there is rarely any way to determine whether there was or was not continuity in either of the cases. Regardless of any theoretical justification for this practice, it contributes little to the study of evolution or to stratigraphic studies in geology. We are, as paleozoologists, forced to document our hypotheses by morphology. In evolutionary studies it is morphological change that is basic to ideas stemming from paleozoology. There seems little reason for inferring genetic change in the absence of supporting evidence other than the supposed lack of continuity of two populations. Most samples of fossils thought to represent a single species include non-contemporary individuals drawn from a time continuum over which there can be found no significant shift in morphology. The samples used in our studies fall into this category and have been considered valid samples of species if no significant shifts in the means and variances of the characters studied were detected among the temporal subsamples.

The difficulties of clear-cut decisions with respect to specific homogeneity of fossil samples must be recognized, and studies based on the species as a unit must be considered with these in mind. It was this consideration, along with others, that prompted the use of living species in our exploratory studies of the validity and power of a number of the concepts presented in this book.

One final note is necessary concerning species and other population units that are used in our studies. These units provide the point of departure and are predetermined by whatever means are most appropriate. The complex and laborious techniques of multivariate analysis can be, and have been, used for testing morphological homogeneity of series of samples and for purposes of discrimination. In general, simpler and more direct methods are as effective as the more complicated techniques. The objectives of the studies reported here are highly varied. In each of the types, however, the problems involve an extensive array of measures that cannot be tested by simple techniques. Where this is not true or where large numbers of variables are not *required*, we feel that it is a mistake to apply complex methods. Only problems that cannot be solved by the use of simple, standard procedures, or which yield richer rewards when treated by complex methods, fall within the province of the type of work with which we are concerned in this book.

The steps leading to the abstraction that is the object of study by types of analysis used later in this book may be summarized as follows: Samples are procured which are appropriate to a study with respect to size, composition, and association in some population. An array of measures suitable to the study and available on each individual is established. Each specimen is measured. The individual measurements of each character provide the basis for estimation of appropriate parameters of their universes. These parameters are calculated. The network of all measures is considered an abstract representation of the population as it has been defined in a particular study. Interrelationship of the measures by simple correlation provides a representation of the population that can be meaningfully studied in its own right or compared with commensurate abstractions of other populations.

QUANTIFICATION AND STATISTICAL METHODS

The fact that statistical studies require quantification has given rise to a feeling of uneasiness and suspicion among many morphologists about the results of statistical tests. The form and beauty of a fine dissection or a careful preparation are likely to be lost as an animal is reduced to a set of numbers. The intuitive feeling of rightness or wrongness of conclusions is sacrified to answers that lie coldly in numbers, far removed from their source. The aesthetic value of an elegant bit of mathematical manipulation can hardly substitute for the loss and is more likely to engender than to assuage suspicion. The feeling that statistics arrive at an answer the hard way is rather widespread and not without justification in many cases. Subjective bias of measuring is suspected of weighting answers in the direction subconsciously desired.

Only if quantitative treatment can probe areas not otherwise accessible, provide a basis for repeatability by different observers, and maintain proper objectivity in accumulation and treatment of data, can its use be justified. Only if complex statistical treatment can do a job where simpler methods fail and accomplish work that is important does it become indispensable in the search for knowledge. We believe that all the criteria can be met and hope that the later parts of this book will make our case evident. If our judgment is justified, we are faced with the consequences of quantitative studies regardless of their scope and nature. Losses that may occur must be absorbed. Implicit in our own faith in complex approaches is the belief that loss is far from sufficient to reduce results to insignificance.

The Consequences of Quantification and Statistics

One evident consequence of measurement is simplification. Theoretically, as we have noted, the complete morphology of an animal can be expressed by counts and measurements; any organ or system, whether fixed or varying in form, can be so described. Practically, this is impossible. Thus something of what the morphologists see or what the artist draws is almost certain to be omitted from numerical description. That this is also true of verbal description is beside the point.

A muscle, for example, can be weighed; its cells can be counted; its many dimensions (with their maxima and minima as a basis for an integral expression of all conditions) may be expressed; even the color may be quantified. By use of a fixed plane of reference, the relationships of this muscle to all other parts of the body could be expressed quantitatively. Granted that the resulting array of data would be utterly incomprehensible, it could be assembled, given enough time. The result would be more or less equivalent to a complete description by a morphologist but in a form of little or no value. The concept of the totality of measures leads to this absurdity if carried to an extreme. Somewhere, if the concept of the use of multiple variables is valid and if information that leads in the direction of totality is productive of important results, there must lie a middle ground. This must be below the level of confusion and above the level where character-by-character analysis is effective. Wherever it is, simplification of the totality is involved, and some loss occurs.

The majority of studies on morphology that have employed some form of multivariate analysis have carried synthesis to a high level and have produced results in the form of a discriminate function or a series of primary factors. The studies by Wright on path coefficients and applications of Thurstone's factor analysis fall in the second category. Theoretical considerations by Fisher (1946) and Rao (1948) provide means of discrimination in which the variables can be weighted and the number of variables necessary for discrimination greatly reduced. The classical study by Barnard (1935) is an example of this type. Burma (1953) has applied the Hotelling t test to samples of the extinct blastoid *Pentremites* in an effort to base discrimination on a large suite of characters. The studies in the present book and other general approaches that have involved the use of multivariate analysis are based upon an entirely different philosophy. Retention of the totality is here considered essential for insight into the internal characteristics of a population and for analysis of the biological relationships between populations. Operational loss of information is minimized under this type of approach.

An early presentation of the concept of association of characters in pairs is to be found in the study of angiosperms by Sinnott and Bailey (1914). Sporne (1945, 1948, and 1954) and Stebbins (1951) have made use of correlation to express the association of characters of plants in an interesting and instructive way. The approach is very different from that used in this book but, in a very general sense, is analogous in its application of the concept of interaction of characters. De Beer (1954), in a discussion of transition stages in evolution, lays stress on what he terms "mosaic evolution." The role of many characters that act as a mosaic of static ancestral and evolving descendant characters provides an interesting approach to the concept of multiple characters in evolution. Each of these studies is, in one way or another, directed toward the use of the association of characters in morphology and evolution and offers a possible alternative approach to the general problem of morphological integration. Schaeffer (1956) has analyzed evolution in subholostean fish by treatment of association of pairs of characters by χ^2 tests. The use of regression lines in analysis of multiple characters is found in studies by Reeve (1940) and Olson (1951).

Somewhat closer to our approach are the various studies based on correlation in multivariate analysis and factor analysis. The psychometricians have been by far the most active both in the development of factor analysis and in its applications. A general survey of their work is found in Cattell (1952). Essentially, factor analysis can be classified as using either "R-techniques" or "Q-techniques." The "R-techniques" utilize in one form or another a measure of correlation between variables over a number of individuals. In this notation, our independently developed "ρF" method would fall in this class. The "Q-technique" measures the correlation between pairs of individuals over a number of variables—just the converse of the previous technique. An interesting application to problems of taxonomy by use of the Q-technique is to be found in Michener and Sokal (1957).

The loss of information accessible to the descriptive morphologist is serious or trivial, depending upon the point of view and the objectives of a particular study. Quantitative comparison of a homologous muscle in individuals of two closely related species, for example, will usually not include all aspects compared in a qualitative study. If, however, only certain aspects are to be compared—for example, the weight or the force exerted in the lever system—quantitative data are far superior to qualitative comparative terms such as "heavier" or "stronger." In such a case a single pair of numbers can tell more than paragraphs of description or a page of measurements of dimensions.

The questions of what to measure, how to measure, and how many measurements to take are fundamental in all studies that involve a large number of variables. Economy dictates the use of as few measures as possible, and a desire for coverage encourages the use of the largest suite that is practical. A limiting factor in paleozoological studies is provided by the fact that only hard parts are preserved. This and the practical need of economy have both led to what we may term the "principle of representative measures." Subconsciously this principle is in constant use; in our studies it has become of primary importance. The idea is simply that one measure may stand for many. In a description, in a search for internal relationships of systems in organisms, or in a comparison, one measure can function in place of the suite of measures that it represents. The illustrative example of the muscle, used earlier, may be carried over in explanation. Say that the interest is in the force that the muscle exerts in its lever system. It appears, from various empirical studies, that weights of comparable muscles in similar lever systems give good approximations of the force they can exert, and for illustration we shall assume this to be true. It often happens that direct measurement of the force exerted is impossible. The measurement of weight can be used as a substitute measurement to express force. Let us further assume that weight is highly correlated with the areas of origin and insertion and that these two likewise are highly correlated with each other. To continue further, assume that it has been found that length and width measurements of the area of origin are highly correlated with each other and with the measurements of area and that the same applies to comparable measurements of the insertion. Without any greater extension, it becomes evident that any one of these measures could serve as a valid expression of all the others. To the degree that correlation holds for all measures of a system, a single measure or a limited series of measures can represent a system for analysis. If there are two or more systems represented by such measures, the relationships of the systems may be studied by use of the representative measures. A series of assumptions has been introduced in the example, but these do not alter the basic idea. The validity of the general line of reasoning was first explored in the unpublished doctoral thesis of Robert Miller (1950). The intimate relationships between mathematical association and biological relationships of measures is considered in chapter vi.

The importance of the generalization of this concept for economy and in inferences to measures not directly available should be evident without further explanation. The value for probing relationships within and between systems to arrive at a better understanding of biological features of extinct organisms is perhaps less clear. This use of multiple variables is one of the most important yet developed, but it is also one of the most difficult to grasp. It is a primary center of interest in later parts of this book and will be considered only briefly at this time.

An expansion of the organism that approaches complete coverage by measurements would entail numbers of measures far beyond the bounds of practicality. The use of representative measures for a particular system, however, can reveal the underlying order and give a basis for understanding intersystem relationships. Conversely, associations of measures in the absence of knowledge of their biological relationships can lead to recognition of systems and provide clues to their origins and functions as well as to their changes during evolution.

Measurements and Samples

The unit of measurement must be appropriate to the materials and objectives of a study. First, the absolute size of maximum and minimum dimensions must be considered, to provide a suitable scale that will include both without being cumbersome. The unit used must be small enough not to mask the variance of the smallest measure but should be sufficiently large that it is greater than the errors of observation. The method of measurement should ideally be the same for all materials that are to be treated together. Data gathered for the various studies in this book came from direct measurements by calipers on prepared skeletal materials, on fossils, and on cleared and stained specimens and from measurements made with a micrometer ocular. When a sample includes a very wide range of size-measurements, it may be necessary to employ two or more techniques of measuring. Whether the same or different methods are used throughout, the errors in measurement must be considered and their effects taken into consideration. Excessive error, approaching the order of magnitude of the variance, requires that the measure be discarded, unless more accurate means of measurement can be developed.

To this point, the general need for quantification, the problem of losses and gains of information, the means of attaining necessary economy, and problems of measures and measurements have been considered. The efforts toward economy represent attempts to solve problems in sampling of measures. When an animal is expressed by measurements, its morphology is in effect expanded, and when manipulations are performed to reduce the measurements to a workable unit or series of units, the expansion is, so to speak, contracted or collapsed. Economy in sampling, under the general principle of representative measures, is, of course, reflected directly in the representation of the morphology in its contracted form, and this must be kept in mind when the initial steps of economy are undertaken. The principle of representative measures involves inference from few measures to many but in terms of relationships only. The existence, for example, of a high degree of association between two or more measures is taken to imply the existence of some biologically integrated system. It may be inferred that other measures of this system will show a high degree of mathematical association with those known to be highly associated. There is, however, no basis for an inference, from the fact that some measures taken are strongly associated, that other associated measures do in fact exist or, if they do exist, what they may be. Outside information is necessary to proceed to this level. In other words, if all possible measures of some defined unit be considered the universe, the sample taken does not provide a basis for estimation of this universe. It indicates, when properly arrayed, and when elements are brought into association, the pattern of association of measures of the universe so far as the measures taken are concerned. If at least two measures from each system, or subset, of the universe are present, the totality of the system will be *represented*. If this is not the case, the missing systems cannot be inferred. In no case can the measures not taken be inferred from the mathematical associations alone.

Other sampling problems arise when the universe of each measure is considered. Here problems are strictly statistical and subject to treatment by statistical techniques. A sample of some measure, such as head length in a species of reptile, can be used to estimate the statistical parameters of the universe. The universe in this case would be the collection of all head lengths of this species of reptile. The average head length in the sample would be used to estimate the average of all head lengths in the species. This is a necessary step in the statistical procedures that lead to the abstraction defined as the basic unit in our treatments of multiple variables. Actually, of course, samples are dealt with in both qualitative and quantitative work, and the problems of sampling and samples are no different. Quantitative studies that involve statistical inference, ostensively at least, place restrictions on the nature of samples that are not necessarily pertinent for qualitative work.

Sample size is one of several aspects of samples to be studied by statistical means that must be given careful consideration. The problems in this area fall into two general categories: the determination of an upper limit of size of sample and the determination of a lower limit. The first problem exists in cases in which a large number of individuals is present, and it is necessary to reduce sample size to some smaller number for practical reasons. An effort is made to arrive at a sample size that is sufficiently large to give reliable results, which, of course, involves some knowledge of a lower limit but is not so large that the work of collection, preparation, and measurement is prohibitive. This type of problem is frequently encountered in studies of contemporary species and also occurs for some fossil species, more among the invertebrates than the vertebrates. The problem of an upper limit does not, of course, exist when the number of individuals is small, as is frequently the case in samples of fossils, but the problem of the lower limit is intensified. To some extent the problems of limits of size depend upon the statistical procedures that are to be applied to the materials and must be considered specifically with respect to these procedures and to the reliability of estimates desired.

In the context of this book, with reference to problems of sample size, we are primarily concerned with the hypothesis that the population correlation coefficient, ρ, is equal to or greater than some arbitrarily fixed value. As described in detail in the formal model (chap. iii), we go through the following procedure:

1. A confidence level is selected. For example, it may be decided that we wish to incur a risk of being wrong only five times in one hundred.

2. Then, for a fixed value of ρ (expressed as $|\rho|$) and the fixed confidence level, we seek the range within which the sample correlation coefficient $|r|$ must lie if the sample is drawn from a population with parameter $|\rho|$. Since the confidence level and $|\rho|$ value are fixed, the range within which $|r|$ must lie depends upon sample size, n.

Reference to David's tables (David, 1938) provides an example of how the range of $|r|$ varies relative to the sample size. Let us suppose that $|\rho|$ has been set at .80 and the confidence level (the risk of rejection when the hypothesis is true) is .05 (two-sided). Table 1 gives some ranges of $|r|$ for various values of sample size, n.

It will be observed in the table that, as sample size increases, the range for $|r|$ becomes smaller. For very small samples, such as $n = 4$, the required value of $|r|$ may be anywhere within the whole range from .000 to 1.00. Obviously, in this case nothing is learned, since any value of $|r|$ satisfies the requirements for $\rho = .80$ and a confidence level of .05. It will also be noted that the reduction of the range becomes relatively less as sample size is increased. A basis is thus established for the determination of both the upper and lower limits in terms of the conditions set forth in the preceding paragraphs.

Samples of twenty to forty have been found generally satisfactory in our work, from the standpoint of both economy of effort and the extent of information gained. When less than twenty individuals have been available, smaller samples have been used, for some information is available and this, after all, is better than none. It is important to realize, in the evaluation of results, that there is a very rapid decrease in accuracy of estimation as sample size is reduced below the level of twenty. The basis for discrimination between samples becomes increasingly less sensitive, and errors of inclusion are increasingly probable.

The problems of randomness and bias and the form of size frequency distributions are more critical and less easily solved. Reliable estimates of universe parameters require randomness in samples. Excessive bias can lead to completely erroneous interpretations. The problem of determination of the nature of samples is less difficult in neozoological materials than in those of paleo-

TABLE 1

SAMPLE SIZE AND RANGE OF $r_{sig} > 0$

n	Approximate Range in Which $|r|$ Must Lie
4	.000–1.00
10	.400–1.00
20	.575–1.00
25	.600–1.00
50	.675–1.00
100	.725–1.00
400	.760–1.00

zoology, for proper design of sampling in the former can result in random and unbiased samples or at least in sampling in which departure from randomness may be directly evaluated. Fossil materials, however, are drawn from preserved assemblages accumulated by natural processes and are rarely random or unbiased with respect to the living population. For the most part paleozoological studies are directed toward aspects of a once living population rather than toward the fraction of it that is the preserved assemblage—the fossil population. No matter how excellent the design and how meticulous the sampling, the defects introduced by the initial natural sampling cannot be remedied. To be sure, a satisfactory sample of the *fossil assemblage* may result, and this is useful for many purposes. The types of studies that we have undertaken, however, are directed toward the living population, and the problem of the relationship of the sample to this population is present throughout. We are continually faced with the difficulties inherent in possible biases.

Two different concepts expressed by the word "bias" have particular importance here; one has a strictly mathematical connotation, and the other concerns the departure of some property of a sample from the conditions of the universe. The first involves the idea that if a certain population parameter is to be estimated by a sample statistic, the following criterion must be satisfied in order that the statistic be unbiased:

Repeated sampling from the universe and computation of the statistic for each sample will result in a frequency distribution of values of this statistic. If the statistic is unbiased, the mean of its frequency distribution should coincide with the value of the parameter it seeks to estimate.

Now consider a common situation in paleontology. The investigator desires to estimate a certain parameter of the universe, for example, the average head length of a fossil species. In accord with good statistical practice, he chooses the statistic that is least biased, in this case the mean ($\bar{x}$), which is an unbiased estimator of the universe mean (μ). At this point difficulties arise that are independent of the statistical consideration and relate to the second concept of bias noted above. A sample of the fossil species, for example, may have been collected in such a way that it includes only large forms, or a sample may have been drawn from a fossil assemblage in which selection by natural processes resulted in preservation of only one size group; that is, sorting dynamics, chemical processes, and so forth may have operated selectively on the sample size distribution. This sort of thing is bias just as surely as the formal notion of bias involved in mathematical considerations of statistic-parameter relationships. Both types must be considered in studies of paleontological materials.

There appear to be various ways of resolving the problems introduced by the second type of bias, that introduced by selective preservation of the materials from which samples of fossils are drawn. The intricacies of solution of such problems are not pertinent to this general discussion and need be mentioned only briefly. A general solution requires information that cannot be supplied by measurements on the individuals of a sample. The two principal sources of this information are the biological nature of the sampled population and the nature of occurrence of the sample in the sediments from which it was collected. From the first it is possible to estimate the probable size frequency distribution of various measures in the theoretical death assemblage of the population, the assemblage in which every individual that could have been preserved was preserved. From this the size frequency distribution of the living population can be estimated, as it existed over time sufficient to reduce short-term fluctuations to ineffectiveness. An understanding of the nature of preservation and the effects of diagenesis must come from geological criteria. Knowledge from these sources can be variously interrelated to provide an estimate of the total bias of the sample with respect to its universe. Once this has been done, it may be possible to redefine the universe in such a way that some biologically meaningful part is represented with little bias by the sample. Such solutions have been worked out (Olson, 1957). The samples of fossils used in our studies have been considered in the light of these solutions, and the general extent of bias is understood and taken into consideration in the conclusions reached.

The simplest illustration of the use of biological knowledge is found in cases of non-growing structures, of which mammal teeth provide a good example. A sample of first lower molars of a species of mammal in which there is no tooth growth after eruption can be considered a proper sample so far as dimensions are concerned, if no bias has been introduced in collecting. This interpretation requires the knowledge that such a structure does not grow and that the size range of the tooth is such that there is little chance of introduction of bias during accumulation and preservation. There may, of course, be difficulty in that only a local population and not the full population was sampled in the course of preservation, and this may seriously affect the estimates of the means and variances. Appeal to both biological and geological knowledge is necessary for a decision on this point. Mammal teeth can also provide an illustration of a way of arriving at a restricted, but biologically meaningful, concept of the universe. The full dental array generally provides a key to age (relative, at least) and makes it possible to consider some particular age group. We may know, for example, that the individuals of a sample were adults, in which growth commonly does not occur. The chances of excessive bias are thus greatly reduced. Measures on teeth or any other part of the animals may be safely used to estimate their universe parameters, if it is proper to conclude that the full species population of adults was sampled. By more complicated methods somewhat similar redefinition becomes possible with growing structures as well.

A critical aspect of the general problems of randomness and bias involves the form of the frequency distribution of a measure. Measurements lend themselves well to arrangement in size frequency distributions, and estimates of universe parameters are generally directed to such distributions. The form of the size frequency distribution of a measure on a sample of fossils rarely approximates the distribution of the universe if growth is involved. The critical factors in the development of the form of size frequency distributions of the death assemblage of a population, accumulated over a period of time that exceeds the span of normal fluctuations in population structure, are the rate of growth and the rate of mortality relative to the maximum life-expectancy of individuals. These factors provide a basis for estimation of the form of the frequency distribution for measures in this assemblage and for inferences on the distribution in the life-assemblage. From this theoretical construction it is possible to evaluate the departure of the sample distributions from that of the hypothetical universe and, in many cases, to make the corrections necessary to provide a workable situation.

Studies of living populations and theoretical constructions for extinct populations show that distributions of measures of structures that grow during ontogeny are not normal. This poses a serious problem which has been too little recognized in quantitative work in paleozoology. The form of the universe distribution in paleobiometric studies is very commonly assumed to be normal.

If departures from normality in the universe frequency distribution are not serious, we feel fairly well justified in making the required assumption. In many empirical studies in this work, tests of goodness of fit have confirmed this. Certainly we are not alone among workers in biometry or other fields in the feeling that even a weak basis for assumption of normality is better than abandonment of statistical analysis altogether in a given problem. The use of transformations to induce normality is recognized but has not been applied in these studies.

In recent years non-parametric methods have been developed, methods that do not require assumptions of normality. In such cases estimation of parameters and tests of hypotheses do not require a knowledge of the form of the distribution function. Basically, the techniques involve an ordering by size or value of the variable and utilize the following property of order statistics:

The area under the probability density function between any two ordered observations is independent of the form of the density function.

Such tests can be used to produce valuable results. A study of quantitative proportions of reptilian vertebrae by Johnson (1955) is an excellent example of the use of non-parametric techniques. He tested hypotheses concerning vertebral form in relationship to taxonomy and mode of life in a series of families of snakes. There was good reason to believe that the size frequency distributions were not normal, and there was no way to determine the true nature of the distributions. By direct comparisons of the frequency distributions, the relative importance of taxonomic association and mode of life in the vertebral morphology was clearly defined. Non-parametric tests are in general less efficient than tests based on normal distributions, but they are coming to occupy an important place in statistical inference in areas where distributions are unknown.

When structures that are treated do not grow during ontogeny or when restricted age groups are used, the assumption of normality seems justified theoretically and has been confirmed in various empirical studies. In other cases there is less basis for the assumption, but, for reasons discussed more thoroughly in chapter iv, the studies reported in this book have been based upon this assumption. For the present we shall note only that the biological consistency of the results, where normality was assumed without strong a priori reasons, suggests that errors introduced were insufficient to obscure orderliness and meaning. Such reasoning, however, may seem to have the flavor of circularity, for it appears to presuppose the nature of answers and to validate the methodology and procedure by the fact that these answers were forthcoming. It is, however, general orderliness and conformity to biological principles rather than specific answers that are presupposed. It seems highly improbable that an array of erroneous conclusions would produce an orderly series of results that were biologically meaningful. The experiments designed to test the hypothesis that groupings of measures developed mathematically (ρ-groups) implied biological groups (F-groups) are pertinent. In these studies qualitative biological groups were established, and groups were formed independently by correlation. Striking coincidence of the groups formed in these two ways was found. It appears highly improbable that this could have occurred in complex situations that involve thirty to fifty variables if the estimates of population parameters were seriously in error.

By no means all types of problems that arise because of sample size, randomness, bias, and form of frequency distributions have been satisfactorily resolved at the present time. Inaccuracies still enter into results because of inadequacies of the knowledge of the relationships between samples and the universes from which they were drawn. As problems are extended into the realm of more complex operations, difficulties in the nature of distributions tend to become increasingly troublesome. There exists an interesting and fundamental field of research in this general area.

The Model Concept

Several problems that arise in quantification in general, and in the treatment of arrays of many variables in particular have now been considered. The most difficult and in some ways the most important—the plan of operational procedure—has been touched only lightly to this point. Plans of procedure that have been found suitable for the purposes of our studies may, in a broad sense, be considered as models. They are general, applicable to a great number of similar problems, and so constructed that theoretical studies may be carried out within their general framework by modification and manipulation of the variables that enter into the construct. The need for a formal framework for operation became apparent early in our studies. The problems considered were such that they defied treatment by simple concepts that involved only one or a few variables. Several of the analyses were motivated by our desire to learn something about a specific group of animals, some taxon, series of taxons, some adaptive array, or temporal assemblage. Other studies were concerned with the exploration of some concept, and the materials selected for study were chosen because they were appropriate to the investigation of the concept.

The initial study on ρF was undertaken to arrive at a better understanding of certain Permian captorhinomorph reptiles. This led to the development of several concepts, particularly the formalization and extension of the concept of the relationships of association of measures by biological and mathematical means. Subsequent studies on *Rana pipiens* and *Sciurus niger* (chap. vi) were designed to explore these concepts. The confidence in the relationships of biological and mathematical associations and an understanding of at least parts of their meanings led to the study of a second concept: that changes in groups of associated measures could lead to new understanding of certain aspects of evolution. This was investigated in studies that used *Pentremites* and *Hyopsodus* as subject matter (chap. viii), not so much because of specific interest in the species as because the materials were particularly suited for study. Problems of the role of ontogeny in the formation of ρ-groups arose in the course of a number of the studies. This prompted a special study on the ontogenetic development of ρ-groups in rats. The work on *Hyopsodus* indicated the need for an analysis of upper and lower dentitions simultaneously, and this was carried out on a living species of the monkey *Aotus trivirgatus*. The important aspect of this series of studies and the motivation of one by another, from the standpoint of statistics and quantification, is that these tools were used in two distinct ways. One is the usual use of statistics to test hypotheses; the other is the use of the procedures and the framework, or model, as an avenue to the development of of new concepts. In the latter instance, quantification and statistics play a fundamental role because they provide a way of exploration and thinking by use of a logical, often complex, construction or model. Throughout the sequence of studies there was an intimate interplay of these two areas of investigation—exploration and the testing of hypotheses. At some stage or stages of such work, the investigator arrives at a stopping point, when his hypotheses have been tested. At this point subjective interpretation is necessary if the results are to have meaning in the framework of biology, regardless of the degree of objectivity that has gone into the studies that preceded this stage. The important difference between the type of study described above and qualitative studies, or less ambitious studies using statistics and quantification, is not that one type is wholly objective and the other wholly subjective, for this is not true, but rather in the level at which subjective interpretation takes place.

The complexity of many zoological problems makes them particularly suitable for study by the use of models. The great power of the stochastic model has been repeatedly demonstrated in neozoology in such fields as genetics, population studies, and ecology. The constructions represent some situation and include the variables that are pertinent, entered in such a way that they may be manipulated to test their effects upon the results expressed in terms of probability. That similar approaches have found less use in paleozoology stems in part from the persistence of a typological approach to many phases of the work. Some problems cannot be treated otherwise because of practical considerations of the materials available. However, while much has come and will continue to come from this direction, there are many areas that are closed to study unless data are viewed from a population point of view under general probability theory. A most powerful extension of this approach is found in the model concept.

A real, sometimes critical, problem in the development of models is the necessity for simplification of the complex processes or situations that the models are established to represent. Critics have not been slow to point out this difficulty in both the physical and the biological sciences. The attempt to reduce complexity to a comprehensible level may oversimplify the problem to the point that some critical variables play no part in the solution and reality is lost. This, of course, can and does occur, and the criticism is frequently valid. The effectiveness of the model, in the last analysis, is a reflection of the judgment, insight, and ingenuity of the one who formulates the construction. Whether or not it performs the task for which it was designed is not primarily a function of its mathematical elegance or its use of clever manipulation but rather of whether all essential elements have been entered in proper and realistic relationship, with proper weighting, without producing a structure so complex and cumbersome that it defeats its own purposes.

CHAPTER II

Areas and Methods of Study

Statements about the areas to which the studies reported in this book are directed cannot at present be fully definitive or inclusive. We are interested primarily in stimulating further work along the lines that have been investigated and in areas as yet unexplored. Without doubt, ideas and methods that have not been recognized at the present time will develop with more extended investigation. Some areas as shown in the chapters that follow have yielded tangible results. Answers to questions in others are known to exist but have not as yet been fully formulated. It is clear, for example, that corollary studies in sedimentation and environments of deposition will produce results that are of interest in themselves and doubly important as they can be related to information gained from studies of fossil organisms. On the other hand, some of the speculations and theoretical extensions from the work already done may well prove to be overly optimistic.

Our major area of research is the study of evolution in the perspective of geological time. We trust that our preoccupation with this field will not be taken to indicate that we feel that the type of study we propose is limited to this area, for without question some phases of neozoology can yield significant results when attacked under the concept that we have applied in our own limited province. We prefer, however, to leave specific research in these areas to those more qualified in the particular disciplines.

Paleozoological studies derive their basic information about organisms from morphological, spatial and temporal relationships, and the texture and composition of the rocks in which the fossils occur. We have already considered some of the limitations imposed by morphological studies plus the additional restrictions inherent in the use of fossils. Areas are, of course, limited, but certain types of analysis point the way to enlarging the amount of information available from fossils and thus increasing the scope of justified inference. Important among these, we feel, are studies that explore the evolution of animals as total organisms as we have attempted to do through the concept of the animal as an abstraction based on the association of measures. The mechanical procedures involved in the development of the abstract units depend in the sequential application upon basic concepts and do not depart from orthodox considerations. The product, however, permits a fresh point of view and casts various phenomena of evolution in a new light. The importance of this change or how radical it may appear in retrospect we cannot judge. Some of the possibilities are examined later in this section, with particular emphasis upon the role of the concept of morphological integration in various phenomena of evolution.

The two principal objectives of the statements on areas and methods are, then, (1) to relate the studies that involve morphological integration to the current synthesis and (2) to consider the broader aspects of association and the integrated units without reference to a particular framework. First, however, it is important to examine more thoroughly the consequences of the fact that the basic data for the development of the abstractions are derived from morphological studies based on skeletal anatomy.

MATERIALS AND CONSEQUENCES

The section on units in chapter i outlined a step-by-step formalization of the process that led to the abstraction considered profitable for study. The source of data is morphological, characters are expressed as numbers, and studies are based on complexes derived from these numbers. Many other sources of information were used, but knowledge so gained was directed to an understanding of the meaning of quantitative morphology. All quantification involved measures that could be taken upon fossil animals, whether living or extinct populations were sampled.

The extent to which generalizations pertinent to evolution can derive from our studies is a function of the importance of the phenotype as a direct source of information about the factors and phenomena of evolution and also as a basis for inference of factors and phenomena not directly revealed. With reference to the first point, organisms themselves provide the documentation of what has existed and does exist, and, when arrayed phylogenetically in time, either observed or inferred, they constitute a record of change—the amounts, the kinds, and the rates. As to the second point, there exists the chance of obtaining specific information related to ontogeny and growth, function, adaptation, and ecology and to more general concepts such as biological uniformitarianism or the continuity of life. Within these inferential categories, the role of analogy is extremely important in the study of fossils. Were it not possible to know the relationships of morphology to other aspects of biology, which must come in large part from living populations, inferences from the morphology of fossils would be pitifully weak. Several of our studies, in recognition of this fact, are devoted to a search for a more sufficient basis for inference. The morphological abstraction based on association of measures seems destined to be particularly important in this regard. If the associative abstraction can be thought to reveal, through its approach toward totality of the animal, things not evident in characters taken one at a time or in limited relationship, it is logical to suppose that inferences drawn from it will provide a basis for deeper penetration into known phenomena of evolution and, perhaps, reveal other factors and phenomena not evident at less complex levels of relationships of characters.

The fact that use of skeletal parts restricts the taxonomic scope of the work hardly needs mention, for this is true of the great majority of studies on fossils. A further taxonomic limitation, less obvious and more difficult to evaluate, arises from the requirement that measures have evident biological meaning. There is, for example, a practical limit to measures available on many micro-fossils, either because their hard parts are very simple or because their size makes accurate measurement of their characters impractical. Colonial animals, such as corals, bryozoans, or graptolites, while measurable, do not generally yield the type of measures necessary for inferences for which multivariate analysis is best suited. In general the more complex the individual skeletons and the more closely soft anatomy is reflected in hard structures, the better suited the organisms are for our studies.

Change may be documented regardless of the materials that are used and the measures that are taken. If the meaning of change is sought—and to us this seems the important aim—then the characters that change must provide a basis for inferences of causation. Hard parts that are intimately related to soft systems of the body or reflect something about the impact of external environment provide the best bases for inferences essential to a search for causal factors. They allow for broader application of the principle of representative measures and thus a closer approach to an understanding of the totality which is vital to our efforts.

Table 2 presents an evaluation of taxonomic groups at appropriate categorical levels. While the approximation is rather crude and individual exceptions can readily be found, the general areas of the animal kingdom to which our studies are best adapted are made clear. The assignments, however, give no weight to the problems of procurement of samples. It is evident that the groups listed in categories 1 and 2 of the table vary widely in this respect. Appropriate samples of fossil insects, for example, are extremely rare, whereas the opposite is true for many species of trilobites, brachiopods, or Foraminifera. This has a direct bearing upon the utility of any quantitative work that involves inferences from samples of fossils. Such methods cannot be applied to a great many known species of extinct animals.

The types of restrictions cited apply in some cases to fossils only and, in others, to both extinct and living organisms. The seriousness of limitations of taxonomic categorical areas depends upon the objectives of study. If the purpose is an investigation of principles and phenomena of evolution, as we believe proper for our studies, materials are available for investigation of a large segment of the Metazoa and limited suites of the Protozoa. Within these arrays of materials, the full skein of principles, modes, and phenomena of evolution is surely represented. If, however, there is thought of using such methods as a general way of study, many areas are closed to investigation.

APPROACH TO TOTALITY OF THE ANIMAL

In the annals of evolution two groups of animals exceed all others in prominence—the horses and the fruit flies. *Drosophila* has contributed to a knowledge of animal genetics to a degree that far overshadows the total contributions from all other animals. Studies of the equid evolution have been primarily the province of paleontology, and from this source have come a

TABLE 2

EFFECTIVENESS OF TREATMENT BY MULTIVARIATE ANALYSIS—ρF—ON VARIOUS ANIMALS SELECTED FROM FOSSIL POPULATIONS

Highly Effective (1)	Moderately Effective (2)	Ineffective (3)
Foraminifera-complex tests, espec. Fusilinidae	Foraminifera (most)	Most Protozoa
Cephalopods (ammonoids)	Cephalopods (most nautiloids)	Foraminifera (very simple tests)
Arthropoda (most) (see cols. 2, 3)	⎰Arthropoda ⎱Ostrocoda (most)	Radiolaria Porifera
Cystoids	Edrioasteroids	Hydrozoa
Blastoids	Agnatha-vertebrates (most)	Graptolites
Crinoids		Scyphozoa
Echinoids		Anthozoans
Asteroids		Bryozoans
Ophiuroids		Worms, etc.
Gnathostome vertebrates (all except some Chondrichthyes)		Cephalopods (dibranchiates)
		Arthropoda; Ostrocoda (very simple test); Copepoda; Cirripedia

knowledge and understanding of many of the fundamental factors and phenomena of evolution through geological time. We know a great deal about horses, from eohippus to *Equus*, but the potential of the group is far from fully realized.

Equids are more pertinent to our direct efforts than fruit flies; therefore, an examination of methods and results in this group best provides a perspective for consideration of the relationships of studies of morphological integration to thinking in evolution based on fossils. The most general, and generally recognized as most authoritative, studies of "four-dimensional biology," are those of George G. Simpson. The important role of the equids in modern concepts of the animal in time is shown in several of his books, *Tempo and Mode in Evolution, Meaning of Evolution, The Major Features of Evolution,* and *Horses.* It seems proper to conclude that this best known of all families and the studies of it by a leading authority provide the best chance of understanding how animals have been considered.

There is no doubt that, in a broad sense, the horse has been considered as a total animal and that to some degree its evolution has been evaluated in this light. The synthesis, however, has been accomplished by collection of detailed analyses of the several parts of the structure, especially the limbs, the teeth, the skull, and the brain. The concept of the animal as an integrated whole is clearly expressed by Simpson in his book *Horses* in the introduction to the chapter entitled "How Horses Changed." The key to thinking upon the whole animal seems to be expressed in the following statement quoted from this introduction: "It would be a lengthy affair and of little interest except to the specialist if we were to follow through in all detail all parts of the body."

Were such a plan followed to the extent that material permits, there would still remain the prodigious task of integration, a task that is inconceivable from the qualitative point of view or from a quantitative point of view that does not incorporate some simplifying mechanism for expression of association of characters in some workable unit. Recognizing the difficulty and the general audience at which *Horses* is aimed, Simpson considers important anatomical features separately with a cautioning remark: "Remember, however, that the whole animal was evolving, and that the animal is an organism. It is called that because its parts are organized into a unit, the whole body, and no part can operate in complete independence of any other part. At all stages in the history of the horse family the animals as a whole were integrated and the animals as a whole evolved."

In these words and thoughts the stage for treatment of animals as integrated, functional units is set, and the need for approaching totality in studying evolution is made evident. When, however, such features as rates of change are considered, it is necessarily the rate of change of a character or a structure that forms the basic datum, and comparisons are of relative rates of different parts or of the same parts in different phyla. The question of change of integration, which is fundamental in so many aspects of evolution, can hardly be touched even in the most general terms. One way of avoiding the problem by use of a substitute proposition has been to employ the species or the genus as a point in the time series. Integration is implicit, since a succession of functioning, adapted stages is treated.

Quite a different aspect of integration in evolution is covered extensively by Schmalhausen in *Factors of Evolution*. Comparative embryology forms an important part of the base from which the thinking of this author stems. He is particularly concerned in the later phases of the study with the increasing complexity of correlation mechanisms, because there is increase in complexity of organisms, and especially with implications of the regulatory functions of morphogenetic correlations in ontogeny. He states very directly: "All forms of

interdependence invariably become more complex as evolution progresses. . . .
The greatest complexity is attained by the system of morphogenic correlations
which is the fundamental apparatus of individual morphogenesis. Furthermore,
all systems of interdependence, in particular morphogenetic and ergontic correlations, acquire increasingly regulative character. Thus the growing complexity
of correlation mechanisms is one of the directed processes of progressive
evolution."

A cogent statement of his concern with the total animal is found in the conclusion of this book, as follows: "Progressive growth in complexity of the
organization, centralization of the nervous system of animals, and general increase in activity and individual adaptability of both animals and plants are
the most important factors determining success in the struggle for existence."

Concern with pleiotropy of single gene effects and multiple gene effects brings
the concept of totality into genetics. Epigenetics, as shown particularly in the
studies of Waddington, must be concerned with more than single characters. It
is hardly necessary to document further the fact that the idea of the whole
animal permeates the thinking of many outstanding evolutionists today. Yet
it seems in large part true that, when we get down to cases, little specifically is
done to bring this problem to a working basis at which study of the whole
animal as it evolves through time can be handled with meaning that is subject
to concise analysis.

The idea of integration extends beyond the organism and the species population and is expressed variously in ecological studies involving interspecies
populations and in ecosystems that include the totality of interactions of
organisms and their environments. The vast complexities of this whole area
are somewhat staggering, and a reader of attempted verbal discussions is certain to be lost in a welter of detail that obscures the heart of the problem. Even
when a single local population is considered in detail with respect to its environment, for example, in *A Herd of Mule Deer* by Linsdale and Tomich
(1953), the minutiae are necessarily stultifying. That the aspects of the ecosystem are critical in evolution can hardly be denied. Even treated in a crude
fashion, as in "The Evolution of a Permian Vertebrate Chronofauna" (Olson,
1952*a*), they reveal information that sets in context evolutionary phenomena
that have little meaning when viewed as a sequence of events in a single
phylum. It appears highly probable that it will become possible to extend the
quantitative concepts of association and integration into this general area and
to study more effectively the complex systems from this point of view.

The cited writings and many more that give a place to integration in evolution make clear two general thoughts in this area. Schmalhausen, in particular,
has specifically treated both. On the one hand, increasing complexity of organisms is a general directional trend in evolution. As complexity increases, integration becomes an increasingly important problem. On the other hand, regard-

less of the level of complexity, as specifically noted by Simpson in the quoted passages, an animal is an integrated unit. There is intended no implication that all animals are at a maximum level of integration or sufficiently integrated to perform their multiple function with the highest possible efficiency; but some degree of integration must exist if the animal is to function as an individual.

This duality poses some extremely interesting questions about evolution. We may ask, for example, does the intensity of morphological integration tend to increase as more complex forms evolve, or does it, in view of increasing complexity, become reduced relative to the totality? Is intensity of integration in a system directly related to the degree of adaptation of this system? Are highly integrated systems in an organism also highly integrated with each other? What is the role of integration during speciation? Or a much more general question should perhaps be the first objective of study: Does integration evolve? Clearly there appears to be an affirmative answer to the last question, but there can hardly be a specific answer beyond the fact that change occurs, unless some quantitative basis for measurement of integration is developed. When this is done, the measurement of integration can be extended into the more specific areas mentioned above, as well as many others. The significance of integration in evolution can then be explored thoroughly.

The assumption was made in our early studies that mathematical association, expressed by ρ, implied biological integration. Tests of this assumption gave strong support, and a model was designed to provide a means of arriving at an associative abstraction that could be translated into meaningful patterns of morphological integration. Various extensions of this model deal with special problems of the meaning of groups of measures and the factors basic to them. Some inroads into the types of questions posed above have been made. These, however, have touched on only a few problems of single populations and individual phyla. There has been no direct approach to the problems of interspecies populations or the relationships of populations to their physical and biological environments as these change through time. Much more study is needed before any attempt at such a synthesis will become a possibility.

In summary, morphological integration is directly related to the concept of the totality of the animal, the species population, or some more complex unit. The ideas of totality are found throughout the history of the study of evolution, couched in various terms and applied in many ways, and it is self-evident that it is the whole organism that is involved in evolution. Quantitative association of measures appears to offer a means of expression of morphological integration at population levels and presents the opportunity to explain more fully the importance of integration in evolution. With this general idea in mind, it is of interest to examine in somewhat more detailed form how integration is related to other factors of evolution and how, as a factor, it may lead to a better understanding of some of the observed phenomena of evolution.

MORPHOLOGICAL INTEGRATION IN STUDIES OF EVOLUTION

Studies of evolution that are based on data obtained from organic materials generally involve the following steps: (1) description of materials; (2) classification of biological events, taxonomic and, where possible, temporal; (3) description of phylogenetic events; and (4) search for causation.

Description of materials is not necessarily formal but involves procurement of appropriate data from the materials and some process of recording, either mental or written. Taxonomic ordering, or classification, necessarily comes after this first step and, since taxonomic position is generally conceded to have phylogenetic implications, normally precedes the third. Temporal ordering is often a direct process in paleozoological work, but this is not the case for modern materials. Here the actual specimens are contemporary, and ordering involves inferences about the relative times of appearance of the entities involved in the study. A knowledge of phylogenetic events is implied in such ordering, and this step, for modern materials at least, is not clearly separate from the third. In some instances actual temporal ordering of the individuals found in the fossil record is ignored, for series are arrayed in an inferred succession of morpho-types, even though the specimens do not actually occur in this temporal order. The process and reasoning are much the same as for the arrangement of modern materials.

Ordering provides the basis for interpretation of evolutionary or phylogenetic events. This step is, in essence, a description of what has occurred. It is at this stage that many studies are concluded. The answers to questions of what has happened and when it happened, relatively at least, are now at hand. The "what" may, of course, be broad enough to involve elements of interpretation such as the matter of rates of change.

The goal of the study of evolution, at least to many students, is an understanding of the "why" of observed phenomena. The search for cause is basic to all syntheses of evolutionary theory and is at once the most stimulating, the most difficult, and the most speculative step in the analysis. It is essential that knowledge from many fields of biology be incorporated into interpretations. It is to this end that the modern synthesis of evolution has contributed so richly, for until recently there has tended to exist a dichotomy in paleo- and neozoological studies at the point of transition from description to analysis of causal factors. This is by no means obliterated even now, but the overlap of interests, knowledge, and understanding has been greatly increased.

The steps leading to the study of integration by use of the concept of quantitative association of measures follow the general pattern of procedure in evolutionary study as outlined above. The primary materials are animals, and, on the basis of observations, these are arranged taxonomically and temporally. The descriptive data applied in these steps are no different from those used in other studies. Quantitative methods may be involved in description and taxonomic ordering, but they are not themselves necessarily directly involved in the investigation of association of measures and integration. The same general considerations apply in the process of phylogenetic ordering.

The steps that led to the study of integration in *Pentremites* (chap. viii) provide an illustration of these procedures. The specific goals of the study were to analyze changes of morphological integration through time and to gain an understanding of what was behind any changes that might be found. Samples were collected with this in mind. Primitive descriptive studies in this case consisted merely of character analysis sufficient for proper taxonomic placement of individuals. The basis for taxonomic assignment was developed without reference to the study of association of measures. Temporal ordering was provided by the stratigraphic field relationships of the samples. A discriminant function was used to aid in ordering specimens, but this was merely to insure a sound basis for further work. To this point, then, standard procedure was used. This fact is emphasized to point out what we consider most important, that the work involved in study of integration is based on procedures of arrangement and basic interpretations that are standard and that studies of integration are not proposed as substitutes for these generally effective methods.

At the stage of description of phylogenetic events, the situation is quite the contrary. These events are studied under the concept of association. This first involves a new or greatly elaborated description of materials in which measures form the basis of description. In order that these measures may be as meaningful as possible, anatomical studies are carried out to the extent that is practical, prior to the measuring. Measurements are taken, in most studies, with this information as a guide. In the *Pentremites*, for example, the preserved parts of the anatomy were studied by careful and delicate "dissection" that involved use of thin sections as well as delicate preparation after use of a softening agent. All measures that were reasonable, that could be accurately made, and that did not appear to be directly duplicative were then taken. By correlation, an associative unit was derived for each sample. Phylogenetic events were then described in terms of the integrations revealed in the various populations as ordered in time.

The procedure differs from standard types of study in methodology only because there is a different series of objectives. A direct parallel, for example, may be drawn in the case of an array of temporally successive samples in which variability of some character is to be analyzed.

Regardless of what steps are taken to reach the level of description of phylogenetic events, the search for causation must bring into play information beyond that supplied by the samples themselves. As the events differ and present different arrays of evolutionary phenomena, so do the methods for search of causation differ. The study of morphological integration, through knowledge of associations of measures, presents us with a series of phenomena that, broadly, may be subject to interpretation under some of the current concepts of evolutionary change. It presents these, however, in a light for which there is relatively little precedent for interpretation. It has been necessary, for this reason, to make preliminary studies to determine the biological meaning of the phenomena observed. Much more than has been done to date along this line must eventually be undertaken before complete formulation will become possible. Again we may take *Pentremites* in illustration. It was found possible in this group to assemble considerable information about several systems—digestive, respiratory, nervous, and so forth—and to express these systems by measures. A basis for interpretation was established. But this could have meaning only if it were possible to have information from other sources upon the precise meaning of integration, expressed mathematically within a species. A corollary study on *Rana* and *Sciurus* provided a pilot investigation in this area. With the information that such studies give, the fossil assemblage was analyzed with more confidence concerning the biological meaning of changes of integration in the animal as a whole and in particular systems. An approach to an understanding of the causal factors in this particular instance, of course, requires additional information on a multitude of other aspects which must come from the biological and geological aspects of the sites of deposition and knowledge of areas of zoology such as genetics and ecology. To the degree that such information is available and applicable in a given situation, there may be hope of arriving at a comprehensive and accurate interpretation.

The power of morphological integration studies in evolution can be measured by the amount of information that is subject to meaningful interpretation after the basic analyses. We feel that knowledge of this area, as shown in later parts of this book, provides a potent wedge to enter into recognized fields of evolutionary study and to open the way to discovery of principles and generalizations not heretofore subject to recognition or evaluation.

The index of integration, described in chapter vi, is a case in point. Modification of the index through time has been explored in a series of samples of *Pentremites* (chap. viii). It has been found that the index is high in the "parent species," low in the two developed populations but at different levels, and then slightly increased in succeeding populations. We can only raise the question, without hope of answer until more studies are completed, whether this is a typical situation. That some general principle of this sort can arise from such studies seems probable, but no definitive formulation is now possible. We do know, even with this one series, that much of the animal is presented as a single number and that the indexes are subject to meaningful interpretation through a study of how different systems have contributed to the increase or decrease in integration. With knowledge of the biological meaning of these changes, which can be ascertained in part at least, the hope for basic analysis of causes of change is greatly enhanced.

Finally, in this evaluation of the general role of integration in the study of evolution, we raise again the potentiality of extension beyond the species level to interspecies integration, to integration of interspecies populations and the physical environment, and to the flow of biophysical complexes through time. There seems little doubt that such studies can be carried out, for the elements are being developed in parallel studies on the biological and physical phases. The synthesis will undoubtedly be highly complex, for establishment of necessary relationships must involve complicated models that follow logically from the model through which the elements are related. The complexity seems inevitable, since the situations to be evaluated are in themselves so intricate that they are difficult to grasp. The eventual synthesis, however, may be such that the complexities will not confuse the essential patterns but will bring us closer to a more complete understanding through recognition of essentials in an integration where the part that each ingredient plays is subject to evaluation.

THE PHENOMENA OF EVOLUTION

The final consideration of morphological integration in this introductory section will involve an effort to relate this factor to phenomena that are recognizable in fossil materials or that may be related to them through inference rather than direct observation. It will be evident from the considerations to this point that by no means all these phenomena have been subjected to firsthand study in the perspective of integration. For the most part we shall be considering possible rather than demonstrated relationships and theoretical rather than empirical results. The format of treatment can be much the same as it would be for evaluation of the relationship of any factor to phenomena and to other factors. We are interested in exploring how various phenomena can be viewed from the perspective of morphological integration and how the meanings of other factors can be evaluated within this framework. The treatment is topical for simplicity and clarity, but there are, of course, broad areas of overlap between the different topics.

Directional Evolution

The phenomenon of persistent trends in evolution has been variously called "orthogenesis," "rectilinear evolution," "straight-line evolution," and so forth. The reality of such evolution, without regard to the fine shades of meaning in the various labels, hardly seems open to question in either a gross or a refined sense. Case after case can be found in the fossil record. Continued size increase, progressive development of hypsodonty in teeth, reduction of lateral digits, and increase in size and complexity of the brain have been amply demonstrated in the evolution of the horse. Successive stages of flattening in the evolution of amphibian skulls in widely divergent lines and increase in the complexity of sutures in ammonites represent cases at somewhat higher categorical levels. The increase in "mammalness" in various skull characters of the mammal-like reptiles illustrates the existence of multiple trends in lines related only at ordinal and subordinal levels.

It would appear that in some cases of limited scope only one character or a small array of characters was involved. For example, in the lower dentition of *Hyopsodus* there is progressive loss in successive species of the "paraconid." This loss, as seen superficially at least, gives little indication of relationship to changes of other characters. Many similar cases of appearances or disappearances of dental characters in lines of mammals are known. Frequently they provide an excellent basis for differentiation of species. The simplest and most evident cases of directional evolution are of this sort. Single gene mutation has been suggested as an explanation of modifications of a character in the phenotype (for example, see Kurtén, 1955). There exists real question whether or not such an explanation is adequate to account for the observed changes.

It seems important to consider whether such a single character change can and does occur without a relationship to other changes. If a multiple gene effect is, in fact, involved, it is unlikely that a single change will be the only effect of modification in the gene complex, especially in view of known pleiotropic effects of even single gene changes. From qualitative studies of various lines in which a simple change has been observed, it is evident that other changes in dentition and other features of the hard anatomy do occur simultaneously. Contemporaneity is of course no criterion for causal relationships of changes. There are many ways to attack such a problem with modern materials, which are subject to experimentation, but the problems with fossils are more difficult, since they are not subject to experimental study. If, however, one of our goals is knowledge of causal relationships in evolution, such problems must be solved. One course that might be followed involves the use of analogy, but the difficulty of the time factor continually plagues such efforts and raises serious doubts about equating observations in modern and fossil materials.

The use of integration offers some possibilities for determination of causal relationships between contemporaneous events. Correlation, of course, does not indicate causal relationship in itself. Some of the studies to date, however, have shown that mathematical correlation of measurements on organisms does have biological meaning and that in tested cases associated measures are related to common causal factors in development and function. High correlation of elements in morphogenetic fields suggests relationship of association and genetic heritage, although at the present stage the evidence is admittedly tenuous. From what has been done up to now, it appears that it may be possible to investigate causal relationships of minor concurrent changes through studies of integration of systems or the animals as a whole.

Three studies have been conducted in part along these lines. One is the study of *Hyopsodus* in which a series of associative patterns of lower molars have been studied. A second concerns upper and lower dental patterns in *Aotus*. A third involves the search for underlying factors of change in amphibian skulls (Olson, 1953). In the last it is evident that many minor changes which might be considered single differences between two species of *Diplocaulus* are related to a single factor of change that is adaptive in nature. As noted in this study, similar suggestions have come from qualitative studies in other groups of organisms. Such studies involve the concept of integration without the precision or inclusiveness possible in quantitative studies.

More difficult problems of directional evolution arise when several characters show similar trends over time, either in a single line or in a series of related lines. An excellent example, but only one of many in which the phenomenon is observed, is found in the mammal-like reptiles as in several lines they approach more and more closely over a period of time the skeletal condition we call "mammalian." It is extremely difficult to give a definitive answer to the question of what is back of such a phenomenon except in the most general and vague terms in our present state of knowledge. A search for common factors such as was carried out for the amphibian skulls would appear to offer some chance of simplifying the search by directing attention to one or a few basic relationships that could be the primary subject of study. Given the proper materials, such a study could be made by analysis, not of single characters, but of integrated patterns of measures. As is so often the case, practical difficulties are enormous. At present such a study would have to be restricted to skulls, and it would be difficult to assemble suitable samples even for this part of the skeleton. The problem does suggest, however, a way that the concept of integration could serve to unravel a problem too complex for treatment by qualitative or simple quantitative methods.

Finally, there is the point, discussed briefly above, that integration itself seems to evolve from simple to complex in the broad course of organic evolution. Rather than an increase of morphological integration *in toto*, as expressed by the coefficient of integration, it seems probable that new integrated systems will appear. A fundamental suite of interesting problems that could be studied involves the questions whether total integration is increased or decreased with the addition of such systems; whether there is an increase in integration of multiple systems as evolution proceeds after new systems appear; and whether integration is less related to phylogenetic position and more to the specificity of individual systems, on the one hand, and their co-operative features in the total body, on the other. Were definite answers to these questions available, an entering wedge would be driven into many of the broad questions of evolution concerned with the relationships of specific adaptations to perfection of general body systems in reorganizations that occur as new high categories arise. Possibly, also, the relationships of such modifications to the vast array of data on evolution at the species and subspecies level might be clarified.

Rates of Evolution

We have alluded to rates of evolution at various places in this section and have indicated briefly some aspects of the concept of integration in this area of study. These remarks have been directed primarily to a representation of the whole animal as it changed in time and to a measure of the rate of change by use of an index of integration. This limited use of the concept as a direct indication of rate has interesting extensions when the contributions of the various parts of the organism are considered. Each system may be studied separately and the effects of change in each related to the part that the distribution of bonded measures plays in determination of the index.

Relative rates of change of different systems in evolving lines pose extremely interesting problems, as brought out by various studies, particularly on horses, by comparisons of single characters. The potential for a more complete understanding of some aspects of this problem exists in the study of systems and related integrated units. Not only can the degree of concurrent change be evaluated, but the degree of interdependence and interrelationship is available for analysis.

Related to the subject of rates of change is the question of cyclical change through time. Evolution in some major categories, such as the vertebrates or the brachiopods, appears to proceed in a series of surges of major reorganization that alternate with times of strong adaptive radiation. Similar phenomena are apparent at lesser categorical levels, as among the mammals or, within the mammals, in the horses, to note one of many possible examples. Whether such a phenomenon appears at still lower levels, as at the species level, is more difficult to detect. It would appear, however, that in some evolving lines there is something of a temporal ebb and flow of speciation. In major categories it is often possible to account for the surges of change and periods of adaptive radiation through their relationships to major physical and biological upheavals. At lower levels the evidence for fluctuations is less precise, and the chance of learning of causation more remote. Under most present concepts of evolution, minor changes are generally conceived to be basic to major shifts and to provide the totality of change which, through temporal telescoping, so to speak, appears major because of its relative rapidity.

Under this philosophy the study of phases through time at the lower levels is important. A delicate indicator of change is needed for such a study. Investigations carried out to date suggest that patterns of morphological integration can provide one such indicator. No time series that are long enough to establish the existence of cyclical change with respect to integration have as yet been studied. Such a task is by no means impractical, for a complete conceptual framework is now available and there are numerous series of fossils in the record suitable for investigation.

Speciation, Selection, and Adaptation

The comments on directional evolution and rates as well as the brief remarks on specific essays earlier in this section have had reference, either direct or implied, to the related phenomena of speciation, selection, and adaptation. Much more must be known about the relationship between integration and speciation before its importance can be evaluated. At present it appears that studies of species populations as associative units can provide insight into a number of aspects of speciation. Minor adaptive shifts, in some cases at least, are clearly indicated in modifications of patterns of integration. These differences can, in part at least, be related to modification in adaptation and provide an avenue of inquiry into selective processes. Work in this area probably should be carried out on modern materials, where factors of selection and adaptation can be better related to slight morphological differences, before any extensive interpretation of the fossil record is undertaken. Potentially, this appears to be a fruitful area for study, but the most valuable analyses—those that consider change with time based on fossils—must depend rather heavily upon studies of modern cases, of which there are all too few at the present time.

Genotype-Phenotype Relationships

One of the great stumbling blocks to synthesis in neo- and paleozoology lies in the difficulties in relating the evidence on phenotypes from the fossil record to the knowledge of the genotype-phenotype relationships available for some living organisms. It is beside the point, for our consideration, that there is equally extensive ignorance of these relationships for most living organisms; for the special problem of the paleozoologist is not to find among the fossils examples that repeat modern situations but to know what has happened and how it happened in the perspective of time. The few efforts to interpret genetic changes among fossils have met with little general acceptance and have, for the most part, been subject to considerable criticism by geneticists. The criticism has not been without justification, but it must be recognized that the task of determining the nature of genetic change behind the events in time series is truly monumental and that even some beginnings in this direction are encouraging. The most extensive work in this area has been done by Kurtén (1953, 1955). Reasons for the difficulties are many. First, of course, is the lack of opportunity for direct observation in modern materials of genetic structure as it has changed over an extended period of time. Second is the fact that only part of the total animal can be observed, so that the total phenotypic effects of genetic change are lost to view. Third, there is the problem of the magnitudes of phenotypic changes seen in modern materials, where the genotype-phenotype relationships can be established, and those seen in fossils, where changes that are recognized are commonly at a much grosser level. The often minute changes traced to genetic effects in experimental studies are usually not evident in fossil materials. A fourth difficulty arises from the necessity of extrapolating from the very few modern groups that are well known to highly varied, often remotely related, fossil groups.

With these, among many, difficulties standing in the way, we may seriously question whether it is even worthwhile to attempt to understand the genotype-phenotype relationships in fossil phyla. Would it not be better to leave well enough alone? In spite of a strong feeling in this direction among many zoologists, we believe that this question should be answered in the negative, for a serious gap in the search for causality in evolution must continue to exist unless some insight into this relationship is possible. Our understanding of some of the phenomena of phyletic evolution, for example, would be much poorer than it now is, were it not for the efforts of such pioneers as Osborn, Robb, or Simpson to draw upon genetic theory in their interpretations, regardless of the credence given some of the results.

If this objective is granted as valid, any approach that may aid in its realization must be given consideration. A principal area of criticism of efforts to give genetic interpretation of observed phenotypic events has concerned the tendency for paleozoologists to oversimplify genetic situations in their interpretations. We cannot quibble with this undoubtedly valid objection. Its sources are manifold. They stem in part from the difficulty that the paleozoologist has in reading, absorbing, and evaluating the voluminous literature in genetics, merely to keep abreast of modern developments. They arise as well, it would seem, from the preoccupation of the paleozoologist with an oversimplified concept of morphological change. This comes in part from the restricted nature of his materials but also from an unnecessary preoccupation with single characters or a few characters as evidence of change. There is nothing wrong with documentation of change by the use of one or a few characters. For discrimination and ordering, it is the most simple, direct, and efficient procedure. The difficulty comes when the character or characters are considered not only as indicators of differences or expressions of differences but, in fact, as expressions of the total differences upon which evolutionary interpretations are based. With this philosophy, it is a natural step in sequence to seek genetic interpretation on the basis of these differences and to make statements concerning adaptation, selection, and so forth based on this oversimplified interpretation of genetic constitution.

The intimate relationships of suites of phenotypic features to the simplest, single gene mutation have been repeatedly documented by modern genetics, and the complex genetic background of even a single character is evident in many cases. In general, it would seem that we should know much more than we do about the related changes in organisms through time if we are to be able to infer, with some hope of validity, the genetic background of their origins. The hope that studies of morphological integration hold in this direction should not require much elaboration. In the first place, studies already made indicate that a way to recognize very small changes exists and that changes of integration expressed as modifications of associations of measures tend to bring us closer to the level of change available to the geneticist. Empirical validation requires analysis of integration in which the genetic situation has been evaluated. This must, of course, be based on modern materials. No such studies have as yet been carried out. The detection of small changes, however, holds the promise of establishing a firmer basis for interpretation of phenotype-genotype relationship in fossils, once the recent situation has been more thoroughly investigated.

The study of morphological integration opens a second opportunity for closer inference from phenotype to genotype by directing attention to a larger part of the animal in terms of relationships of measures of the whole animal and of integration between systems. There exists a better chance to sort out primary and secondary effects and to set any single change more properly in the essential array of total change, which is a full expression of evolution. As studies of integration provide means of gaining insight into the adaptive significance of systems and characters, into fields within organisms, and into factors basic to small character changes, they provide a chance for more specific interpretations of phenotype-genotype relationships by bringing into closer harmony the knowledge of fossil populations and their modern analogues.

Ontogeny and Evolution

The fact that selection is active throughout the full ontogenetic history of organisms is basic to the current concept of evolutionary theory. Ontogenetic events in evolution have been emphasized in expositions of the synthesis of evolutionary theory, particularly by Schmalhausen, and have been the object of detailed studies by many workers, especially Waddington (1940) and Orton (1955). The paleozoologist is in a rather poor position to analyze his materials from this point of view. His samples tend to be poor in the stages of ontogeny when skeletal materials are first being formed and, of course, give essentially no record of any earlier stages. Some efforts to study ontogenetic stages have been made, for example, by Kurtén (1953, 1955), Croneis and Geiss (1940), and Olson (1951). The studies have, for the most part, only minor evolutionary implications. The problem that confronts the paleozoologists is discouragingly difficult; yet it seems certain that various phenomena in evolution that he observes must be importantly related to events of selection and adaptation that occurred during ontogeny. Samples of the great majority of known species are not adequate for even a sketchy knowledge of ontogeny. Collections made with the express purpose of gaining ontogenetic series can improve the situation somewhat. It is, however, too much to hope that any more than a very small percentage of the species can ever be so represented. Study of ontogeny in evolution in time series of fossils cannot become a general procedure. This does not mean, however, that important contributions cannot come from series that can be so studied.

The fact that such series do exist does not mean that we proceed directly to analysis, for there exists an array of baffling problems of age distributions, sample constitution, and methodologies of treatment. The most usual treatment has been by analysis of bivariate regressions, termed "studies of relative growth." Even when results of a large series of such bivariate analysis on a sample of a population are brought together with interesting and instructive interpretations, they do not get at the heart of the importance of ontogenetic change. A study by Lundelius (1957) on *Sceloporus* and one by the senior author (Olson, 1951) show the extent to which interpretation is feasible. In both, some degree of integration is attained, and some indications of changes in rates as they may affect selection are evident. These are, however, at a very gross level.

A study of ρ-groups in ontogeny is presented in chapter vii, based on samples of white rats. Such series can be handled in much the same way as time series in phyletic evolution. Minor and related changes during ontogeny can be detected and studied in relationships to growth rates of the dimensions involved. If the technical difficulties that plague efforts to determine growth stages in fossil materials can be solved, it may be possible to apply such analyses to some fossil species. This hope is highly tentative. If it is demonstrated that important information on ontogeny can come from studies of morphological integration, it still remains to relate this information to events that are important in evolution. Even if these conditions can be fulfilled, the special problems of dealing with extinct populations in time series must be solved. As yet, only a start in this direction has been made.

Hybridization

The evolutionary significance of hybridization has been gaining recognition rapidly during the last decade. This is particularly true for botanical materials. Stimulated by the work of Anderson (1949) and others on introgressive hybridization, research on hybridization as an essential factor in floral evolution has made great strides. Less significance has been accorded this factor in animal evolution, but there is a rapidly growing literature in this area. Whether or not hybridization in animals has played a vitally important role in evolution is still an open question. Regardless of its significance over long periods of time, hybridization in animals is a common phenomenon and cannot be ignored in studies of evolution. The majority of studies of animal hybrids are concerned with characters that are readily accessible to study. Relatively little is known among wild populations about the effects of hybridization on skeletal systems, and it certainly may be questioned whether or not most of the differences between hybridizing animals are sufficient to be recognized in this system. Certainly in the majority of cases a delicate means of detection of the differences between the species and between hybrids and parent species would be required.

The part that morphological integration might play in the detection and evaluation of the effects of hybridization must be purely speculative at present, for no work has been carried out in this area. The use of associations of measures provides an exceedingly sensitive device for detection of minor differences and could, perhaps, aid in recognition of such differences as might be expected. Even if positive results could be obtained in experiments on living species and their hybrids, there would remain difficulties in application to fossil groups of animals. The solution of problems would involve a precise knowledge of geographic and stratigraphic distribution of populations, knowledge that is often most difficult or impossible to acquire. Although this imposes severe restrictions, there do exist a number of situations in which it might be possible to test the effects of hybridization over extended periods of time, particularly in the rather recent parts of the geological record. Studies of species of mammals based on dentitions come to mind as the most likely to succeed. Application of the concept of integration to cusp patterns and interrelationships of other dental characters can be readily carried out. First, however, it would be necessary to run test cases on modern species to determine the nature of changes to be expected and to investigate whether these changes differed in any significant way from those that occur in the normal course of speciation.

It is entirely problematical whether such studies would yield results of real interest, but the possibility does exist. We merely make the suggestion that this could prove an interesting field of study and that the concept of association of measures as an indicator of integration could provide a means of investigation.

Parallelism and Convergence

We shall make no effort to draw a fine distinction between parallelism and convergence but shall consider them more or less together from a morphological point of view as tendencies to attain the same or similar ends by paths that differ with respect to starting points.

As in the case of hybridization, our comments are speculative. There does appear to be a very broad area in which studies of morphological integration might be of great interest. There is, for example, the question whether certain characters have developed in a parallel or convergent fashion or have been inherited directly by different lines from a common ancestry. If there were, in fact, independent development of the features that show a high degree of similarity in the groups in question, patterns of organization would be expected to be different from those in cases in which the converse was true. These differences might lie within the systems that include the characters that show similarities, or they might appear in other systems of the body. In either event or in some combination, studies of morphological integration should provide a broader basis for determination than studies that involve but a few characters.

A problem of this general nature has arisen, for example, in the case of the New and Old World hystricomorph rodents. Studies of the type suggested would undoubtedly have to be very extensive for a major group such as this. Many species would have to be analyzed in order to determine whether the differences between species were greater than the differences between the geographic subdivisions of the hystricomorphs. Any other problems that might be studied would, of course, pose similar problems of design. Regardless of what design might be used, it seems highly probable that the concept of the total organism expressed in patterns of integration would reveal much more than could be expected of methods and concepts less complex and less rigorous in execution.

The fossil record provides numerous cases to which the general plan of attack suggested in the last paragraph might be applicable. There is, for example, the whole problem of the sources and relationships of the actinopterygian fish grouped together as subholosteans (see, for example, Westoll, 1944; Schaeffer, 1956). The complexities of parallelism and convergence in their evolution are such that little headway has been made toward a thorough understanding of the major patterns of change. Somewhat similar, at least in the complexity of the problem, is the study of the evolution of sutural patterns in the various stocks of ammonite cephalopods. The concept of integration surely could aid in solving some of the problems that have not yielded to simpler analyses.

Another category of problems occurs where there is no question of ancestry but where particular structures show strong resemblances in different phyla. The questions that arise involve, for example, the functional significance of similar, but independently derived, systems. There are many examples in this general area, but we shall cite only two from the field of paleomammology, a field in which the suggested methods might be particularly useful. The striking adaptive similarities of the limb structures of the single-toed horses in North America and the advanced litopterns of South America are a case in point. While clearly from very different ancestral stocks and widely different in many characters, these two groups have evolved highly similar adaptive counterparts so far as the limb structure is concerned. It would be extremely interesting to study the patterns of integration of the locomotor structures from a more comprehensive point of view than has been possible to date. A second example involves the highly similar dentitions in a number of groups of early Tertiary mammals—primates, condylarths, and some insectivores, among others. Detailed species differences, within and between these categories, are evident, but placement of species, genera, and even families in the proper higher categories has been a matter of great difficulty. Placement has been based upon the recognition of general patterns, characteristic of the major categories, and the existence or tendency toward this pattern in the problematical groups. But such patterns, conceived and expressed qualitatively, must involve, if nothing more, difficulties in connotation and, more seriously, suffer from lack of comprehensiveness and precision. It appears probable that each major group, by the time it can be considered an entity, would have established a basic pattern of integration of dental and cusp characters distinct from the patterns of the groups that it resembles. These patterns, provided that they exist, should be subject to determination from species whose assignment is unquestioned and then serve as a basis for judging the position of forms that can be assigned with less certainty.

To the degree that problems of parallelism and convergence involve many characters of organisms, the solutions must be sought in concepts that encompass systems or the totality in their framework. Since the concept of morphological integration expressly treats just such problems, it should in theory prove of great value. Actually, if the theoretical value be granted, there will always be serious limitations to its applicability because of the nature of samples that are available for study in many critical areas. Samples of teeth of *some* primates and condylarths, for example, are suitable, but, at present, samples for many species are inadequate. We must stress again that the concept and methodology are best suited to the investigation of principles and less to the solution of particular cases. The lack of suitability of samples of many species is probably unavoidable, for the sparseness of representatives suggests that even exhaustive collecting would not yield enough. It is possible in many cases, however, that recognition of the need for larger samples will motivate collecting that will provide adequate samples for many species now represented only by inadequate materials. We know from our own field activities that we frequently have failed to give proper consideration to sample design and to the care necessary to obtain proper materials.

ECOLOGY AND EVOLUTION

To this point our discussions have dealt in large part with the general concept of integration in categorical units that are based on taxonomic or mode-of-life considerations. Most attention has been paid to the species population. We have discussed briefly how the various factors of evolution and the phenomena that result from them may be viewed under the concept of quantitative integration. Morphological integration itself has been suggested as a factor in evolution. The measures that have been considered are based on morphology, and integrations express relationships in morphological systems. Comparisons have had their inceptions in networks of measures based on morphological features, whether the common basis of relationship is at the species or higher categorical level. The phenomena of evolution have been viewed in a phyletic perspective, with the center of interest in the population and what happened to it in the course of time. Outside features have been considered only as the population is related to them—for example, how the elements of the population are adapted to this or that situation. As the concept of integration so viewed adds to knowledge or understanding of each population and each phylum, it contributes to a synthesis that has its roots in the changes of organisms. By such procedures, regardless of the way that knowledge of populations and their changes is gained, a picture of the flow of an associated group of animals through time under some array of physical conditions can be obtained.

Such a study, on the Clear Fork Permian beds of Texas, was described some years ago by the senior author (Olson, 1952a). Even in a study simplified by deletion of poorly known genera and species and lumping of environmental types, the complexities became almost unmanageable. It was necessary to introduce additional simplification by consideration of only two or three related lines at one time. At the time of that study, it was felt that there would emerge more effective ways of analyzing faunal evolution, just as it was believed earlier that methodologies that brought together many characters of organisms could be more effective than analyses of one or a few characters.

The concept of totality applies with equal strength to the animal and to the complex of populations in ecological association. As in the case of the morphology of animals, we may also adopt a point of view of totality with respect to the ecosystem. So far as the organic aspects of such systems are concerned, the populations assume the position of elements in the universe. It seems as improbable that a single population can change materially without affecting other associated populations as it does that a single character change is an isolated event in an organism. With this point of view, it is possible to arrive at a concept in which the change of the whole, the ecosystem, is the center of focus and the changes of the elements fall into context as modifications, not independent of the whole, but dependent upon all other related, or integrated, changes. This is the perspective that recognizes faunal evolution as basic or, as stated in the cited paper, the concept of modifications of the chronofauna.

It is an almost inevitable step to enlarge the general concept of integration to encompass this idea, and we are certainly not alone in taking this step, any more than we were with regard to the concept of integration applied to the organism. There does, however, seem to have been some lack of recognition of the vastly different light in which the concept of faunal evolution casts many of the phenomena of evolution of phyla. This is evident, for example, in the problems that are posed by the major extinctions that have occurred at various times in the geological past. While it has been recognized that ecological factors have been involved, preoccupation with the groups that have become extinct and the environment as related directly to them has, in our opinion, generally failed to provide a grasp of the complexities that are in fact involved. Rather there has been a tendency to search for one or another prime factor, usually with unsatisfactory results. Extinction, viewed from the perspective of the evolving ecosystem, represents a modification in the system that affects and is affected by all other concurrent and earlier events. Both points of view can yield interesting results and neither should be abandoned in favor of the other. The general preoccupation with the first, however, does appear to have obscured a great deal that can come from materials that are available for study.

The analogy of integration in the organism and in the ecological system is only partial, for the latter involves not only the organic relationships of the animals and plants but also the physical aspects of the evolving system. These must be brought into relationship with the populations and the relationships between populations, in order that the complete integration of critical factors may be included.

Application of quantitative integration to the problems of evolving ecosystems is a natural step from its application in the study of interspecies populations of animals and their evolution under this concept. Tentative models for such studies have been drawn up, and these point rather specifically to the complexities that are involved, to the difficulties in practical application, and to the potential power in such studies. It is not yet possible to offer a practical formalization that can be applied in actual cases. We shall, however, review briefly some of the problems and progress to indicate that there does exist a stimulating but difficult area of study in the extension of quantitative integration to the field of evolution of ecological systems of the past.

Considerable progress in the analysis of quantitative association as a basis for understanding physical environments has been made by the junior author. Correlation has proved a powerful tool, as shown in two published reports (Miller, 1954; Miller and Olson, 1955). In the second paper the concept of representative measures is developed for the analysis of environments of sedimentation, stated as the key facies measure. The necessity for reduction of complexities by meaningful representation is as important in environmental analyses as it is in the morphological studies cited earlier in this section. The problems of sampling, samples, and universes are in general similar, and the use of analogy plays an equivalent role. The way seems clear, in an extended program of research now being conducted by the junior author and from studies of others in the field, to push forward to the construction of a conceptual framework based on associations of measures that will serve for the analysis of the nature and evolution of many types of sedimentary environments.

The problems of the biological aspects of ecological systems of the past fall into two general categories, one of practical problems and the other of more strictly theoretical problems. The first group arises from the fact that by no means all of any evolving biological system is preserved in the record and that what is preserved does not necessarily maintain in the record the spatial and detailed temporal relationships that the elements had in the living system. We can be quite sure that important elements are absent, but the only means of evaluation of the nature of the missing parts is through recourse to analogous systems living today. Success in interpretation is thus likely to diminish as time from the present increases, for the use of analogues is less and less satisfactory and the organisms involved are more remotely related to those alive today. The problems of the determination of the spatial, temporal, and biological relationships of the elements are practical difficulties, and the answers must come from field observations, morphological interpretations, and, again, recourse to modern analogues.

The more theoretical problems involve the determination of what measures are best used to express relationships of the elements of the system. Knowledge of the elements themselves comes primarily from morphology, providing the array of known populations to be considered. Quantitative expressions of morphological characters of the populations—for example, means or variances of single characters, or expressions of integration—can provide single numbers whose relationships can be studied. Alone, a suite of such individual numbers may mean little with regard to the role of a population in the system. Considered in relationship to other measures, however, they may take on real meaning.

A more detailed understanding of the basic features of morphological integration than is given by the general statements in the first two chapters of this book is necessary for the comprehension of a framework for the study of ecological evolution in which morphological integration is one of the fundamental aspects. In particular, the formal model developed in chapter iii is critical, for any more comprehensive model which would include both physical and biological aspects of an evolving system must be consistent with this model. Furthermore, an intimate knowledge of what is revealed by application of the concept of morphological integration to single populations and to sequences of populations in evolution is an essential prerequisite. Chapters vi, vii, and viii provide this background. Consideration of this area, which we feel to be the most important objective of our work, is deferred until the final chapter of the book in order that we may draw freely upon all that has been revealed in the studies made to the present time.

The ρF-Model, Modifications, and Extensions

There have been three rather distinct phases in the development of the model that we now consider suitable for studies of morphological integration. The first includes the preliminary work and synthesis that led to the formulation of the ρF-model, first presented in 1951 (Olson and Miller, 1951b). The second phase includes a series of studies on fossil materials and, resulting from the need for better control, a series of studies on modern materials. The third involves the formulation of the model and interpretations of various experiments that became possible after its formulation. The studies of two extinct amphibians, *Trimerorhachis* and *Diplocaulus*, were undertaken in an effort to determine the nature of integration of the dermal patterns of the skulls and to find factors basic to the derived patterns (Olson, 1953). Several modifications of the original model were found necessary in the course of this study and were stated in the short report of this work cited above. When additional studies of *Rana* and *Sciurus* had been completed, it was apparent that a more suitable model was needed in order that continued research could be cast into a stable framework. The reformulated model is presented below.

BASIC FEATURES OF THE MODEL

The model, which we present both verbally and formally in this section, provides a mathematical basis for operations that involve biological theory and interpretation. These operations, however, are carried out beyond the formal framework of the model, in contrast to the original ρF-model (see appendix to this chapter). Ambiguities and confusion that have resulted from the introduction of both quantitative and qualitative considerations into a single framework are thereby eliminated.

A full understanding of the technical details of the model is not essential to a comprehension of the results of the experimental studies presented in later chapters. We shall attempt to present in this section a verbal and somewhat loosely drawn exposition of the salient features for readers whose interests lie primarily in the biological significance of these studies. Formal exposition is deferred to the following section.

A quantitative characteristic of a species, such as head length, will be called a "measure" in the following discussion. The degree of association between measures is considered to be a sensitive expression of biological processes. A useful and convenient quantitative expression of the degree of association is found in the coefficient of correlation, in which values range from zero for no correlation to ± 1 for complete correlation. In this study the correlation between any pair of measures, for instance, head length versus body length, is

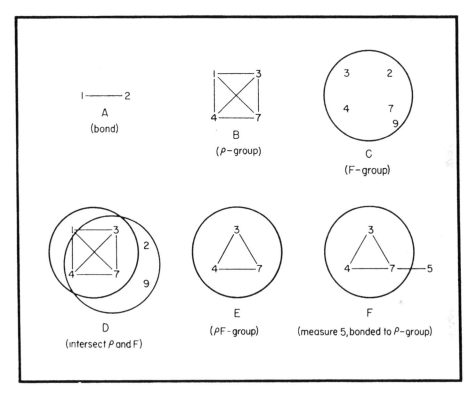

F_IG. 1.—Illustration of definitions of terms in the ρ F-model, based on revised definitions. *A* to *F* as indicated by captions and described in text.

expressed as ρ. An abstract, quantitative representation of a species is thus provided in convenient form, if the correlation between all possible pairs of measures is computed. The resulting numerical pattern is assumed to be unique for the morphology of a given species at a given time, provided that the totality of all possible measures is so treated (see Fig. 1).

Analysis of the numerical pattern of ρ values is effected by a subdivision depending on the value of ρ, as, for example, the class of all values greater than 0.80 (see Fig. 2, based on data of Table 3).

Definition: The Term "ρ-Group"

Consider the class of all pairs of measures with ρ greater than some selected level, such as 0.80. Form all maximal classes of measures such that within each class all pairs have ρ greater than the selected level. These classes are defined as "ρ-groups." They form the fundamental basis for biological interpretation in this study.

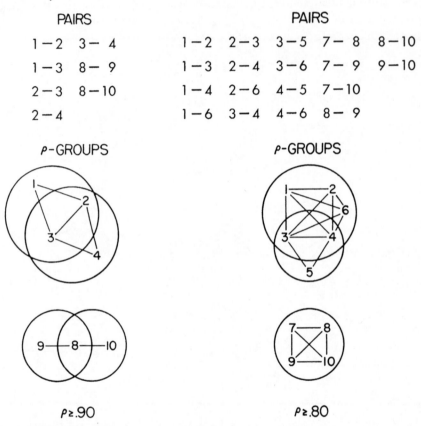

PAIRS				
1 — 2	3 — 4			
1 — 3	8 — 9			
2 — 3	8 — 10			
2 — 4				

PAIRS				
1 — 2	2 — 3	3 — 5	7 — 8	8 — 10
1 — 3	2 — 4	3 — 6	7 — 9	9 — 10
1 — 4	2 — 6	4 — 5	7 — 10	
1 — 6	3 — 4	4 — 6	8 — 9	

P-GROUPS

P-GROUPS

ρ ≥ .90

ρ ≥ .80

FIG. 2.—Bonded pairs of measures and ρ-groups, based on data of Table 3, formed at levels of ρ ≥ .90 and ρ ≥ .80, respectively; ρ-groups inclosed in circles.

Since a given measure may be correlated with several other measures at the required level, it is clear that a measure may be in more than one ρ-group and that ρ-groups may overlap. In practice, biological interpretation is often greatly hampered by complex overlaps (intersects) of ρ-groups. Reduction of overlap is often important and will be considered later in this section.

In actual studies the sample correlation coefficient, r, is obtained rather than ρ. To derive a ρ-group (when dealing with r values) which is comparable to that previously developed for the population, some level of ρ greater than zero is again selected, ρ with a fixed value being expressed as $|\rho|$. All values of r

consistent with those $|\rho|$ values equal to or greater than the arbitrary level of $|\rho|$, at some selected probability level, are accepted as belonging to the ρ-group. Determinations may be read directly from the tables compiled by David (1938).

Definition: The Term "Bonded"

The term "bonded" is used when the correlation between a pair of measures is greater than or equal to the value $|\rho|$ used in forming a class of ρ-groups. Suppose the ρ-group-forming level is $|\rho| \geq 0.90$. Then all pairs of measures whose correlations are ≥ 0.90 are bonded, and all other pairs are not.

TABLE 3

HYPOTHETICAL MATRIX OF ρ VALUES

(Lower Limit of Band)

	2	3	4	5	6	7	8	9	10
1........	.90	.97	.84	.53	.81	.73	.61	.54	.19
2........		.91	.94	.61	.83	.40	.36	.78	.41
3........			.94	.82	.84	.61	.51	.72	.68
4........				.81	.86	.64	.61	.69	.73
5........					.31	.32	.51	.78	.41
6........						.71	.79	.69	.37
7........							.84	.84	.81
8........								.91	.90
9........									.84

Definition: "Basic Pairs"

Subsets that consist of pairs of measures may be derived by successively raising the arbitrary level of $|\rho|$ by small increments and retaining, irrespective of the level at which they occur, each pair that is formed. This procedure, which is an extension from the procedure that produces ρ-groups at some fixed level of ρ, results in pairs of measures that are hereafter called "basic pairs." Basic pairs have the characteristic that each measure is more highly correlated to the other than to any other measures. In practice the values of the correlation coefficient are entered into a matrix in which the rows and columns are defined by the measures taken. Basic pairs may be readily discerned by scanning of this matrix. If the highest value for each pair of measures is at the intersection of the corresponding row and column, these two measures form a basic pair.

A principal function of basic pairs is to reduce intersects that occur in most initial groupings. In this role the basic pair serves as a "group-former." The process of group formation is as follows:

1. All ρ-groups from the quantitative study, where ρ = some level appropriate for the study, are listed. In some cases several levels are listed.

2. All ρ-groups that do not contain both measures of at least one basic pair are dropped.

3. All measures that have a higher correlation with at least one member of a basic pair not in ρ (from step 2) than with either measure of the basic pair(s) in ρ (from step 2) are

eliminated from the ρ-groups being considered. Note that this automatically removes any single measure of a basic pair when the other measure of that pair does not occur in the considered ρ-group. (This step may be carried out rapidly, using r if desirable, by inspection.) The theory, however, is precisely comparable with that described for the formation of basic pairs by raising the level of ρ by small increments, the difference lying only in the substitution of the bond between one measure of a basic pair and one other measure not in the basic pair.

4. Any measure that now remains in more than one ρ-group, except in intersects that involve basic pairs, does so by virtue of the fact that its bonds with one or more measures of basic pairs in each of the ρ-groups involved are identical. This is almost certainly an artificial situation, resulting from rounding of the values of r. The treatment of "ties" is handled as follows: (1) To make an assignment to one ρ-group under this circumstance, examine the

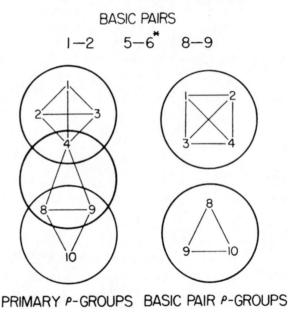

BASIC PAIRS

1—2   5—6*   8—9

PRIMARY ρ-GROUPS   BASIC PAIR ρ-GROUPS

Fig. 3.—Effect of basic-pair analysis, based on data of Table 2 ($\rho \geq .80$). ρ-groups circled. Note development of two discrete groups, one for basic pair 1–2 and the other for basic pair 8–9. Basic pair 5–6 occurs below level $\rho \geq .80$ and thus does not form a group at or above this level.

bonds of the measure in question to the other members of the basic pairs involved; (2) the measure is assigned to the ρ-group in which the highest bond to these measures in the basic pair occurs. In effect, this brings into play the summation of intensity of association of the measure in question to each measure of the two or more basic pairs.

A simple, hypothetical example of the formation of basic pairs is shown in Figure 3, based on the data in Table 4. The procedure demonstrated is mechanically effective in the reduction of intersects, for all groups so formed are mutually exclusive and contain one or more basic pairs. Single measures of a basic pair cannot occur separately in two groups, since the chance of intersection has been eliminated. The procedure can be considered biologically meaningful, however, only if the basic pair and the measures that group with it have biological significance. That ρ-groups have biological meaning was assumed early in the studies, and the assumption was verified by experiments discussed in detail later in the text. It follows logically from the assumption that statistical association has biological meaning that the highest level of association, as in the basic pair, should indicate the highest intensity of biological relationship, in other words, the highest level of morphological integration. Furthermore, it is to be expected that the measures that cluster around the basic pair on the basis of level of bonding should also be highly integrated.

TABLE 4

HYPOTHETICAL MATRIX WITH r AND ρ

		2	3	4	5	6	7	8	9	10
1.......	r	.97	.81	.87	.21	.09	.47	.43	.16	.41
	$\rho \geq$	.99	.92	.95	.63	.56	.78	.76	.61	.75
2.......	r		.69	.70	.01	.31	.27	.49	.17	.02
	$\rho \geq$		.88	.89	.50	.70	.67	.71	.62	.51
3.......	r			.86	.33	.19	.20	.31	.33	.17
	$\rho \geq$			.94	.72	.62	.62	.69	.72	.61
4.......	r				.29	.20	.14	.61	.73	.12
	$\rho \geq$				.68	.62	.59	.84	.90	.58
5.......	r					.41	.21	.14	.10	.01
	$\rho \geq$					.75	.63	.59	.56	.50
6.......	r						.41	.39	.17	.17
	$\rho \geq$						.75	.74	.62	.62
7.......	r							.28	.41	.11
	$\rho \geq$							.68	.75	.56
8.......	r								.94	.81
	$\rho \geq$								.97	.93
9.......	r									.93
	$\rho \geq$									.95

An example taken from the study of *Rana pipiens* will serve to illustrate the empirical validation of this reasoning. The measures of *R. pipiens* were taken to cover the entire skeleton. Fifty measures were used for this purpose (for details, see chaps. v and vi and Appendix A). These were grouped into a series of functional arrays (F-groups) on the basis of information gained from dissections and studies of functional behavior of living animals. A number of F-groups was derived. We shall consider for illustration only the measures that pertain to the anterior and posterior locomotor structures—the limbs and girdles. While both are involved in locomotion, they function differently with respect to the various actions and have very different subsidiary functions. For these reasons they were grouped into two distinct F-groups. The F-groups included the measures presented in Table 5 (for full definition see chap. v and Appendix A).

At the high level of $\rho \geq .91$, these two groups showed considerable intersection, which was to be expected from their common functional attributes. The following is an example of one ρ-group at this level:

$$Ta_l\text{–}Fe_l\text{–}Fe_w\text{–}Hu_l\text{–}Hu_w\text{–}An_w\text{–}Epic_{pe}\text{–}St_p\text{–}Epic_{de}\text{–}Hu_{cr} \,.$$

TABLE 5

MEASURES OF *Rana pipiens* IN LOCOMOTOR SYSTEMS

Abbr.	Description
Forelimb Array	
Co_l	Length of coracoid
Hu_l	Length of humerus
Hu_w	Width of humerus
An_l	Length of antebrachium
An_w	Width of antebrachium
Mc_l	Length of 2d metacarpal
$Epic_{pe}$	Length of mid-line of epicoracoid
$Epic_{de}$	Length of deltoid insertion on epicoracoid
St_p	Length of osseous sternum
$Epst_d$	Length of episternum
Hu_{cr}	Length of crista ventralis of humerus
Hind-Limb Array	
Pp_l	Posterior process of ilium, length
Cr_l	Maximum length of crus
Cr_w	Minimum width of crus
Tl_l	Length of 2d tarsal
Tl_w	Width of 2d tarsal
Fe_l	Length of femur
Fe_w	Minimum width of femur
Mt_l	Length of 2d metatarsal
Ac_w	Width of acetabulum
Crh_w	Width of head of crus

Intersects occurred to the level of $\rho \geq .95$, but only pairs were present above this level. Basic pairs that pertain to these groups were isolated and found to be as follows:

F-forelimb
$Hu_l\text{–}Hu_w,$

F-hind limb
$Cr_l\text{–}Fe_l.$

At the level $\rho \geq .91$, grouping by basic pairs resulted in the following arrays:

F-forelimb
$Hu_l\text{–}Hu_w\text{–}An_l\text{–}An_w\text{–}Epic_{pe}\text{–}St_p\text{–}Epic_{de}\text{–}Hu_{cr},$

F-hind limb
$Cr_l\text{–}Ta_l\text{–}Fe_l\text{–}Mt_l.$

These two groups include only measures in one or the other of the two F-groups. They do not, to be sure, include *all* measures allocated to the F-groups, which would hardly be expected at this high level, but the important facts are that they are discrete and that neither includes any measure from the other F-group. This has been found to be the case for other systems in *R. pipiens*, as well as for other test cases. Most significant is the fact that F-groups have been derived by purely mathematical procedures. Merging occurs in these two groups only at levels of ρ where the basic pairs form a quadruplet, this being a level where the common function is the dominating factor. At this level, a larger functional complex than the one designated by the F-groups as formulated is indicated.

Partial Correlation Extension

Partial correlation may be performed within the framework of the revised model. One or more measures are "held constant" to eliminate their effects upon the association of the other measures. The derived correlation coefficients may be treated in the same manner as those that result from simple correlation. A principal use of this extension is the detection of factors important in integration of measures. The mathematical and biological aspects of this process are rather complex, and detailed consideration will be deferred to chapters iv and vi.

THE FORMAL MODEL

We have now discussed most aspects of the model which will be used throughout the rest of this book. Detailed work with the model and its extensions, as well as a framework for the derivation of other extensions, requires a formal development, which is presented below.

Consider the morphology of an animal to be completely represented by a series of linear measurements. The animal would then appear as a complex network of linear dimensions. In the ideal case the *total* morphology is abstractly and objectively recorded. A time sequence of abstractions of the type just described thus records abstractly and objectively the *total morpho-change* through the time sequence. Analysis of the model is effected by partitioning the collection of morphological measurements into classes based on the degree of association. It is a fundamental assumption that underlying orders of biological significance can thus be ascertained, and in this way the model is used as more than just an abstract, objective way to record morphology.

If $X_1, X_2, \ldots, X_p$ are random variables, corresponding to the characteristics of a species (or other unit), they may be treated as if they followed a multivariate distribution. In the following, quantitative characteristics of this sort will be called "measures." If X_1 is such a measure and is taken for all the members of a species, the values $X_{1(1)}, X_{1(2)}, X_{1(3)}, \ldots, X_{1(N)}$ result, N representing the number of individual animals.

Consider the set M of all such measures on the species. From M form a product space, Q, whose elements are all possible pairs (X_iX_j). Then Q is mapped into A; where $|\rho_{ij}|$ is the image of (X_iX_j); A is the closed interval, $[0, 1]$; and $|\rho_{ij}|$ denotes the correlation coefficient between a pair of measures X_i and X_j.

Now consider a subset, S, of A, where S is defined to be

$$\{ |\rho_{ij}| \, \epsilon[0, 1] \, \big| \, |\rho_{ij}| \geq \text{ some arbitrary value of } \rho \}.$$

This subset forms the basis for analysis from a biological point of view.

	X_1	X_2	X_3	$\cdots$	X_p
X_1	$|\rho_{11}|$	$|\rho_{12}|$	$|\rho_{13}|$	$\cdots$	$|\rho_{1p}|$
X_2	$|\rho_{21}|$	$|\rho_{22}|$	$|\rho_{23}|$	$\cdots$	$|\rho_{2p}|$
X_3	$|\rho_{31}|$	$|\rho_{32}|$	$|\rho_{33}|$	$\cdots$	$|\rho_{3p}|$
.	.	.	.		.
.	.	.	.		.
.	.	.	.		.
X_p	$|\rho_{p1}|$	$|\rho_{p2}|$	$|\rho_{p3}|$	$\cdots$	$|\rho_{pp}|$

Fig. 4.—Matrix representation of correlation coefficients. Entry $|\rho_{ij}|$ is in ith row, jth column; X_i represents morphological measurement i; X_j represents morphological measurement j. For example, $|\rho_{23}|$ represents the correlation between X_2 and X_3.

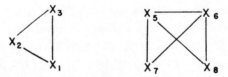

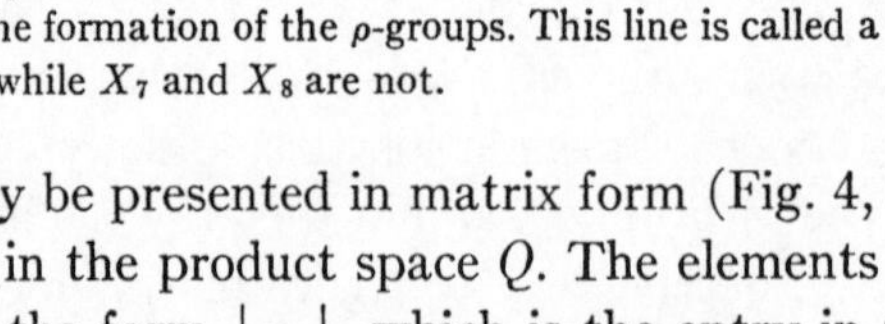

Fig. 4a.—Illustration of the formation and identification of maximum non-contained ρ-groups. The line connecting any pair of X's indicates that the correlation between this pair exceeds the minimum ρ level set for the formation of the ρ-groups. This line is called a "bond." Thus X_5 and X_8 are bonded in (2) above, while X_7 and X_8 are not.

The foregoing may be presented in matrix form (Fig. 4, a). The row-column X's locate an entry in the product space Q. The elements (X_iX_j) are mapped into A and take on the form $|\rho_{ij}|$, which is the entry in the ith row and the jth column of the matrix.

The subset S is then represented in matrix form (Fig. 5) by deleting from the above matrix all $|\rho_{ij}| < $ arb. ρ. Since the matrix is symmetric and $|\rho_{ij}| = |\rho_{ji}|$, we also drop out everything below and including the principal diagonal

$|\rho_{ii}|$. From the above array, all possible non-contained[1] intercorrelated groups of measures are formed. These groups may range in size from pairs to p-tuplets.

It is clear that many of these groups overlap. This overlap introduces severe difficulty in biological interpretation. Thus a primary consideration is to reduce the overlap of the groups to a minimum. Before consideration of various methods to reduce the overlaps in the groups in a biologically meaningful way, sample-population relationships are discussed.

In the actual analysis the sample differs from the population in two ways: (1) The number of animals involved in a given measure represents a sample (n)

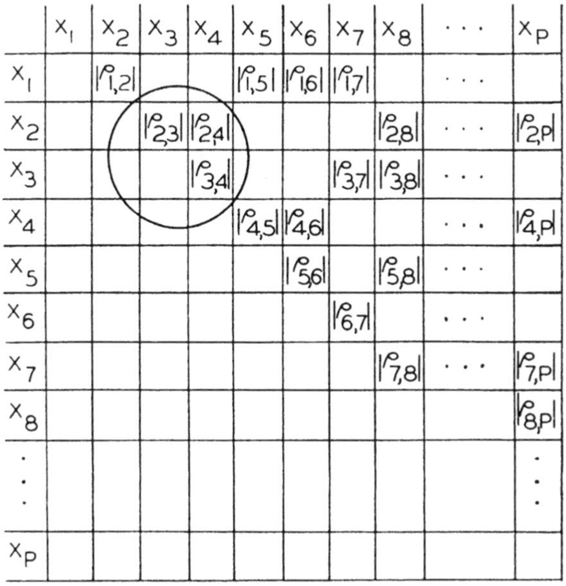

	X_1	X_2	X_3	X_4	X_5	X_6	X_7	X_8	$\cdots$	X_p
X_1		$\lvert r_{1,2}\rvert$			$\lvert r_{1,5}\rvert$	$\lvert r_{1,6}\rvert$	$\lvert r_{1,7}\rvert$		$\cdots$	
X_2			$\lvert r_{2,3}\rvert$	$\lvert r_{2,4}\rvert$				$\lvert r_{2,8}\rvert$	$\cdots$	$\lvert r_{2,p}\rvert$
X_3				$\lvert r_{3,4}\rvert$			$\lvert r_{3,7}\rvert$	$\lvert r_{3,8}\rvert$	$\cdots$	
X_4					$\lvert r_{4,5}\rvert$	$\lvert r_{4,6}\rvert$			$\cdots$	$\lvert r_{4,p}\rvert$
X_5						$\lvert r_{5,6}\rvert$		$\lvert r_{5,8}\rvert$	$\cdots$	
X_6							$\lvert r_{6,7}\rvert$		$\cdots$	
X_7								$\lvert r_{7,8}\rvert$	$\cdots$	$\lvert r_{7,p}\rvert$
X_8										$\lvert r_{8,p}\rvert$
$\vdots$										$\vdots$
X_p										

Fig. 5.—Matrix of Figure 4, after deletions. Deletions include (1) all $|\rho_{ij}| <$ arbitrary ρ; (2) the entries in the principal diagonal; and (3) all entries below the principal diagonal.

of the number of animals (N) in a species. The sample n is usually very small relative to the total N, and N is not known. (2) The number of measures X_1 in the sample is much smaller than the total number of measures p.

Item 1 is treated on a statistical basis. The sample correlation coefficient, r_{ij}, is used as a statistical estimate of the population value, ρ_{ij}. Item 2 is not treated statistically. It represents an unavoidable difficulty, in that it is necessary to infer biological meaning from part to whole when relationship of part to whole is not clearly understood. However, a priori biological knowledge,

[1] By "non-contained" we mean the largest completely intercorrelated group which can be formed. Thus in Fig. 4a, X_2–X_3 form a group of two; similarly, X_3–X_4 and X_2–X_4. But these pairs are contained in the triplets consisting of X_2–X_3–X_4, and, by definition, this is the maximum non-contained group. Figure 4a shows another case in which two maximum non-contained ρ-groups are formed. These are the triplets X_5–X_6–X_8 and X_5–X_6–X_7.

combined with current experimental work described later in the text, seems sufficiently strong in many cases to lend confidence to the interpretation.

The sample matrix is in the same form as the matrix shown in Figure 6. However, a given measure X_1, which takes on values $x_{i(1)}, x_{i(2)}, x_{i(3)}, \ldots, x_{i(N)}$ corresponding to the total N individuals in the population, now takes on

	$\hat{x}_1$	$\hat{x}_2$	$\hat{x}_3$	$\cdots$	$\hat{x}_k$
$\hat{x}_1$	$\|r_{1,1}\|$	$\|r_{1,2}\|$	$\|r_{1,3}\|$	$\cdots$	$\|r_{1,k}\|$
$\hat{x}_2$	$\|r_{2,1}\|$	$\|r_{2,2}\|$	$\|r_{2,3}\|$	$\cdots$	$\|r_{2,k}\|$
$\hat{x}_3$	$\|r_{3,1}\|$	$\|r_{3,2}\|$	$\|r_{3,3}\|$	$\cdots$	$\|r_{3,k}\|$
$\vdots$	$\vdots$	$\vdots$	$\vdots$		$\vdots$
$\hat{x}_k$	$\|r_{k,1}\|$	$\|r_{k,2}\|$	$\|r_{k,3}\|$		$\|r_{k,k}\|$

FIG. 6.—The sample equivalent of the matrix of Figure 4, in which the entries $|\rho_{ij}|$ have been replaced by sample correlation coefficients $|r_{ij}|$; x_i of the ith population is now X_i.

	$\hat{x}_1$	$\hat{x}_2$	$\hat{x}_3$	$\hat{x}_4$	$\hat{x}_5$	$\hat{x}_6$	$\hat{x}_7$	$\hat{x}_8$	$\hat{x}_9$	$\hat{x}_{10}$
$\hat{x}_1$		.68	.82		.91		.87		.66	.73
$\hat{x}_2$			.80		.77	.78		.75		
$\hat{x}_3$					.68	.86				
$\hat{x}_4$					.74			.84	.96	.95
$\hat{x}_5$							.92			
$\hat{x}_6$								.89		
$\hat{x}_7$									.95	
$\hat{x}_8$										
$\hat{x}_9$										.83
$\hat{x}_{10}$										

FIG. 7.—Sample equivalent of the deleted matrix of Figure 5

values $x_{i(1)}, x_{i(2)}, x_{i(3)}, \ldots, x_{i(n)}$, where n is usually much smaller than N. The number of measures, X_k, is also much smaller than the theoretically complete number of measures, X_p, in the model. The sample matrix appears in Figure 7. The sample equivalent to the subset previously shown in matrix form is computed as follows:

We wish to test the hypothesis that $|\rho| \geq |\rho_0|$ *versus* the alternative that $|\rho|$ is lower, where $|\rho_0|$ is some arbitrarily selected value of $|\rho|$. We reject $|\rho| \geq |\rho_0|$ when $|r| < c$ (lower limit of confidence interval for ρ_0). Let the sample size be n and the level of significance a. Then the probability $P_{|\rho_0|}\{-c \leq r \leq c\} = a$ is given in table form in David (1938). (For example, if $n = 25$, $a = .05$, and $\rho_0 = 0.8$, then c from David's tables is 0.645.) Suppose we wish to form the subset S such that all $|\rho_{ij}|$ are $> |0.8|$. We then accept all r_{ij} greater than or equal to 0.645. The form of the sample matrix for the subset analogous to Figure 6 is illustrated in Figure 7.

The Basic Pair

Suppose the level of $|\rho_0|$ which forms the lower limit of a subset is raised in an analysis of a sample matrix. The matrix is reduced in entries as the value of c increases; and ρ is raised until a single entry is left in the ith row. If that same entry is also the highest in the jth column, then the measures X_i and X_j, which determine that single entry $|\rho_{ij}|$, form a basic pair. It is clear that several basic pairs may be found in a single matrix.

An equivalent definition is: Consider the sample matrix of Figure 7. If the highest $|r|$ entry in the ith row is also the highest $|r|$ entry in the j column, then the two measures determining $|r_{ij}|$, namely, X_i and X_j, form a basic pair.

The basic pair may be thought of as an application of a limiting process. As the level of $|\rho|$ approaches the neighborhood of zero as a limit, all measures become bonded in all possible ways. On the other hand, as the level of $|\rho|$ approaches $|1|$ as an upper limit, all bonds between measures will vanish, for in the biological context of this study we expect no correlations equal to $|1|$. If the level of $|\rho|$ is raised toward the upper limit until only pairs are bonded in the sense defined in the previous two paragraphs, the result is the formation of basic pairs.

APPENDIX

THE ORIGINAL ρF-MODEL AND ITS PRESENT MODIFICATIONS

The ρF-model was constructed to bring into relationship measures that were associated by both quantitative and qualitative considerations. The intersects of groups of measures formed, on the one hand, by mathematical procedures and, on the other, by qualitative biological considerations were found to be significant units for the study of certain problems of evolution. The formal development of this model is as follows:[2]

Consider n morphological quantitative characteristics and assume:

1. If $X_1, X_2, \ldots, X_n$ are random variables corresponding to the n quantitative characteristics of a species, then $X_1, \ldots, X_n$ may be treated as if they followed a multivariate distribution. In this case, ρ_{ij} will denote the correlation coefficient between the pair of variables, X_i and X_j.

2. Any subset of X's may or may not be functionally and/or developmentally related. Quantitative characteristics may be called "measures" in the following discussion.

[2] Reprinted from *Evolution*, V (1951), 325–78, by permission of the editor.

Definition.—A ρ-group of the X's is any group, $X_{i_1}, X_{i_2}, \ldots, X_{i_k}$ such that:

$$|\rho_{i_m i_n}| \geq \text{(an arbitrarily selected value)}, \quad m, n = 1, \ldots, k . \tag{1}$$

For any other X_j $(j \neq i_1, i_2, \ldots, i_k)$ there exists at least one $i_m(m = 1, \ldots, k)$ such that $|\rho_{j i_m}| <$ (arbitrary value). $\tag{2}$

Thus (1) insures high intracorrelation in the subgroup, and (2) insures that all highly corre-lated X's have been included. It is clear that the ρ-groups may overlap and a given vari-able may be in more than one ρ-group.

Definition.—An F-group of the X's is any group $X_{i_1}, \ldots, X_{i_n}$ in which all members are thought to be functionally or developmentally related. These groups are chosen independent-ly of ρ-groups on the basis of biological considerations.

Definition.—A ρF-group of the X's is any group of the form $A_\rho \cup A_F$ where:

A_ρ is a ρ-group; $\tag{1}$

A_F is F-group; $\tag{2}$

$A_\rho \cap A_F \neq 0$ (i.e., A_ρ and A_F have at least one common member). $\tag{3}$

The principal differences between the earlier ρF-model and the reformulated model presented in this chapter relate to the use of the concept of F. In the original model ρF-groups were formed by the intersection of a ρ-group and an F-group. Such groups were thought to have biological significance. The ρF-group is not formed in the modified model, for the concept of F has been elimi-nated from the formal framework. This has provided greater flexibility in subsequent mathematical manipulations of the groups, ρ-groups, and follows logically from tests that have confirmed the hypothesis that ρ-groups imply F-groups. The ρF-group remains as an important conceptual unit but one that is beyond the structure of the model.

The original idea of F related it to a function or to a developmental pattern. This concept has now been enlarged and generalized so that F-groups include measures related by some unifying biological factor. This generalization has been necessary so that F may be applied to situations encountered in em-pirical work carried out after publication of initial results.

CHAPTER IV

Correlation and Morphological Integration

The general framework within which our studies have been cast, the model in chapter iii, and the concepts that follow from experiments reported in later chapters depend upon the use of correlation. It is critical that the basis for these procedures be fully understood.

The present chapter is devoted to discussion of the aspects of correlation that are pertinent to our own work. Herein lies the foundation of what we consider a rational synthesis of mathematics and biology, woven into the fabric of the concept of morphological integration. In subsequent chapters associations of measures revealed by correlation are used as the bases for analysis under a wide variety of circumstances. In each case the basic processes are adapted to the particular situation. Fundamentally, however, regardless of the group of organisms treated or the extent of coverage of the individuals by measurements, all studies fall into one of two major categories. One category, which may be considered the general case, comprises studies based on samples made up of individuals drawn from some considerable part of the ontogenetic range of the organism in question. These studies are concerned with growth series. The other, which is a special case, important primarily because of its frequency, includes studies based on samples from a single ontogenetic stage, defined either temporally or physiologically. Within this category fall studies of adults, studies of temporal stages in ontogenetic series, and studies of non-growing structures, such as teeth of mammals.

CORRELATION AND ONTOGENETIC SERIES

The fact that, as an animal grows, the individual components of the body increase at varying rates is revealed by casual observation and has been the center of interest in many and varied types of biological research. The general phenomenon is too well known and understood to require elaboration, but it must be emphasized that it is basic to the considerations in this section and to much of what follows in the later pages of this book. Dimensions of some parts of the anatomy increase in very close correspondence with those of other parts. The other extreme also occurs—a virtual lack of correspondence between developing parts. Between these extremes falls the great majority of relationships of the growth of parts during ontogeny. In order to express the strength of this relationship, we utilize the coefficient of correlation.

In the present section the use of the correlation coefficient is explicitly for expressing the degree of association of morphological dimensions with respect to differential growth, hence change of form, of an animal through his ontogenetic history. It is then only a short step further to show how a sample of individuals of a species, with ages ranging over the entire ontogenetic series, lends itself to the types of numerical analysis that have been used.

Initially we shall be concerned with scatter diagrams, obtained as follows: A random sample is taken from a population of animals or their fossil representatives. For each of the sample individuals two quantitative characteristics are measured without error. The resulting ordered number pairs are then plotted on Cartesian graph paper. The ages at time of measurement are not known. There are, of course, some important simplifications in these statements. First, it is questionable that any paleontological collection can be considered a *random* sample of a population. Second, errors of measurement are inevitable and may sometimes be important. Third, usually more than two quantitative characteristics are of interest. For the moment we shall not be concerned with pursuing these simplifications.

The type of scatter diagram described is frequently used in both zoological and paleontological studies. In the latter the nature of the sample is determined by the death pattern of the organisms in the population, by processes of preservation and diagenesis, and by the luck and skill of the collector. Rarely is anything known about the age of any specimen at death, except perhaps whether or not it was an adult. Such samples thus present a special suite of problems that are important in final interpretations. The model now developed takes cognizance of the facts but cannot in itself offer solution to the problems of sampling.

A simple linear model is first discussed to show the factors that determine the position of a single point on a bivariate scatter diagram. Assume, for simplicity, that growth in time is linear. Consider two measurements on a single individual. The geometric construction necessary to plot a point on the X,Y plane involves a projection of the intersection of planes through $y = g(T)$ and $x = f(T)$ onto the X,Y plane, as shown in Figures 8 and 9.

The variables that affect the position of the observed point (x,y) on the X,Y plane may now be considered. The equation for the curve traced on the X,Y plane consists of three factors: the slope of $f(T)$, the slope of $g(T)$, and t, the time in ontogeny at which the measures were taken. From a biological aspect, the following considerations are of interest: (1) the age of the animal when measured (in paleontological studies the age at death) is a factor that contributes to the position of the point on the X,Y plane; and (2) the two other factors may be considered as rates of growth. The proportion of the rate on y to the rate on x results in the "form" of the adult. The ratio of x to y is thus used as an index in morphological comparisons of taxonomic groups.

We shall now consider the model in terms of empirical curves, fitted to actual observations, and thus replace the simple linear form by more realistic functions. Many studies of growth have been made on the increase in size of a given

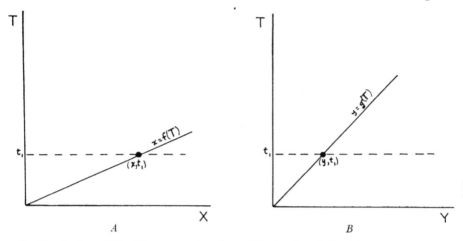

Fig. 8.—A, plane X,T, with construction of $x = f(T)$; B, plane Y,T, with construction of $y = g(T)$. For cases of simple linear growth in time.

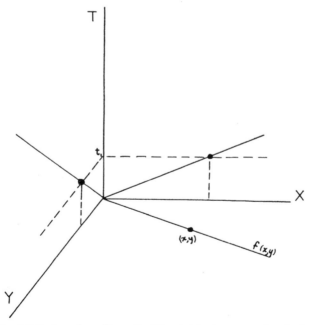

Fig. 9.—X,T and Y,T planes from Figure 8, with point (x,y) on trace of projection of intersection of planes through $x = g(T)$ and $y = g(T)$ on X,Y plane.

measurement of an animal, plotted against time (see, for example, Merril, 1931). The most generally acceptable curve seems to be a sigmoid curve, such as the Gompertz or logistic. We shall assume, for purposes of exposition, that a given morphological measurement increases in size as a sigmoid curve and shall use the Gompertz form. The relative merits of the logistic versus the Gompertz are not important here; nor, for that matter, is the choice of the sigmoid form, other than for illustration.

The Gompertz curve may be expressed as follows:

$$y = a\,e^{-be^{-kt}}, \qquad 0 < t < \infty, \qquad a, b, k > 0.$$

Analysis of the curve:

1. At $t = 0$, $y = ae^{-b}$. The latter is very close to zero in most curves based on biological observations, but the calculated curve actually goes through the

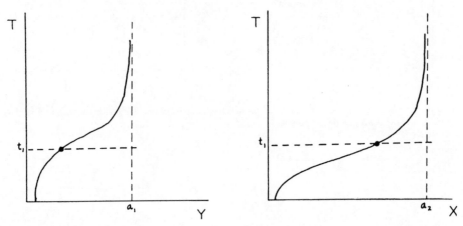

Fig. 10.—First step in formation of a geometric model for development of a point on the X,Y plane, for case of growth curve of Gompertz for y and x plotted at t_1 on a and b, respectively; a_1 and a_2 asymptotes for y and x, respectively.

origin at $t = -\infty$. An unfortunate artificiality is thus introduced, but it is important to understand that the Gompertz function (or any other growth curve) is not a "biological law," but only an empirical curve, a descriptive expression of an array of observations.

2. As $t \to \infty$, $y \to a$. Thus in growth studies a is the asymptotic upper limit of size of the measurement of the animal.

3. The point of inflection is $y = a/e$; $t = \ln b/K$.

Figure 10 illustrates the first step in the formation of a geometric model for the development of a point on the X,Y plane, in which the linear growth curves have been replaced by two Gompertz curves, $y = a_1 e^{-b_1 e^{-k_1 t}}$ and $x = a_2 e^{-b_2 e^{-k_2 t}}$. The resulting function on the X,Y plane takes the following form:

$$x = a_2\,e^{-(b_2/b_1^{k_2/k_1})\,(\ln a_1/y)^{k_2/k_1}}.$$

This curve, traced out on the X,Y plane by $(x,y) = f(t)$, $g(t)$ as the animal grows, may take qualitatively different specific forms, depending on the values of parameters a_1, a_2, b_1, b_2, k_1, k_2. The special forms may be classified as follows:

Class I $(b_1 = b_2,\ k_1 = k_2)$, straight line, $x = \dfrac{a_1}{a_2}(y)$,

Class II $(b_1 = b_2,\ k_1 = k_2)$, simple power function, $x = a_1 \left(\dfrac{y}{a_2}\right)^{b_2/k_1}$,

Class III $(k_1 = k_2)$ giving the form $x = a_2\, e^{-(b_2/b_1^{k_2/k_1})\,(\ln a/y)^{k_2/k_1}}$.

Examination of the literature suggests that Class III curves are prevalent. Some properties of the class III curve are:

Range:

$$0 < y < a_1, \qquad 0 < x < a_2, \qquad a_i b_i k_i > 0\ ;$$

$$k_2 \neq k_1, \qquad a_2 \neq a_1, \qquad b_2 \neq b_1\ ;$$

$$\lim_{y \to a_1} x = a_2, \qquad \lim_{y \to 0} x = 0\ .$$

Rate of change:

$$\frac{dx}{dy} = \frac{b_2 k_2}{b_1^{k_2/k_1}}\, \frac{x}{a_2 y}\left(\ln \frac{a}{y}\right)^{(k_2 - k_1)/k_1}.$$

We can now exhibit the path of a point (x,y) as t varies, on the X,Y plane (Fig. 11). We shall call this curve the "joint ontogenetic path" of an individual animal (expressed, of course, by the two measurements in question). To reproduce the scatter diagram completely, it is necessary to consider a given pair of measurements taken on a number of individuals as illustrated in Figure 12.

To this point we have developed a model which includes the factors underlying a scatter diagram in which the two variables are osteological dimensions. The relationships between the points on the X,Y plane (the scatter diagram) may be analyzed in various ways. Our approach proceeds as follows. We shall suppose that a linear or log-linear equation is representative of the relationships between most pairs of morphological dimensions. This statement is not made for convenience alone. Examination of a large number of scatter diagrams on which pairs of morphological measurements have been plotted has shown that almost all provide a reasonable fit to either a log or a linear function. Thus we conclude that the functional relationship between most variables can be represented adequately by either a linear or a log-linear equation (Olson and Miller, 1951a). This function is represented by a line drawn through the scatter of points. A summary significance is attached to the line. If it represents the points fairly well, the way in which the two proportions will vary with respect to each other is summarized by a line drawn through the scatter of points.

In Figure 13 two lines, one dashed and the other solid, are shown. It is obvious that the dashed line drawn through the scatter of points is not so representative as the solid line. The standard solution to the question of which of the many possible lines that could be sketched through a scatter of points should be used is the *method of least squares* (see, for example, Hoel, 1947, pp. 79, 80). This approach combines the advantage of repeatability with that of providing the line of best fit; in other words, the sums of the squares of the deviations of the points from this line are at a minimum.

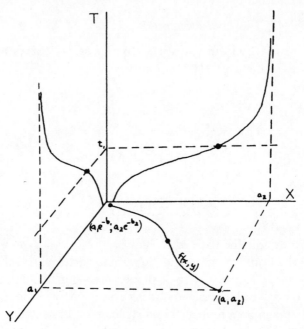

Fig. 11.—Trace of projection of intersections of planes through curves on Y,T and X,T planes (Fig. 10) on X,Y plane, with point x,y plotted at t_1. Based on Gompertz form, $f(x,y)$ is joint ontogenetic path.

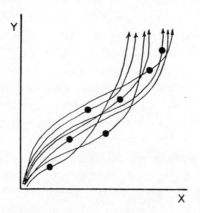

Fig. 12.—Several joint ontogenetic paths based on Gompertz curve for x and y. Points on paths plotted for different values of t.

Were the points on a scatter diagram so distributed that a straight line passed through them all, there would be no need for the least-squares method. However, all degrees of scatter about the least-squares line actually occur, from an almost perfect fit to collections of points so widely dispersed that no demonstrable relationship exists between the two dimensions on the particular scatter diagram. It is both necessary and convenient to take the "degree" of linear association into account. For this purpose the correlation coefficient, r, may be considered as a measure of the degree to which the association between the variables approaches a linear or log-linear relationship (see Weatherburn, 1947, p. 73).

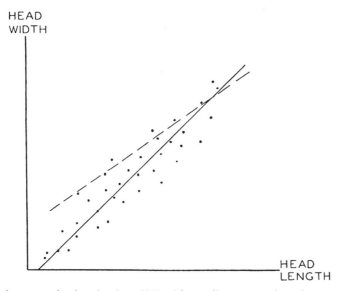

Fig. 13.—A scatter of points in plane X,Y, with two lines drawn through it. Solid line more representative than dashed line.

A straight line is characterized by two parameters, or descriptive properties, namely, the slope and the intercept, which is the point at which the line crosses the co-ordinate. We shall, for the purposes of exposition of the correlation coefficient, confine our attention to the slope of the line of best fit, which will be referred to as the "regression line," in accord with common practice. An additional statement is essential at this time. In computation of the regression line, either y is considered a function of x, or x is considered a function of y. Two lines may thus be computed for each scatter diagram, that of x for a given y and that of y for a given x. The relation between these two lines may be used as a basis for the development of r.[1]

[1] The reader's attention is directed to a different approach to the regression line, in which a single line is developed (see, e.g., Tessier, 1948; Kermak and Haldane, 1950; and Kruskal, 1953). This line appears to be of considerable interest in the treatment of various biological problems but is not pertinent to the present discussion of correlation.

Figure 14 shows the regression lines of y on x and of x on y, based on a scatter diagram of two measures on *Rana pipiens*. It can be seen that the broad scatter of points is reflected in the wide divergence between the two regression lines. Inspection makes it apparent that there is no close linear association expressed between the two dimensions.

Figure 15 shows a high degree of association between measurements of length of pre-maxillary and length of parietal on the individuals of a sample of *Sceloporus olivaceus*. In this case the regression lines approach coincidence. If we use the geometric mean of their slopes as a measure of the divergence between the two regression lines, we then arrive at the desired index of the degree of linear association. The correlation coefficient, r, may be defined as the square root of the product of the slopes (M_y and M_x) of the two regression lines, y on x and x on y (the geometric mean of the regression slopes).[2, 3] Thus $r = \sqrt{M_y M_x}$.

It is important in a consideration of the utility of the correlation coefficient to note that its square has the following useful relationship (see Hoel, 1947, pp. 83, 84):

$$r^2 = \frac{\text{Variance of values estimated by the regression line}}{\text{Variance of the dependent variable}}.$$

Thus r^2 is equal to the percentage of the variance of the dependent variable that is contributed by its relationship to the independent variable. The percentage of variance of the dependent variable that has not been accounted for by the independent variable and is attributed to other factors is represented as $(1 - r^2)$. Thus a correlation coefficient (r_{xy}) of .5 is misleading if it is taken at face value. Actually, only 25 per cent (r_{xy}) of the variance of y has been accounted for by x, whereas $1 - r^2$, or 75 per cent of the variance of y, is assigned to factors other than those represented by x.

[2] Discussion of this definition of r may be found in Fisher (1946), pp. 182–89, or Dixon and Massey (1951), p. 163.

[3] The relationship between the above definition of the correlation coefficient and the conventional form used for computing convenience is as follows:
The slope M_y of y on x is

$$\frac{\Sigma (x - \bar{x})(y - \bar{y})}{\Sigma (x - \bar{x})^2}.$$

The slope of M_x of x on y is

$$\frac{\Sigma (y - \bar{y})(x - \bar{x})}{\Sigma (y - \bar{y})^2}.$$

$$\therefore \sqrt{M_y M_x} = \sqrt{\frac{[\Sigma (x - \bar{x})(y - \bar{y})]^2}{\Sigma (x - \bar{x})^2 \Sigma (y - \bar{y})^2}} = \frac{\Sigma (x - \bar{x})(y - \bar{y})}{\sqrt{\Sigma (x - \bar{x})^2 \Sigma (y - \bar{y})^2}}.$$

Finally,

$$\frac{\Sigma (x - \bar{x})(y - \bar{y})}{\sqrt{\Sigma (x - \bar{x})^2 \Sigma (y - \bar{y})^2}} = \frac{\Sigma xy - \dfrac{\Sigma x \Sigma v}{N}}{\sqrt{(\Sigma x^2 - \bar{x}\Sigma x)(\Sigma y^2 - \bar{y}\Sigma y)}} = r_{xy}.$$

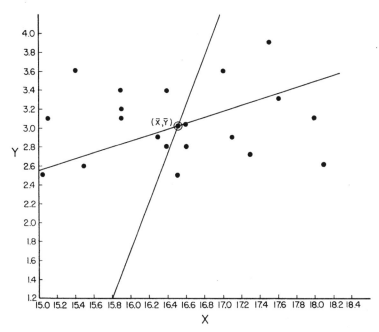

FIG. 14.—Regression line of y on x and x on y based on two measures (x maxillary length and y length of anterior process of squamosal) of *Rana pipiens*. The wide divergence of the lines indicates the broad scatter of points. $r_{xy} = .09$.

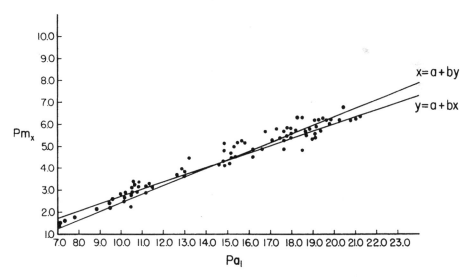

FIG. 15.—Regression lines of y on x and x on y for two measures (x, length of premaxillary; y, length of parietal) of *Sceloporus olivaceus*. The near-coincidence of lines is indicative of the very restricted scatter of points. $r = .998$. (Data from Ernest Lundelius, 1957.)

The sign of the correlation coefficient is dependent upon the direction of the slope of the regression lines. The usual convention in mathematical usage is that (+) is to the right of the abscissa and upward on the ordinate. Thus line ($\overline{oa}$) in Figure 16, which slopes from the origin o in a positive upward direction, has assigned to it a plus value; line ($\overline{oc}$), which slopes downward and to the right, has a negative value. The sense of the slopes of the two regression lines for a given scatter of points will be reflected in the resulting sign of the correlation coefficient. It can readily be seen that a widely divergent pair of regression lines, reflecting a broad scatter or markedly non-linear scatter of points, will give no clear-cut interpretation of sign. In the basic model presented in chapter iii and in much of the work that deals with ontogenetic series in this book, the sign of the correlation coefficient is not taken into account. Use is

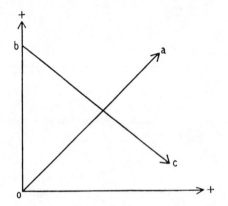

Fig. 16.—Two regression lines, ($\overline{oa}$), which slopes from origin in a positive upward direction and is assigned a positive value, and ($\overline{bc}$), which slopes downward and to the right and is assigned a negative value.

made of the absolute value of r, expressed as $|r|$. Interesting possibilities lie in the utilization of the sign of r as an aid in analysis. Some of these are considered in sections of the book that report studies of samples drawn from single ontogenetic stages.

Negative regression slopes will not occur within the framework of the model presented in this present chapter if (1) a good representation of the points is obtained from all ontogenetic stages and (2) the growth is positive. This holds, no matter what the ontogenetic paths may be. Figure 17 illustrates this point. It can be seen by inspection of this figure that, if points are available in stage I, the earliest ontogenetic stage, the trace of the averages on the X,Y plane must include points in the quadrant adjacent to the origin. If the points are distributed about equally in the various stages, a negative slope for the regression line would not be expected.

The fact is, however, that negative values of r do occur from time to time.

It is necessary to consider briefly the possible meanings of such values. To do this, we direct attention to three aspects of the distribution of points which might be considered pertinent: (1) situations in which the joint ontogenetic path, derived from the distributions, does not pass through, or very near to, the origin; (2) the effects of sampling fluctuations; and (3) negative growth. The pertinent aspects of each of these points are as follows:

1. In many instances the average joint ontogenetic path determined from a scatter of points based on a measurement of a pair of osteological measurements

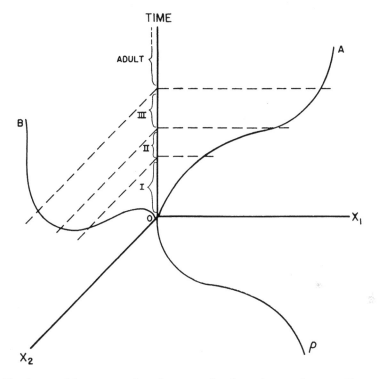

FIG. 17.—Average joint ontogenetic path representing the entire growth range. The time axis is subdivided into hypothetical ontogenetic (growth) stages. Curve OA represents the *average* ontogenetic path for morphological dimension x_1 against time. This average curve is contributed to by numbers of individuals from the various ontogenetic stages. The curve OB represents a similar average ontogenetic path for dimension x_2. The curve OP is the average joint ontogenetic path for x_1 and x_2.

does not pass through the origin ($x = 0, y = 0$). Such a case is shown in Figure 18. This situation can arise under various circumstances with biological materials. In vertebrates, where measurements are made on bones, for example, the initial dimension of the entities measured must be significantly greater than zero in view of the processes of ossification. Readers can readily conceive of many other comparable examples. In such cases, however, even though a marked translation of joint ontogenetic paths may occur during growth as well as change in forms of the joint ontogenetic path, there is no geometric reason to expect negative regression slopes with a resulting negative correlation coefficient.

2. Random sampling fluctuations present different problems. The discussion that follows depends on the assumption that ages at death are not known, as is generally true for paleontological materials. Suppose the variability of the joint ontogenetic paths followed by growing individuals of a species is very large. Then, in a heuristic sense, it is possible that the chance distribution of ages-at-death in our model would result in a completely spurious negative regression and thus a negative value of r. This is illustrated in Figure 19. A high degree of variability and sampling clearly can combine, as illustrated, to produce negative r values, and it seems probable that many that are encoun-

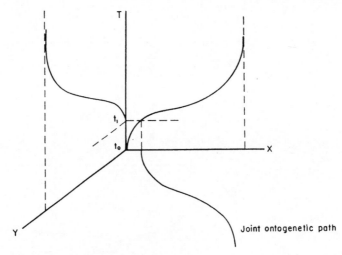

Fig. 18.—The effect of pairing dimensions which have not started growth at the same time. Measurement X started growth at time t_0, while measurement Y started growth at time t_1. A translation of the joint ontogenetic path is the result.

tered fall into this category. We are not at present able to define the limits of the degree of variability of joint ontogenetic paths observed within species or smaller taxonomic units. It is reasonable to suppose, however, that the joint ontogenetic paths will be more variable for some pairs of dimensions of a species than for others. Under given circumstances of sampling, the former may give negative values of r, whereas the latter do not.

3. Finally, a possible source of negative r values is found in cases where negative growth occurs. Such a pattern of growth is more frequent in soft anatomy than in hard structures but is occasionally encountered in growth of hard parts. Such a case was noted in the growth of skulls of the extinct amphibian *Diplocaulus recurvatus* (Olson, 1953). Here, in late stages of development, the posterior width of the skull between the "horns" decreased in absolute width as the length of the skull, measured along the mid-line, increased.

It can be shown that under certain circumstances relating to slope and inter-cept, such as those in *D. recurvatus*, a negative rate of growth for one or the other of a pair of morphological dimensions will result in a negative joint onto-genetic path and hence in a negative correlation coefficient.

CORRELATION AND ONTOGENETIC STAGES

Samples from three more or less distinct situations are included under the general heading of "ontogenetic stages," since they all have the common prop-erty of being drawn from populations which, at the time of sampling, do not exhibit the property of growth. One of these involves animals in which terminal growth prior to death is characteristic of the ontogenetic processes. For sim-plicity, we may define the period of life in which growth (in the sense of normal

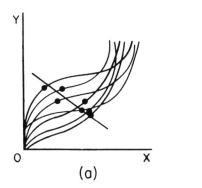
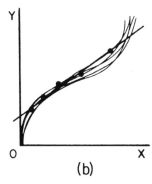

(a)   (b)

FIG. 19.—Two cases illustrating effects of sampling fluctuations upon joint ontogenetic paths of *x* and *y*. (*a*) Points distributed in such a way that they produce a "spurious" negative regression. (*b*) Points distributed to produce a positive regression but one different from that which would obtain were the sampling over a greater range of ontogenetic stages.

and progressive increase in dimensions) does not occur as the *adult* stage. Most mammals and birds are characterized by the existence of such a stage.

A second category comprises structures that do not grow, once the processes of histogenesis that produces the complete morphological entity have termi-nated. The most important examples of this type of structure are found in mammalian teeth, that do not grow after they are fully formed. The indi-viduals in samples of these two categories are not necessarily of the same age, for the criteria that determine their inclusion are basically physiological.

The third category includes samples of individuals of the same age, a tem-poral sample from the ontogenetic series. A sample of fifteen-year-old boys would be an example. The ontogenetic stages of the albino rat considered in chapter vii of this book represent samples of this type that have been used in our own work. In such cases, although physiological "age" may be quite differ-ent, the common property of temporal age lends community to the population sampled. The same type of mathematical treatment can be appropriately applied to each of these types of samples.

Geometric illustration of the type used in the preceding section will be employed here as we now proceed to discuss the use of correlation studies on osteometric properties of ontogenetic stages. For illustration we propose to use adult mammals to avoid the introduction of awkward repetition and phrasing, but examples from the other categories would serve equally well. Adult mammals exhibit a variability for a particular dimension that may be rather large or small. Variability in the height of adult man, for example, is relatively large, whereas variability in the length of one of his molar teeth is relatively small. Some dimensions vary together. A long-legged man usually has long fingers. Others show less correspondence, head width and length of the leg in man, for

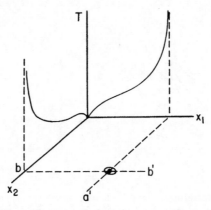

FIG. 20.—The position of point (x_1, x_2) on X_1, X_2 plane when adulthood has been reached (the time of terminal growth). The limit of growth of x_1 is a, and b is the limit of growth of x_2. Point on plane X_1, X_2 at intersections of lines aa' and bb'.

example. In some cases dimensions may show an inverse relationship, as frequently occurs between length of adjacent bones in skulls of vertebrates. These essentially descriptive relationships can be stated analytically, and the degree to which they hold (the degree of association) can be expressed by the coefficient of correlation.

To lay the groundwork for a discussion and analysis of the degree of association between pairs of measures on non-growing structures, with adult mammals used for illustration, we shall start as before with a consideration of the growth of two separate structures plotted against time. The ontogenetic path shown on the T,x plane in Figure 20 is that followed in growth by x_1. As adulthood is reached, the curve approaches the value a on x_1 asymptotically. The value represents size at full growth for the particular osteological measurement x_1. Similarly, b on x_2 represents the limit of growth for the measurement x_2. If the measurements are taken on the animal at a time after full growth is achieved, the result can be expressed by plotting a point on the x_1,x_2 plane at the intersection of the lines aa' and bb', as shown in Figure 20.

Next, a number of individual animals of a given species may be considered, and their ontogenetic paths for the osteological measurement x_1 be drawn, as shown in Figure 21. Added to this diagram is the asymptotic limit for each individual ontogenetic path. It is now possible to illustrate the range and frequency distribution of adult sizes for x_1. These are the values on x_1 at the points where the asymptotes intersect the x_1 axis. The diagram in Figure 21 also includes a construction based on a similar procedure for measuring x_2. We arrive

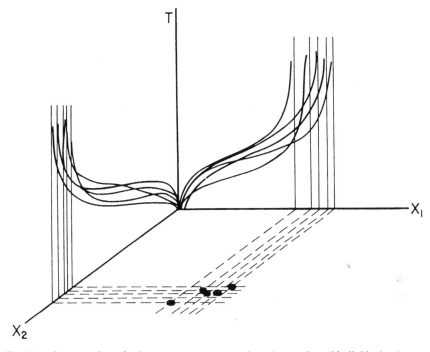

FIG. 21.—Ontogenetic paths for measurements x_1 and x_2 of a number of individuals of a species plus the asymptotic limit for each individual ontogenetic path. Points plotted at intersections of asymptotes of the individual ontogenetic paths on plane X_1,X_2.

at a scatter diagram for x_1,x_2 in which the points are plotted at the intersection of the asymptotes of the individual ontogenetic paths (Fig. 21). The theoretical upper and lower limits (U and L, respectively) for the size of x_1 and x_2 are indicated in Figure 22. If we consider that the height of adult man is the measure x_1, then x_1L would represent the shortest expected adult man, and x_1U would represent the tallest. Similarly, x_2L and x_2U, for another dimension, delimit the theoretical minimum and maximum. In Figure 22 these limits are utilized to mark out the region on the x_1,x_2 plane within which adult measures may be expected to fall. The frequency curves in the figure illustrate an assumed normal distribution of frequencies of adult sizes with few individuals at the upper and lower limits and the greatest number midway between the limits.

Since analysis of scattergrams for two variables automatically removes direct consideration of time in the context of the discussion, we shall dispense with the time axis in the following. In the analysis of various distributions of points in the region shown in Figure 22 in the x_1, x_2 plane, it is assumed that the points form a good fit to a linear or log function. Figure 23 illustrates several possible types of regression lines that may occur. If the general trend of points on the scattergram is from the lower left to the upper right of the adult region, it follows as a rule that when x_1 is small, x_2 is small, and when x_1 is large, x_2 is large. If, on the other hand, the points run almost vertically, x_1 is consistently small, but x_2 is highly variable and may be small or large.

The correlation coefficient in this context is an expression of the strength of such statements. A high negative value indicates a strong inverse relationship, and a high positive value indicates that x_1 and x_2 vary together to a marked degree.

The previous heuristic discussion may be formalized by redefinition of the correlation coefficient in such a way that the relationships described are more clearly shown:

$$r_{xy} = \frac{\sum_{i=1}^{n} (x_i - \bar{x})(y_i - \bar{y})}{\sqrt{\sum_{i=1}^{n} (x_i - \bar{x})^2 \sum_{i=1}^{n} (y_i - \bar{y})^2}}.$$

In words, this equation states that the correlation between x and y is equal to the ratio of the covariance of x and y to the geometric mean of the variances. The covariance is a measure of how x and y vary jointly, and the variance is a measure of how x or y varies independently. If the covariance is approximately the same as the geometric mean of the variances, the two dimensions vary strongly together or vary strongly inversely, depending on the sign of the numerical result.

PARTIAL CORRELATION

As an introduction to the general aspects of partial correlation and our particular use of it in work on morphological integration, we may quote Hoel (1947, p. 120) as follows: "In making statements about the relationship between two variables, it is important to make clear whether it is one that permits the influence of other closely related variables or whether it is one in which the influence of certain of these related variables has been eliminated."

Implicit in our work has been the philosophy that all measurable components of the skeleton interact. No single variable is independent of all others, and, similarly, no given pair of variables interact with each other alone, independently of the rest of the organism. The strength of relationships between any two osteological dimensions—for example, frontal length and parietal length—must, therefore, be interpreted in the light of a pool of underlying factors common to frontal length, to parietal length, and to other dimensions as well.

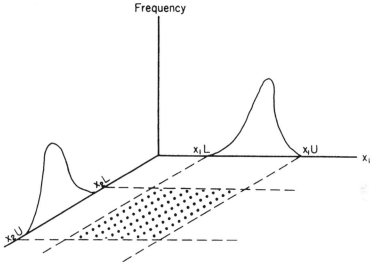

FIG. 22.—Diagram of theoretical upper and lower limits of two measurements on a species, utilized to demark the region on the X_1,X_2 plane within which measures may be expected to fall. X_1L and X_1U represent the lower and upper limits of x_1; X_2L and X_2U represent the lower and upper limits for x_2. Frequency curves based on assumption on normal distribution of x_1 and x_2.

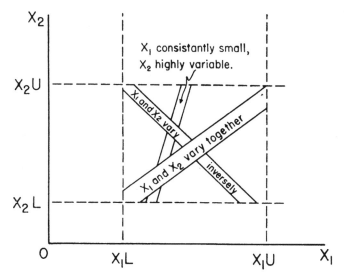

FIG. 23.—Several possible types of regression lines that may occur in cases of a pair of measurements based on adult individuals. Symbols as in Figure 22.

If the complete joint ontogenetic path from very young to full size is considered, the use of partial correlation to remove masking effects of other interacting variables is an intuitively reasonable procedure. The use of partial correlation in the treatment of adult mammal dimensions or, more generally, those of ontogenetic stages requires separate discussion.

The first part of this section is devoted to a general development of partial correlation and a discussion of its application to samples of the full ontogenetic sequence. The second part treats the case of partial correlation applied to samples of ontogenetic stages, with adult mammals used in illustration. The general discussion of partial correlation is based on a derivation by Weatherburn (1947) and on a figure in Hoel (1947, p. 118). The illustrations and the

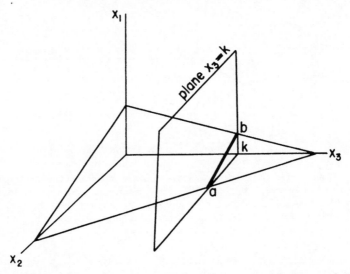

Fig. 24.—Construction in explanation of partial correlation as described in text (see below)

accompanying discussion are intended to present, step by step, the geometric relationships involved in partial correlation in such a way as to be consistent with the development of total correlation (r_{xy}) taken up in the earlier sections of this chapter. The text is written to describe the constructions in Figures 24, 25, and 26 and should be read with direct reference to those figures.

Figure 24: Compute a surface of best fit for x_1 on x_2 and x_3. If the variables are normally distributed, the regression surface will be a plane. Suppose we put a plane at $x_3 = k$, parallel to the x_1x_2 plane. Then the trace of the plane of best fit is the straight line $\overline{ab}$. This line represents the regression of x_1 on x_2 for a fixed x_3. Its slope may be represented by $b_{12.3}$.

Figure 25: The regression plane for x_2 on x_1 and x_3 is computed, as was done for x_1 on x_2 and x_3 in Figure 24. The trace of the plane, with the plane $x_3 = k$ as in Figure 24, is shown by the straight line $\overline{cd}$. This line represents the regression of x_2 on x_1, for $x_3 = k$. Its slope may be represented by $b_{21.3}$.

In the discussion of two-variable correlation earlier in this chapter, the correlation coefficient was described as the geometric mean of the slopes of the two regressions, x_1 on x_2 and x_2 on x_1. As the angle between the two regression lines increased, the correlation coefficient decreased. We may here consider the slopes of the two lines $\overline{ab}$ and $\overline{cd}$ to represent an analogous situation.

Figure 26: The lines $\overline{ab}$ and $\overline{cd}$ represent the regression of x_1 on x_2 for fixed x_3 ($x_3 = k$), and the regression of x_2 on x_1 for a fixed x_3 ($x_3 = k$), respectively. By combination of Figures 24 and 25, the relation between the slopes of the regression lines x_1 on x_2 and ($x_3 = k$), line $\overline{ab}$, and x_2 on x_1 ($x_3 = k$), line $\overline{cd}$, can be illustrated. As the angle between the lines $\overline{ab}$ and $\overline{cd}$ increases, the

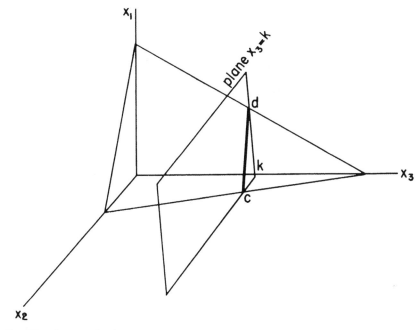

Fig. 25.—Construction in explanation of partial correlation as described in text (see p. 74)

partial correlation between x_1 and x_2 for fixed x_3 decreases. As the angle between the two lines decreases, the two planes approach coincidence, and the partial correlation increases.

We now define the partial correlation coefficient of x_1 and x_2 for fixed x_3 as follows:

$$r_{12.3} = \sqrt{b_{12.3} b_{21.3}}.$$

It can readily be shown (Weatherburn, 1947, p. 251) that this reduces to the familiar form:

$$r_{12.3} = \frac{r_{12} - r_{13} r_{23}}{\sqrt{(1 - r_{13}^2)(1 - r_{23}^2)}}.$$

According to Fisher (1946, p. 190): "If we know that a phenomenon A is not in itself influential in determining certain other phenomena [such as] B, C, D . . . but on the contrary is probably directly influenced by them, then the calculation of the partial correlation of A with B, C, D, . . . in each case eliminating the remaining values, will form a most valuable analysis of the causation of A." On the other hand, "If . . . we choose a group of . . . phenomena with no antecedent knowledge of the causation or absence of causation among them . . . then the calculation of correlation coefficients, total or partial, will not advance us a step towards evaluating the importance of the causes at work."

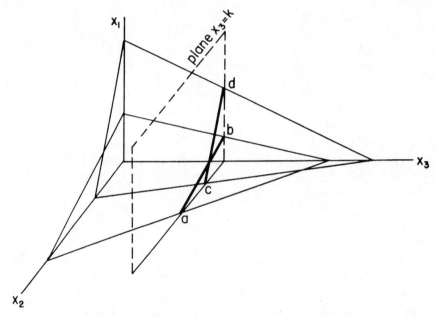

FIG. 26.—Construction in explanation of partial correlation as described in text (see p. 75)

Our own situation does not fit into either of Fisher's two categories. We *do* have some knowledge of the causation underlying osteological dimensions. We are not, however, able to say that a dimension such as head length is not in itself influential in determining other dimensions, such as arm length, length of femur, and length of spinal column; on the contrary, head length is probably directly influenced by these last dimensions.

Our position regarding the use of partial correlation in the specific context of osteometrics is stated in the following paragraph. We shall confine the discussion to three variables in order to be consistent with the previous section, but the principles can be extended to more than three variables with no change in interpretation.

The heuristic derivation of partial correlation which has been presented is based on a comparison of the two multiple regression planes x_1 on x_2 and x_3 and x_2 on x_1 and x_3. The regression planes require further discussion. Ideally, three-variable regression equations involve the estimation of a dependent variable in terms of the two dependent variables. In the present context we are not able to say that any one of the variables is directly dependent on several other independent variables. In place of this statement, we use the premise that all three variables depend upon an underlying set of factors and that, by holding constant one or another of the variables, we may arrive indirectly at two items of information: a better idea of the correlation between two of the dimensions, removing the effect of an underlying factor represented by a third dimension, and an assessment of the importance of this third dimension by noting the change in the correlation of the first two after the influence of this third has been removed.

In our work we have assumed that one underlying factor is over-all increase in size and that another is function. Therefore, instead of indiscriminately computing all permutations of the partials, we have attempted specifically to remove the influence of over-all growth in order better to assess the importance of function. Thus, in a given study, we may compute the partials of a number of pairs of dimensions after removing the effect of such over-all growth dimensions as total body length or total body width. This was done, for example, in studies of a series of *Pentremites*, as taken up in chapter viii. In some instances we have worked with more than three variables by computing the correlation between two after removing the effects of two or more gross size dimensions, such as total length and total width.

In the discussion of partial correlation applied to variables where the sample has been taken over the entire range of growth, the importance of the multiple regression equations has been noted. A rationale has been presented for our particular case, wherein it is recognized that we do not have a dependent variable to be estimated in terms of several independent variables. It is argued that all three variables are affected by an underlying group of factors during growth and that partial correlation will give us an indirect assessment of the external manifestation of certain factors.

To justify the use of partial correlation of adult mammal dimensions in which the sample is taken over the range of fully grown animals only or other cases where samples are from ontogenetic stages, it is necessary to present a biologically reasonable interpretation of multiple regression if we wish to be consistent with the development of partial and total correlation as presented earlier. For example, how shall we interpret the biological meaning of the following statement: Adult dimension x_1 is a function of two other independent adult dimensions? The answer to this question is that we cannot meaningfully interpret such a statement any more than we could when the sample was taken over the entire growth range.

We shall thus refer to the discussion and interpretation of bivariate correlation of adult mammal dimensions presented earlier in this chapter. In this discussion the position was taken that results could best be interpreted in terms of variance and covariance. Thus if covariance, or the way in which the two dimensions varied together, was very large relative to the geometric mean of the variances of the two dimensions, a correlation approaching zero would result. We then inferred that dimension A varied independently of dimension B and that A and B were affected in different ways by underlying factors. To sort out in a crude way the underlying factors, we then proceeded to remove the effect of certain of these by assuming that certain dimensions are most closely related to these factors. This process is best illustrated in the work in this book by its application to molar dentitions of a species of *Aotus* and several species of *Hyopsodus*, reported in chapters vii and viii, respectively. In each of these instances general biological information and the behavior of groupings of measures under simple correlation suggested that over-all size was a dominant factor in the correlations that occurred. This, we felt, was best expressed by such gross dimensions as the total length of the molar series and by maximum length and width of the individual molars. These were held constant, singly and together, in partial correlations in order that their masking effects, representative of those of over-all size, would be removed.

FLUCTUATIONS OF THE SAMPLE CORRELATION COEFFICIENT, r

The sample correlation coefficient, r, is fundamental to the entire structure of investigation discussed in this book. From it the levels of ρ, which form the basis of ρF-groups, are estimated. The structure of the ρ-groups in turn dictates the value of the index of integration, $I\rho$. The basic-pair analysis likewise is built on the estimate of ρ that is given by r.

Discussion in the various chapters on evolutionary sequences and the functional interpretation of ρ-groups begins with likenesses and differences between r values. In what follows we test some of our basic assumptions with respect to r.

ASSUMPTION 1: *Correlations of pairs of dimensions are stable within a natural population, established at the lowest taxonomic level (species or subspecies).*

Clearly, we require that correlation between two measurements must not fluctuate significantly within a homogeneous population if we are to be justified in attaching real meaning to cases in which the correlation differs from one population to the next. To test this assumption, we have repeatedly sampled a breeding population to observe whether the correlation between such a pair of measures as head length and head width changes significantly from one sample to the next or whether, as has been assumed, the r values do not change significantly in sampling of this kind.

The data analyzed were taken from a breeding population of pigeons, as presented and described in Appendix C. To reduce the labor in analysis, which involved five samples of twenty pigeons each, ten of the twenty-six available measures were used in the tests. For each of the five samples the correlation between a given pair of the ten measurements was recorded. This resulted in arrays for correlation between pairs of the included measurements, as shown, for example, by the correlations between measure 2 and measure 6 in Table 6. It is evident from the example in the table that there is a certain amount of fluctuation in the five entries of $r_{2,6}$. We ask the question: Is this fluctuation significantly large; that is, are we justified in assuming that five such different r values could have resulted from sampling fluctuations in a single population, especially in view of the small size of the samples ($n = 20$)?

TABLE 6*

VALUES OF r FOR A PAIR OF MEASURES
FROM FIVE SAMPLES OF
DOMESTIC PIGEONS

Sample No.	$r_{2,6}$
1.	.400
2.	.480
3.	.280
4.	.449
5.	.289

* Measures 2 and 6 used. For identification see Appendix C.

The question is formalized for statistical treatment by erection of the hypothesis that the r values are homogeneous, that they have been drawn from a single population with a fixed value of ρ. Our decision is based on probability considerations. By use of the appropriate χ^2 test (Rao, 1952, pp. 233 ff.), we compute the probability that an array of r values as diverse as the one in the example could have been drawn from a population with a fixed ρ value. In the sets of tests described in this section, the following risk levels have been used consistently:

$\alpha = .05$; that is, the risk of rejecting the hypothesis of homogeneous r values when actually true is no more than 5 in 100.

$\beta = .25$; that is, the risk of accepting the hypothesis of homogeneity of r values when actually false is no more than 1 in 4.

Owing to inherent errors in measurement, including operator variability, we are less concerned about the α error than we are with the β error, since error of measurement would tend to increase the fluctuation of r values.

The results are shown in Table 7. It is quite apparent that we are justified in the assumption that the sampling fluctuations of the correlation between pairs of measurements within this particular adult population are not large. In a strict sense, these results hold good, of course, only for the particular population sampled—the population of pigeons. We feel, however, that it is biologically sound to extrapolate the results for this population to those of other animals where, because of insufficient samples, similar tests cannot be made. All the studies of single populations are based upon this extrapolation, which is certainly to be preferred to the flat assumption of homogeneity usually made and otherwise necessary.

In view of the results of the test of our first basic assumption, we shall now examine the other side of the coin.

ASSUMPTION 2: *The "large" differences encountered in r values in comparisons of several populations for a fixed pair of measurements are not due to sampling fluctuations, such as could have arisen from repeated sampling of a single population, but represent real biologically interpretable changes in degree of association.*

TABLE 7*

HOMOGENEITY OF *r* VALUES IN FIVE SAMPLES FROM A
SINGLE POPULATION OF PIGEONS

Measures	2	4	6	8	10	12	14	16	18	21
2..........		H	H	H	H	H	H	H	H	H
4..........			H	H	H	H	H	H	H	H
6..........				H	H*	H*	H*	H*	H	H
8..........					H	H*	H	H	H	H
10..........						H	H*	H	H	N.H.
12..........							H*	H	H*	H*
14..........								H*	H*	H
16..........									H	H*
18..........										H
21..........										

"H" means that the correlation coefficients are homogeneous at $\alpha = .05$ and $\beta = .25$ levels of significance. "H" means that the correlation coefficients are homogeneous at $\alpha = .05$ but $\beta > .25$. "N.H." means that the correlation coefficients are not homogeneous at $\alpha = .05$. α is the risk of rejecting the hypothesis when it is actually true; β is the risk of accepting the hypothesis when it is actually false.

To test this assumption, we have repeated the procedure outlined in the previous discussion on two evolutionary sequences of *Pentremites* of Mississippian age. The data and discussion of the samples are presented in Appendix F. In this case, the individual entries were arranged so that, for a given pair of measurements, we formed an array of *r* values for each series from each available time level in the evolutionary sequence. The example in Table 8 includes information from each of the two evolutionary lines, godoniform and pyriform, based on measures 3 and 32. The time levels for each line represent separate populations.

Nine measures were taken from the larger collection available. Sample sizes ranged from eighteen to forty. Since the two lines were closely related, the same morphological measurements were used in each case. The arrays of *r* values were tested for homogeneity as in the case of the pigeons. The results of thirty-six such χ^2 tests are given in Table 9, A. Comparisons of Table 9, A and B, show

how strikingly our second assumption has been justified, at least with respect to this particular group of fossil animals. The godoniform (Table 9, A) shows only four homogeneous and three questionable arrays out of a total of thirty-six χ^2 tests. The pyriform (Table 9, B) shows no homogeneous arrays and two that are questionable in the same number of tests.

TABLE 8*

VALUES OF *r* FOR PAIRS OF MEASURES IN
SIX SAMPLES OF *Pentremites*

GODONIFORM LINE		PYRIFORM LINE	
Time Level	$r_{3,32}$	Time Level	$r_{27,32}$
Glen Dean..........	.007		
Golconda...........	.281	Golconda...........	.663
Paint Creek.........	.796	Paint Creek.........	.908
Renault............	.027	Renault............	.928

* Samples from different time levels and in the two lines considered to have been drawn from different biological populations. Measures 3 and 32 used for godoniform line, and 27 and 32 for pyriform line. For explanation of measures see Table 20, p. 105.

TABLE 9*

HOMOGENEITY OF *r* VALUES IN SIX SAMPLES OF *Pentremites*

A. GODONIFORM LINE

	1	3	5	9	14	22	27	31	32
1......		N.H.	N.H.	N.H.	N.H.	N.H.	H	N.H.	N.H.
3......			N.H.	N.H.	N.H.	N.H.	H*	N.H.	N.H.
5......				N.H.	N.H.	N.H.	N.H.	N.H.	N.H.
9......					H	N.H.	N.H.	H*	N.H.
14......						N.H.	H	H*	N.H.
22......							H	N.H.	N.H.
27......								N.H.	N.H.
31......									N.H.
32......									

* See Table 7 for key to symbols.

B. PYRIFORM LINE

	1	3	5	9	14	22	27	31	32
1......		N.H.	N.H.	N.H.	N.H.	N.H.	H*	N.H.	N.H.
3......			N.H.	N.H.	N.H.	N.H.	N.H.	N.H.	N.H.
5......				N.H.	N.H.	N.H.	N.H.	N.H.	N.H.
9......					N.H.	N.H.	N.H.	N.H.	N.H.
14......						N.H.	N.H.	N.H.	N.H.
22......							N.H.	N.H.	N.H.
27......								H*	N.H.
31......									N.H.
32......									

A third study of the fluctuation of r values was made on a series of onto-genetic stages of an inbred stock of the albino rat. The data were taken from materials recorded and described in Appendix G. Ten measurements were selected from the thirty-four taken on each individual rat, to provide the r values used in the χ^2 tests. The measurements were taken on samples of twenty individuals, one sample from each of the following ages: one day, ten days, twenty days, forty days, and adult (250 days).

Interest here lies in the relative stability of r values over the period of growth and development represented by the samples. The procedure involved examination of the r values of given pairs of measurements at the successive ontogenetic stages. An example of an array, using measures 11 and 24, is given in Table 10.

TABLE 10*

VALUES OF r FOR PAIRS OF MEASURES
IN FIVE SAMPLES OF ALBINO RATS

Ontogenetic Stage	$r_{11,24}$
One-day	.097
Ten-day	.172
Twenty-day	.719
Forty-day	.705
Adult	.885

* Measures used are 11 and 24. For explanation of measures see Table 31, p. 141.

The first step in the analysis consisted of a test of the hypothesis of homogeneity for the forty-five r values over the whole ontogenetic sequence. The results are given in Table 11, A, which shows no clear-cut tendency toward either homogeneity or lack of it. It appeared, however, that the ten-day stage, in many cases, was the principal contributor where non-homogeneity was encountered. Table 11, B, shows the results of recomputation of the χ^2 values with the effect of the ten-day stage removed. A clear reduction in the number of non-homogeneous arrays of r values results.

A final test for homogeneity was made on arrays of r values that ranged over only the later periods, from twenty days through the adult stage, with the effects of the one- and ten-day stages removed. Table 11, C, shows that all but two of the arrays that were non-homogeneous in the earlier tests are either homogeneous or questionable when only the later ontogenetic stages are considered. It thus appears, within the limitations of the experiment, that the early ontogenetic stages contribute importantly to the fluctuations in r values within a population.

We do not feel that any obvious conclusions can be reached from this last study. It does appear, however, that these and the results of the first two tests are consistent with the general equation set up and discussed in chapter ix:

$$\rho = \alpha \text{ (function)} + \beta \text{ (growth)} + \gamma \text{ (residual)}.$$

In the context of this discussion we may paraphrase this equation as follows: The coefficient γ is small and is dependent largely upon the ontogenetic range over which the measurements have been taken. This will be expanded and discussed in chapter ix.

TABLE 11*

HOMOGENEITY OF *r* VALUES IN FIVE SAMPLES OF ALBINO RATS

A. ALL FIVE STAGES INCLUDED

	1	4	8	11	14	18	24	27	29	34
1		N.H.	H	N.H.	N.H.	H*	N.H.	H	H*	N.H.
4			H*	H	H	H	N.H.	H*	H*	H*
8				H*	H*	H	H*	H*	H	H*
11					H	N.H.	N.H.	H	H	N.H.
14						H*	N.H.	N.H.	N.H.	N.H.
18							N.H.	H*	H	H*
24								H	H*	N.H.
27									H*	H
29										H
34										

* For explanation of symbols see Table 7.

B. TEN-DAY ONTOGENETIC STAGE REMOVED

	1	4	8	11	14	18	24	27	29	34
1		H*	H	N.H.	H	H*	H*	H	H*	N.H.
4			H*	H	H	H	N.H.	H*	H	H
8				H	H	H	H*	H*	H	H*
11					H	N.H.	N.H.	H	H	H
14						H*	H	H	H	H*
18							N.H.	H*	H	H*
24								H	H*	N.H.
27									H*	H
29										H
34										

C. ONE- AND TEN-DAY ONTOGENETIC STAGES REMOVED

	1	4	8	11	14	18	24	27	29	34
1		H*	H	H	H*	H*	H	H*	H*	H*
4			H	H	H	H	H*	H*	H	H
8				H	H	H	H*	H*	H*	H
11					H	H	H*	H*	H	H
14						H*	H	H*	H	H*
18							H	H*	H*	H
24								H*	H*	N.H.
27									N.H.	H
29										H
34										

CHAPTER V

The Qualitative Aspects of F-Groups

The general concept that measurements taken upon a part of the body that has some biological unity bear meaningful relationships to each other seems to us to have an impelling intuitive rightness. The organism, however, is the product of processes dependent upon the actions of a co-ordinated genetic system and the environment in which the animal develops and lives. During development there occur continual adaptive responses, both to internal and to external factors, limited and directed by genetic constitution. Individual adaptation is essentially a compromise between a perfection of function at each developmental stage and the potentialities of the genetic constitution. Modifications beyond genetically imposed limits cannot occur, and genetic change between parent and offspring cannot be so great that the developing offspring finds the environment to which it is committed intolerable. In less drastic genetic modifications, however, is found the source of change of the limits of developmental potentialities. Through selection, changes that are favorable in the environment of development and adult life are fixed and are, to a greater or lesser degree, reflected in morphological features of particular systems and, with the spread of genetic effects, in the relationships between systems.

F-groups, which represent the systems, provide excellent materials for studies of the details of evolution, for they portray the results of the complex processes of morphogenesis. As F-groups are compared among members of different species, they give information about similarities and differences related to particular functional and adaptive aspects of the organisms. As the comparisons are extended to many systems and information upon interrelationships of the various systems is added, a significant step toward the ideal of analysis based on the total organism is taken.

The constitution of an F-group, with respect to the measurements included, can be ascertained either by qualitative or by quantitative methods. It has been necessary in preliminary phases of our work to depend upon qualitatively formed F-groups to test the proposition that ρ-groups $\Rightarrow$ F-groups. The remainder of this chapter is devoted to this end, first, in a theoretical consideration of the problems and, second, through illustration of the qualitative development of some of the F-groups that provide the basis for the analysis of the experiments presented in chapter vi.

BIOLOGICAL PROBLEMS

If the definition of an F-group as a subset of all measures on a species related by some unifying biological factors is to fit the concept of morphological integration, which involves quantitative assessment of relationships, the unity dependent upon the biological factor must be the unity that is expressed mathematically. The number of possible biological factors that might result in association is large—how large no one can say at the present time with any hope of accuracy. At a specified level of integration, only one F-group can fulfil the requirements given above, and, of course, it may be that none does. It is surely pertinent to ask what chance there can be of selecting qualitatively *the* proper factor and including only the measures appropriate for the level of intensity set quantitatively. Any answer must first take cognizance of what is measured, the amount of detail that is expressed in the measurements, and the objectives of the study, which will in large part direct the selection of factors. In the present studies, by virtue of their concern with the measurements available on fossil materials, measurement is confined to hard parts, and information is obtained only about the skeleton and that part of the soft anatomy that can be inferred from the skeleton. Within this context we erect the following equation, which we believe encompasses all essential features required for analysis:

Morphological integration = α (function) $+ \beta$ (development) $+ \gamma$ (residual) .

In the present context—the consideration of an F-group—the term F can be substituted for "Morphological integration" in the equation. If it be assumed that γ is small, the factors to be considered in the formation of F-groups are α (function) and β (development). These, of course, are extremely gross, and both are dependent upon a suite of biological processes which might in themselves be considered critical centers of focus within some other framework. We have preferred in our exploratory studies, however, to assume that these two are the essentials to be considered in the formation of F-groups and to search out the underlying processes in light of information gained through analysis of function and development.

The problem of the individual F-group is part of the broader problem of the relationships of patterns of integration within a system and between systems and their relationships to the total integrative pattern of the organism. It seems inconceivable that a set of, say, fifty measures on a skeleton can be neatly arrayed in discrete groups without overlaps between groups and without internal differences in intensities of integration within F-groups. Even were "intragroup" intensities of integration reasonably consistent, which seems doubtful, there still would remain the problem of placement of "intergroup measures." Simple, precise solutions cannot be expected for groups formed either qualitatively or quantitatively. The studies of quantitative groups in later chapters will show that any attempts at simple ordering are doomed to at least partial failure. The fact is that the situation is just not that clear-cut.

A general basis for assessment of intensity of integration from a qualitative point of view is essential, in spite of the evident difficulties of precision. Within the limits imposed by the use of skeletal measures, it has seemed to us logical to suppose that measures of the elements that participate in some intricate function, in ontogeny and in the adult, would exhibit a high degree of morphological integration. This proposition was used as a guide in the measurement and formation of F-groups in the study of *Rana pipiens*, as detailed later in this chapter. The systems measured were all involved in dynamic functions in which actions were directed to cope with circumstances provided by the external environment. In addition to measures within systems, some measurements that crossed systems were made, as well as a gross measure—head length—to serve as a guide to the over-all size of the animals.

By no means all the skeletal elements of vertebrates are involved intimately in dynamic functions. Many parts of the skull in most tetrapods, for example, play no explicitly dynamic role. The bones that encase the brain fall into this category, although some are to a degree related to dynamic functions as they serve for insertions and origins of axial, masticatory, and appendicular muscles. Predominantly, however, the bones of the brain case play a static role. We may raise the question of the degree of integration that such elements possess, whether there is a nice mutual adjustment, as assumed for structures involved in dynamic functions, and whether whatever integration may exist is due to functions or to some other dominant factor. In all structures, but especially in skulls, where elements lie close together, mere proximity may have a strong effect upon relative changes of dimensions during growth. This is evident in skull bones that grow in contact with each other. Is there, then, an integration that depends primarily on proximity, and how does it compare, if present, with integration between other structures less closely associated but participating in a common function? Another series of questions can be raised about the integration of the components of systems that attain mature expression relatively early in ontogeny, such as the brain and cranial nerves. It seems reasonable to think that a high degree of integration exists at this stage of development; whether or not this persists—assuming that it was present—during later ontogeny, as more intimate contacts with other structures takes place, remains an open question. Morphological integration is certainly related in part to a factor of gross size relationships of the parts of single systems and parts of diverse systems, for most structures tend to contribute to total body size by increasing in dimensions during ontogeny. It is of interest to determine whether this influence is dominant, overshadowing the effects of function and other possible factors.

A second study was designed to seek answers to some of these problems. The skull of *Sciurus niger* was used, for it offered the opportunity to construct groups whose nature was reasonably well understood. The procedure involved tests of efficacy of ρ-groups as predictors of the F-groups that were established. The construction of the F-groups for this species is presented later in this chapter, and the tests are described in chapter vi.

It will be noted that in both the studies the questions asked refer to "factors" that are gross and related either to functions or growth. The approaches used in the formation of groups were those of anatomical and functional studies. They do not, by any stretch of the imagination, get to the heart of the problems of more intricate factors involved in the analysis of morphogenesis. The need for full information to provide F-groups in which the grouping was reliable dictated this policy; F-groups that were based on guesswork were not admissible. It is our opinion that validation at this rather gross level provides a basis for the inference that similar relationships hold where more delicate and intricate structures and factors are involved.

Many biological problems were necessarily left for study by strictly quantitative methods, pending the outcome of tests at the levels noted. Prominent among such problems, as they related to the skeleton, are those that pertain to integrative systems that develop without performing a dynamic function during growth but, when fully developed, are intimately engaged in such a function. The dentition in many mammals, where growth of the teeth does not occur after full formation, provides excellent examples of this type of system. Mammalian teeth have proved of such great value in studies of evolution that an understanding of their integrative characteristics is of paramount importance. More generally, the type of system that they represent offers somewhat more direct evidence of the part that genetics may play in the development of integration, since the important factor of function during development is absent. The futility of efforts to establish F-groups by qualitative means for such systems is too evident to need special comment.

Another large suite of problems relates to the ontogenetic development of F-groups. Do the same groups exist at various ontogenetic stages; does the intensity of integration increase or decrease in successive stages, or is there no general rule about changes; how do different types of systems differ with respect to their ontogenetic history; and what differences occur between results obtained when a major part of an ontogenetic system contributes to the sample and when the sample includes only one ontogenetic stage? These and comparable problems cannot be adequately answered by qualitative assessment. They are taken up from a quantitative point of view later in this book.

In the course of studies of *R. pipiens* and *S. niger*, it was found that a notation, more detailed than that used in the original study, was an aid in quick interpretation of the meaning of F-groups. F' is used to show that a sample rather than a population is being considered. The nature of a particular F-group will be indicated hereafter by a lower-case letter. A group that owes it existence predominantly to some function, for example, will be indicated as either "Ff" or "F′f." If a size relationship independent of any other recognizable dominant influence is involved, the designation "Fs" or "F′s" will be used, and so forth. The fact that there is a multiplicity of factors, and that many undoubtedly have not been recognized, makes it impossible, at present, to establish a system of reference in which the letters have specific meaning in all experiments. For the time being at least, the designations will apply only to the particular case under consideration. Finally, where necessary, a descriptive subscript will be used to indicate something about the nature or location of the group. Thus a functional group that involves mastication will be designated as "Ff_m" or "$F′f_m$" and a forelimb group as "Ff_{fl}" or "$F′f_{fl}$." The subscripts will be given a specific meaning only for the sample under consideration and will not necessarily be carried over into samples of other populations.

EXAMPLES OF QUALITATIVE FORMATION OF F-GROUPS

Empirical studies of the proposition that ρ-groups imply F-groups necessitated the formation of F-groups by qualitative means. A reversal of this process was later carried out, in the study of the fossil blastoid *Pentremites*, to check the accuracy of F-groups formed by quantitative means by estimation of the reasonableness of the results obtained. The species upon which these formulations were made are considered below in some detail, since the methods of formulation of the F-groups and the reasoning that led to the final results are most important in evaluation of the validity of the tests presented in the next chapter.

Rana pipiens

A sample of twenty adult and subadult individuals drawn from a breeding population constitute the materials used in this study. Fifty measurements were used to cover the major dynamic, functional systems of the body which involve the skeleton. The measurements are shown in Figures 27 and 28 and described in Table 12. Each measurement was repeatable with an error no greater than ± 0.05 mm. The measurements furnished the raw data for calculations of the correlation coefficients as used in chapter vi, but our present interest is in the way that they were arrayed into F-groups.

Before the measurements were taken, a series of observations was made to aid in development of a meaningful network. Five specimens of *R. pipiens* were dissected with particular attention to the relationships of the musculature and the osseous skeleton. Areas of origins and insertions of muscles were studied to determine which could be accurately measured on prepared skeletons. The functional behavior of *R. pipiens* was then studied in a sample grown in the laboratory and in individuals in their natural habitat. From these observations on anatomy and function, it was possible to arrive at satisfactory conclusions about the significance of many of the possible measures, and from this series fifty measures were selected to represent several functional systems of the body.

Each of the F-groups in *R. pipiens*, as represented by the measures, is considered to be a group in which the integration is dominated by function. Each,

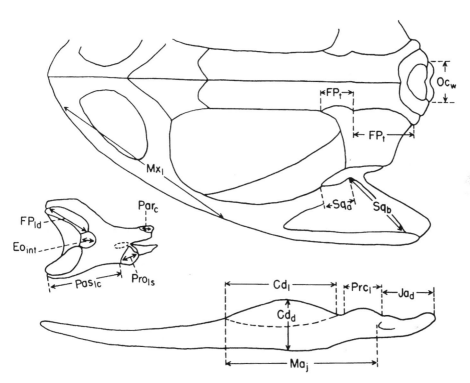

Fig. 27.—Measurements of the skull and jaws of *Rana pipiens* used in studies throughout this book. (For explanations of measurements see Table 12.)

of course, is represented only by a sample of the total possible measures. The following F'f-groups were determined (for abbreviations, see Figs. 27 and 28):

1. F'f$_j$ (JAW MOVEMENT F-GROUP)

The measures in this group were taken on structures that are related to movements of the jaws. These movements in *R. pipiens* are primarily associated with feeding. The jaws and teeth play a role that is somewhat secondary to that of the tongue or, more precisely, the hyoid musculature. Opening of the mouth serves, in essence, to provide an avenue for entrance of food into the body, and teeth and jaws are used, for the most part, to orient and hold large particles. These structures are not highly adapted to special feeding processes, as is the case in many vertebrates, and there would be little, if any, selective advantage in high integration of the parts involved in the composition of the jaws and in jaw movements. It is to be expected that quantitative studies would show a morphological integration lower than that in some of the other systems noted below.

All parts of the body are, of course, involved to some degree in food procurement. The hyoid structures, which are intimately involved, could not

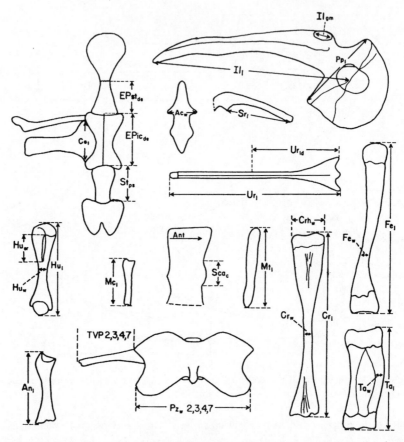

FIG. 28.—Measurements of the post-cranial skeleton of *Rana pipiens* used in studies throughout this book. (For explanations of measurements see Table 12.)

be measured accurately on the prepared specimens, and thus do not figure in the formation of the F'f$_j$-group. Once the animal is near some item of food, the forelimbs and structures related to head orientation play a role in procurement. Somewhat less directly related but still of some importance are structures of the axial system. The forelimbs, in addition to their part in orientation, are used on occasion to aid in altering the position of large particles grasped in the teeth and to dislodge oversize and excess fragments. Functions of procurement and handling of food are relatively crudely performed, and there is no reason to

TABLE 12

MEASURES OF *Rana pipiens*

Abbr.	Description
Mx_l	Chord length from suture of maxillary and jugal, at ventral skull margin to maxillary-premaxillary suture
Oc_w	Maximum distance between lateral margins of the occipital condyles
$PZ4w$	Maximum distance normal to the mid-line of the vertebral column between lateral margins of posterior zygapophyses of vertebra 4; site of attachment of *M. longissimus dorsi* and poorly differentiated *M. intercrurales*
Ur_l	Maximum length of the urostyle; origin of *Mm. coccygeo-iliacus, coccygeo-sacralis* and *longissimus dorsi* (part); the muscles serve to fix and adjust the position of the urostyle and pelvis and urostyle and trunk
Pp_l	Length of pelvic plate from *Spina pelvis* anterior to *Spina pelvis* posterior; a measurement expressing one dimension of the area of origin of *Mm. triceps femoris* (part), *sartorius, adductor longus, adductor magnus, gracilis major, gracilis minor, pectineus,* and *rectus abdonis;* areas of origin of individual muscles cannot be made out on the plate
Il_l	Length of ilium from anterior tip to suture in center of acetabulum; directly involved in locomotor function of hind limb as supporting structure between trunk and hind limb
Cr_l	Maximum length of crus (tibio-fibula)
Cr_w	Minimum width of crus
Ta_l	Maximum length of anterior tarsal element (tibiale)
Ta_w	Minimum width of anterior tarsal element
Fe_l	Maximum length of femur
Fe_w	Minimum width of femur
Mt_l	Maximum length of 4th metatarsal
Co_l	Maximum length of coracoid, measured parallel to axis
Hu_l	Maximum length of the humerus
Hu_w	Minimum width of the shaft of the humerus
An_l	Maximum length of the antebrachium (radio-ulna)
An_w	Minimum width of the shaft of the antebrachium. See An_l
Mc_l	Maximum length of 2d metacarpal
Ac_w	Distance between the dorsal rims of the acetabuli, measured normal to the axial plane
Sr_l	Length of the sacral rib measured along the posterior edge; site of origin of *M. ilio-lumbaris;* site of insertion of *M. coccygeo-sacralis*
FP_{pt}	Length of origin of *M. pterygoideus* on fronto-parietal
FP_t	Length of area of origin of *M. temporalis* on the fronto-parietal
Sq_a	Length of the anterior process of the squamosal; the site of origin of *M. masseter minor*
Sq_d	Length of the descending process of the squamosal; site of origin of part of *M. masseter major*
Ma_j	Length of area of insertion of *Mm. masseter major* and *minor* on the lower jaw; the single areas of insertion of the two muscles cannot be differentiated on the bone
Cd_l	Length of coronoid process, the place of insertion of *M. temporalis*
Pre_l	Length of pre-articular prominence on the lower jaw; used for an estimation of the strength of insertion of *M. pterygoideus*

TABLE 12 (*Continued*)

Abbr.	Description
Ja_d	Distance from posterior end of the jaw to the posterior end of the pre-articular prominence; this measure gives a general estimate of the leverage involved in opening the jaw and thus is related to the function of *M. depressor mandibuli*
FP_{ld}	Width of area of insertion of the *M. longissimus dorsi* on the fronto-parietal
EO_{int}	Width of scar on exoccipital lateral to foramen magnum; this scar marks the area of insertion of *M. intertransversarius capitis superior*
Pro_{ls}	Width of scar on pro-otic medial to the stapes; this scar marks the area of insertion of *M. levator scapulae superior*
Pas_{lc}	Lateral extent of the parasphenoid wing from mid-line, marking the extent of the insertion of *M. levator scapulae inferior*
Par_c	Width of scar on end of paroccipital process (opisthotic), marking the area of insertion of *M. cucullaris*
Sca_c	Width, along anterior margin of scapula, of prominence that marks the extent of the origin of *M. cucullaris*
$Epic_{pe}$	Length of the mid-line ridge on the cartilaginous epicoracoid; this is the area of origin of *M. pectoralis pars epicoracoidea*
St_{ps}	Length of osseous part of sternum, which marks the place of origin of *M. pectoralis pars sternalis*
$Epst_{de}$	Length of osseous part of episternum, marking extent of origin of *M. deltoideus pars episternalis*
Hu_{cr}	Length of *crista ventralis* of humerus; this crest is the site of insertion of a large series of upper-limb muscles and tendons, including *M. pectoralis* (3 parts), *M. deltoideus pars clavicularis* and *scapularis*, and *M. coracobrachialis brevis*; these muscles function to produce various motions of the humerus; their areas of insertion cannot be distinguished on the bone
TVP3	Length of the transverse process of vertebra 3; the site of origin of *M. intertransversarius 2–3*
TVP2	Length of transverse process of vertebra 2; origin of *M. intertransversarius capitis* and insertion of *M. intertransversarius 2–3*
TVP4	Length of transverse process of vertebra 4; origin of *M. intertransversarius*, including special branch 3–4
TVP7	Length of transverse process of vertebra 7; insertion of intramuscular tendon of trunk muscles, especially *M. longissimus dorsi*
Ur_{ld}	Length of scar of origin of *M. longissimus dorsi* on urostyle
PZ3w	Maximum width normal to the mid-line of the posterior zygapophyses of vertebra 3
PZ2w	Maximum width normal to the mid-line of posterior zygapophyses of vertebra 2
Il_{gm}	Length of scar on superior process of ilium, marking the area of origin of *M. gluteus maximus* (*C. posticum* of *M. triceps femoris*)
Cr_{hw}	Width of head of the crus; area of insertion of various tendons of upper limb and origin of tendons to lower-limb muscles; individual tendons, partly capsular in their attachments, cannot be differentiated on the bone
Cd_d	Depth of coronoid process of lower jaw, marking the area of insertion of *M. temporalis*
PZ7w	Maximum width normal to mid-line of postzygapophyses of vertebra 7; no specific muscle origins or insertions; related to long axial muscles by intramuscular septum and dorsal fascia

expect that the systems that enter into this process in a secondary capacity are highly integrated with the structures that are primarily adapted to it. The measures in $F'f_j$ (jaw movement) with brief notes on their functions are given in Table 13.

There is no reason to suppose that all the measures in Table 13 portray the same degree of adjustment to jaw function or that there is an equal intimacy

TABLE 13

MEASURES IN $F'f_j$ (JAW MOVEMENT) IN *Rana pipiens*

Abbr.	Description
Mx_l	A measure that expresses the length of the maxillary dentition, which is used for grasping and orienting large items of food
Fp_{pt}	Length of the origin of *M. pterygoideus** on the fronto-parietal; this muscle plays a role in maintaining and controlling both lateral and vertical positions of the lower jaw relative to the upper
Fp_t	Length of the area of origin of *M. temporalis* on the fronto-parietal; this muscle is a principal adductor of the lower jaw
Sq_a	Length of the anterior process of the squamosal (tympanic of Ecker, Wiederscheim, and Gaupp), the site of origin of *M. masseter minor;* this muscle functions in adduction of the lower jaw
Sq_d	Length of the descending posterior process of the squamosal (tympanic), the site of origin of part of *M. masseter major*, which functions to close the lower jaw and to move the tympanic annulus
Ma_l	Length of the insertion of *M. masseter major* and *minor* on the lower jaw; the insertions are so intimately associated that they cannot be distinguished on a prepared lower jaw; function of *M. masseter* as given above
Cd_l	Length of coronoid process; the extent of the length of insertion of *M. temporalis*, function as above
Pre_l	Length of pre-articular process of lower jaw; used to estimate the strength of the insertion of *M. pterygoideus* on the lower jaw; the measure is only an approximation, since the total process is not involved in the insertion
Ja_d	Distance from the posterior end of the jaw to the posterior end of the pre-articular prominence; this measure gives an estimate of the leverage involved in opening the jaw and is thus related to the function of *M. depressor mandibuli*
Cd_d	Depth of the coronoid process, marking the dorso-ventral extent of the insertion of *M. temporalis*, function as given above

* Terminology is that used in *Anatomie des Frosches* by Ecker, Wiederscheim, and Gaupp (1896) unless exceptions are noted.

in their interrelationships. They do, however, all act in performance of a single function. Some, of course, are related to other functions as well.

2. $F'f_{ho}$ (HEAD ORIENTATION F-GROUP)

This group consists of measures that are involved in movement and orientation of the head. Orientation is related to several important functions of the animal in relationship to the external environment. Since a series of closely related structures performs the movements necessary to these functions, they may be considered to form a single functional complex. The relationship to feeding has already been mentioned. The senses of sight, hearing, and balance are variously brought into play by positioning of the head in various circumstances. There are mechanical relationships to other parts of the body, particularly to the forelimb and the axis, both in resting postures and in locomotion. Movements of the head are rather limited but quite specific, and a high integration is to be expected. There would also appear to be a high probability of strong overlap of integration with axial, forelimb, and feeding systems.

TABLE 14

MEASURES IN $F'f_{ho}$ (HEAD ORIENTATION) IN *Rana pipiens*

Abbr.	Description
Fp_{ld}	Width of insertion of *M. longissimus dorsi* on fronto-parietal; this muscle is involved in both vertical and lateral head movements
Eo_{int}	Width of scar on exoccipital lateral to foramen magnum; this scar marks the area of insertion of *M. intertransversarius capitis superior;* the muscle retracts and gives lateral movement to the head
Pro_{ls}	Width of scar on pro-otic, medial to the stapes; the area of insertion of *M. levator scapulae superior;* this muscles serves to bring the scapula forward and for various adjustments of head position relative to the axis
Pas_{lc}	Lateral extent of the parasphenoid wing from mid-line, marking the extent of insertion of *M. levator scapulae inferior;* this muscle serves to move the shoulder girdle forward and to adjust the lateral position of the head
Par_e	Width of scar on end of "paroccipital" process (opisthotic); the area of insertion of *M. cucullaris;* the muscle functions to move the shoulder girdle and to pull the head backward and laterally
Sca_e	Width, along anterior margin of scapula, of the prominence that marks the extent of the origin of *M. cucullaris;* function as above

3. $F'f_a$ (AXIAL F-GROUP)

This group is involved in movements and orientation of the axis of the body. There is a definite, although limited, pattern of movement of the axis of *R. pipiens* in the course of its various activities. Differential movements of the osseous parts occur primarily in the pre-sacral area, and it is to this region that the group refers. Movement is largely confined to dorso-ventral flexure, and there is only very limited lateral adjustment. The axial muscles, through their attachments to the vertebrae by fibers, tendons, and intramuscular septa, accomplish the movements. The axial complex is in part integrated by the strong dorsal fascia. For example, it is only by means of this fascia and the intramuscular septa that the medial axial muscles are attached to vertebrae from the sacrum to the fourth post-condylar vertebra. More laterally the intramuscular septa alone perform the attachments.

Axial movement occurs during feeding, in locomotion primarily on land, and in mating behavior. Differentiation of the structures involved from other systems in the body is clear in the posterior parts of the complex, but more anteriorly there is rather intimate relationship both with the system involved in head orientation and with the forelimb. This is well shown, for example, in the continuity of the long medial axial muscles with muscles involved in head movement, in the relationship of *M. latissimus dorsi* and the dorsal fascia, and in the continuation of the dorsal fascia into the fascia of the posterior part of the head. The *M. depressor mandibuli* also arises in the dorsal fascia in close association with the axial muscles. Some of these interrelations are undoubtedly reflected in the morphological integration of the skeletal parts to which the muscles of the different systems are attached. Accurate measurements of the axial system are rather limited in *R. pipiens* and in the present case involve only the vertebrae and gross dimensions of the transverse processes and the ribs. The measurements and their abbreviations are shown in Figure 28, labeled *TVP* and *PZW*. No special discussion is necessary, since relationships are clear in the figure. The parts measured are related both to the intimate contacts of osseous parts of the skeleton (in the posterior zygapophyses *PZw*) and to the axial muscles in these processes and the transverse processes (*TVP*). Over the same area *PZ2w* to *PZ4w* and *TVP* are sites of origin of muscles that insert on the head, but posteriorly the vertebrae are related to the musculature only through the intramuscular septa.

4. F′f_{fl} (FORELIMB LOCOMOTOR F-GROUP)

This is the forelimb group. It consists of measures of structures that have specific functions, intimately related to external environment. A high level of integration presumably exists. Relationships to systems already described have been noted briefly. The primary function is that of locomotion. In slow movements on land the actions of the forelimb and hind limb are closely co-ordinated and rather similar. In other types of locomotion, such as jumping and swimming, the actions are very different. These are the predominant phases of locomotion. The forelimbs are little used in swimming, being held, for the most part, passively pressed to the body. In jumping actions they provide a cradle for the anterior part of the body, both in the maintenance of orientation at the initiation of a jump and, often ineffectively, in absorbing some of the shock of landing. This differentiation of functions between the forelimbs and hind limbs suggests that a different organization exists and that the internal integration of each system would be greater than that between elements of the two systems. One other function of the forelimb is of some importance, that of its use by the male in clasping during mating. There is evident sexual dimorphism in *R. pipiens* related directly to this function. The measures taken were not designed to reflect this dimorphism, and the measurements of the two sexes are pooled in our studies.

The measures in forelimb locomotor group are shown, with their abbreviations, in Figure 28 and listed in Table 15.

A complex system like the forelimb or the hind limb presents special difficulties in decisions on measurements, since it allows almost unlimited assessment of detail. The suite of measures used has been selected to represent only the grosser aspects of this system, in order that the picture may not be unduly confused by the introduction of detailed intrasystem differentiation. In this way, it is believed, a high level of over-all integration is represented; but, in all probability, extremely high integration such as might be expected in the small elements of the wrist and foot has not been recorded.

5. $F'f_{hl}$ (HIND-LIMB LOCOMOTOR F-GROUP)

This group includes measures on structures of the hind limb and pelvic girdle. The hind limb is, of course, very highly specialized for the function of hopping

TABLE 15

MEASURES IN $F'f_{fl}$ (FORELIMB LOCOMOTOR) IN *Rana pipiens*

Abbr.	Description
Co_l	Length of coracoid
Hu_l	Length of humerus
Hu_w	Width of humerus
An_l	Length of antebrachium
An_w	Width of antebrachium
Mc_l	Length of 2d metacarpal
$Epic_{pe}$	Length of mid-line ridge on the cartilaginous epicoracoid; the area of origin of *M. pectoralis pars epicoracoidea*
St_p	Length of osseous part of sternum, which marks the site of origin of *M. pectoralis pars sternalis*
$Epst_{de}$	Length of osseous part of episternum, marking the extent of origin of *M. deltoideus pars episternalis*
Hu_{cr}	Length of *crista ventralis* of the humerus; this crest is the site of insertion of a large series of upper forelimb muscles and tendons, including *M. pectoralis* (3 parts), *M. deltoideus, pars clavicularis* and *scapularis*, and *M. coracobrachialis brevis;* these muscles serve in a complex way to perform various movements of the humerus

and highly adapted to provide a powerful organ of propulsion in both jumping and swimming. Both the limb proper and the pelvis share in this adaptation. The functions are specific, and a high integration of parts is to be expected. There is relationship to the forelimb, as discussed above. Interaction of the hind limb and axial structures occurs through the pelvis, but there is only a general functional relationship, and it is improbable that measures of the two systems are highly related. The pelvis, however, poses two difficulties in measurement. First, it bears an intimate relationship to the two systems, axial and hind limb, and many of its structures appear to participate about equally in the actions of the two systems. Second, the origins of most of the muscles that act in movements of the hind limb are closely grouped around the acetabulum and not only are difficult to separate in dissection but do not leave sufficient evidence on the bone for measurements of their individual origins.

The problem of integration of structures that function in two systems is not one upon which a judgment can be made from qualitative data. Three measures that fall into this category were taken, and the integration of these is considered in the quantitative discussions in chapter vi. These inter-F-group measures are given in Table 16.

Measures of the hind-limb complex, as in the case of the forelimb, include a number that express gross length and width and a few that pertain more directly to special musculature. They are shown in Figure 28 and Table 17.

The difficulties in definitive measures are shown in the composite nature of several that have been used. There exists a question about the propriety of con-

TABLE 16

INTERGROUP MEASURES IN *Rana pipiens*

Abbr.	Description
Ur_1	Maximum length of urostyle; the site of origin of *Mm. coccygeo-iliacus, coccygeo-sacralis*, and *longissimus dorsi* (part); these muscles serve to fix and adjust the position of the urostyle, the pelvis, and the urostyle and trunk, operations important both in the attitudes of the hind-limb complex in locomotion and in the relationships of the trunk to the pelvis
Il_1	Length of ilium from anterior tip to suture in center of acetabulum; directly involved in locomotor function of hind limb and as a supporting structure between the hind limb and the trunk
Sr_1	Length of sacral rib measured along the posterior edge. Site of origin of *M. iliolumbaris*, and of insertion of *M. coccygeo-sacralis*

sidering them as adequate representatives of the integrative pattern of the hind limb. This system is, on the whole, less well represented than are the others described above.

Sciurus niger ruviventer

Measurements taken on a sample of skulls of *Sciurus niger ruviventer*, from the collections of the Chicago Natural History Museum, provide an example of the problems of grouping where dynamic function is not the only dominant element in the relationship of parts of a system. The measures used were selected from a series of 235 made by the junior author on a sample of fifty individuals. These specimens were all from a local population, assuring a high degree of homogeneity. Forty-five measures from the total array were selected as representatives, and N was reduced from fifty to thirty by utilization of only those individuals for whom the suite of forty-five measures was available. The measures are shown in Figures 29 and 30 with their abbreviations and are described in Table 18.

In this instance all measures were taken on skulls and jaws, so that the situation is very different from that in *R. pipiens*, in which the whole skeleton was

TABLE 17

Length and Width Measures of the Hind-Limb Locomotor Group

Abbr.	Description
Fe_l	Length of femur
Fe_w	Width of femur
Cr_l	Length of crus
Cr_w	Width of crus
Ti_l	Length of anterior tarsal element (tibiale)
Ti_w	Width of anterior tarsal element (tibiale)
Mt_l	Length of 4th metatarsal
Pp_l	Length of pelvic plate from *Spina pelvis anterior* to *Spina pelvis posterior;* a measure that expresses one dimension of the origin of *Mm. triceps femoris* (part), *sartorius, adductor longus, adductor magnus, gracilis major, gracilis minor, pectineus,* and *rectus abdonis;* the functions of these muscles are the same, in that they play a role in motions of the hind leg, but very different in the particular motions that they impart
Il_{gm}	Length of scar on superior process of ilium, marking the area of origin of *M. gluteus maximums* (*C. posticum* of *M. triceps femoris*)
Cr_{hw}	Width of the head of the crus; the area of insertion of various tendons of the upper part of the limb and origin of tendons to the lower limb muscles; individual tendons partly capsular in their attachments and lacking demarcations on bone

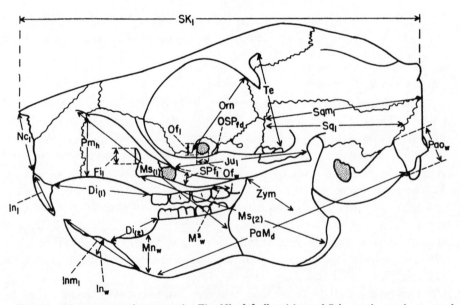

Fig. 29.—Measurements (part; see also Fig. 30) of skull and jaws of *Sciurus niger ruviventer* used in studies. (For explanation of measurements see Table 13.)

covered. Detailed dissection of the head preceded the measurement and served as a basis for selection and interpretation of the measures. Formation of qualitative groups, on the basis of the biological associations of measures, presents greater difficulties than were encountered in *R. pipiens*. In the latter, most measures were on structures that were directly involved in some dynamic function. This is not equally true for measures that adequately represent a mammalian skull, such as that of *Sciurus*. Many of the structures measured maintain a fixed relationship to each other, and relationships are static rather than dynamic. The measures of *Sciurus* directly related to movement of the jaws express a dynamic situation, and this is true as well for some of the meas-

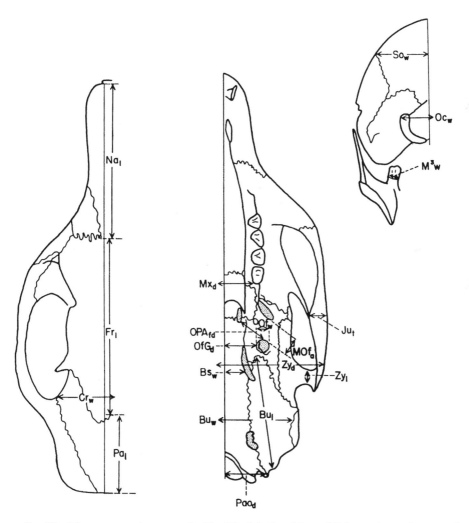

Fig. 30.—Measurements (part; see also Fig. 29) of skull and jaws of *Sciurus niger ruviventer* used in studies. (For explanation of measurements see Table 13.)

TABLE 18

MEASURES OF *Sciurus niger ruviventer*

Abbr.   Description

Oc_w........Greatest width of occipital condyle at right angles to plane of symmetry

Pao_d.......Distance from lower tip of paroccipital process to mid-line at right angles to mid-line

Pao_l.......Length of edge of paroccipital process

So_w........Greatest width of supraoccipital taken normal to mid-line

Bu_l........Length of bulla parallel to mid-line, from ventral aspect, suture to suture

Bu_w.......Greatest width of bulla from suture to suture normal to mid-line

Sq_l........Greatest anterior-posterior length of squamosal plate from suture to suture

Zy_l........Antero-posterior length of zygomatic processes of squamosal taken at point of greatest constriction of zygomatic process of squamosal

Pa_l........Length of parietal along mid-line (including interparietal when present)

Sqm_l.......Distance from supraoccipital-squamosal-parietal to parietal-alisphenoid-squamosal junctions

Fr_l........Antero-posterior length from parietal-frontal at mid-line to nasal-frontal at mid-line

Na_l........Distance from nasal-frontal mid-line junction to nasal-frontal–premaxillary

Or_n........Distance from palatine-orbitosphenoid-frontal to vertex of angle made by horizontal and vertical components of the frontal-parietal suture, just behind the post-orbital process

Ju_l........Length of jugal from posterior extremity of maxilla-jugal suture, to where suture runs normal to plane of symmetry on ventral surface of zygomatic arch (measured on lateral surface of arch)

Ju_t........Greatest thickness of jugal measured approximately normal to length of jugal on arch, normal to greatest width

Pm_h.......Width of premaxillary from point of junction of masseteric ridge with premaxilla–maxilla suture, to point where premaxilla–maxilla suture goes over into flat, ventral position of premaxilla just posterior to incisive foramen

Bs_w........Width of basisphenoid, measured along basisphenoid-basioccipital suture to mid-line (when suture ankylosed, use double width/2 at approximate point of juncture of the two bones)

M_w^3........Width of M^3 measures below crown, normal to mid-line

M_l^3........Length of M^3 normal to width

$Di_{(l)}$.......Length of diastema in upper tooth row

In_l........Length of incisor from angle made by surrounding premaxilla on anterior side, to cutting edge of incisor

Sk_l........Skull length along dorsal mid-line from lambdoidal crest to anterior end of nasal

Pao_w.......Distance between ventral extensions of paroccipital processes

Zy_d........Distance between zygomatic arches, measures in lateral sides at point where squamosal process first appears, on anterior margin

Cr_w........Width of cranial vault at widest point, taken on squamosal-parietal suture

Mx_d.......Distance between M^3, right and left, in the maxilla, at medial posterior edge where tooth enters alveolus

Mn_w.......Width of mandible just posterior to incisors, measured vertically, running through the mental foramen

TABLE 18 (*Continued*)

Abbr.	Description
$Ms_{(1)}$	Distance from anterior edge of masseteric ridge on maxilla to point of inflection of anterior curve of masseteric ridge, on lateral side of mandible (NOTE: All measures between cranium and mandible were taken with upper and lower cheek teeth in occlusion.)
$Ms_{(2)}$	Distance from point of origin of *masseter superficialis* to posterior point of inflection of angular process of lateral side
Zy_m	Distance from approximate center of origin of *zygomaticomandibularis* on ventro-medial edge of zygomatic arch, to approximate center of posterior edge of fossa on mandible
Te	Distance from tip of coronoid process to a point on temporal ridge just posterior to the post-orbital process
Pam_d	Distance from anterior tip of paroccipital process to point of junction of the two sides of mandible posteriorly (anterior end of insertion of *digastricus*)
$Di_{(2)}$	Length of diastema from P_4, anterior edge, to posterior edge of incisor at point it enters bone
Inm_l	Greatest length of incisor measures from anterior insertion to cutting edge
In_w	Width of incisor normal to length
Of_d	Distance from medial edge of foramen ovale to mid-line, normal to mid-line
Orf_w	Greatest width, normal to length, of foramen ovale
OPA_{fd}	Distance from anterior edge of foramen ovale (shelf over foramen ovale) to palatine foramen
Nc_l	Vertical length of nasal cavity
Fi_l	Vertical length of anterior *foramen infraorbitalis*
$OSPf_d$	Distance from nearest edge of optic foramen to sphenopalatine foramen
SPf_l	Length (long axis) of sphenopalatine foramen
Of_l	Long axis of orbital foramen
Of_w	Short axis (normal to long axis) of optic foramen
MOf_d	Distance from masticatory foramen at posterior edge, to anterior edge of foramen ovale

ures in the occipital region, which bear directly upon head movements. The latter system, however, is only partially represented, since measurements of the vertebrae and shoulder girdle were not included. One dynamic complex comparable to those in *Rana* is thus present in the measures—the masticatory complex—which we shall designate as $F'f_m$. The partial group involved in head movement will be designated as $F'f_{ho}$. There is no assurance that this incomplete group includes the most highly integrated structures of the system.

The residue of measures, not in $F'f_m$ or $F'f_{ho}$, cannot be related directly to a specific, dynamic function, although they do play a secondary role in head dynamics. Some measures of this residue express gross dimensions of structures that owe their associations to topographic proximity as elements of a common part of the body—the skull. The consistent relationships of the gross aspects of these structures, manifest in the fairly constant shape of the skull during late ontogenetic stages, arise primarily from their harmonious patterns of growth. Dynamic function, while exerting some influence during later growth stages, cannot be considered a dominant factor in their integration. The suite of measures so related we designate as $F'g_{sk}$ to indicate that co-ordinated growth in topographic proximity is the dominant influence in integration. Growth, to be sure, is a controlling factor of the measures in Ff-groups, but superimposed upon the general growth pattern of topographically associated structures is the effect of the dynamic circumstances in which they exist and grow.

A number of measures in the total array are related to dimensions and positions of cranial foramina through which the cranial nerves pass from the cranium. The central nervous system, to which these nerves relate, is laid down and rather fully developed early in ontogeny. It seems reasonable to assume that integration is established early and that the bones which develop later around the exits of the nerves arise in at least partial conformity with this established pattern. The interrelationships between the bones and their subsequent growth are, however, but little influenced by this pattern. Neither dynamic function nor growth, in the sense designated as Fg, is a dominant factor in establishment of the basic pattern of the cranial nerves. The size and relative positions of the cranial foramina change with increase in skull size. Bone growth, however, is by increments at the sutures and by surface apposition. Resorption and deposition around the foraminal margins may occur during growth, but there is no necessary relationship to changes in shape and size of the surrounding bones. Thus there exists a complicated situation which makes it difficult, if not impossible, to evaluate the validity of a grouping of measures of cranial foramina. These measures have been assigned to a single group, designated as $F'o_n$, with the thought that their relationships to the cranial nerves may provide some degree of association. Because of its uncertain nature, however, this group cannot be used, as can the F'f and F'g groups, for tests of the value of ρ-groups in predicting F-groups in the work taken up in chapter vi.

There is no apparent biological basis for assignment of a few of the forty-five measures of *Sciurus* to any one of the groups cited above. Such measures appear to be more or less closely related to two or more groups. These are called "mixed" measures to indicate this characteristic. The qualitative grouping of the measures of the skull and jaws of *S. niger* is shown in Table 19.

Pentremites

The extinct genus *Pentremites*, a blastoid, serves as a third example of certain of the aspects of morphological integration as witnessed from a qualitative point of view. Six samples of this group, representing two phyletic lines in time, were studied and are taken up in detail in chapter viii with respect to their bearing upon problems of the evolution of morphological integration. The inclusion of *Pentremites* in this section is primarily for illustration of the ways in which biological considerations play a role in interpretation, even when there is no intention to form F-groups qualitatively. In the study of this genus the principal interest was in the change of integrated groups through time. The same measures were taken on the members of each sample, in order that commensurates could be compared. Although some effort was made to assure reasonably complete coverage, the actual composition of groups was not a primary concern so long as the changes were subject to interpretation. Quantitative groups were formed, but no effort was made to arrive at qualitative groups for comparisons. At several stages of this study, however, recourse to general biological relationships that apply in the formation of qualitative F-groups came into play. Since this is necessary in most studies, it is of some importance to consider the implications as illustrated in *Pentremites*.

There are no close modern analogues to *Pentremites*, for there are no living blastoids. The crinoids and echinoids give the best basis for interpretation of

TABLE 19

QUALITATIVE GROUPING OF MEASURES OF *Sciurus niger*

$F'f_m$			$F'f_{ho}$	$F'g_{sk}$		$F'o_n$	"Mixed"
Zya_1	$Di_{(1)}$	$Ms_{(2)}$	Oc_w	So_w	Fr_1	Fi_1	$PaM_d(F'g_{sk}$ and $F'f_m)$
Or_n	In_1	Zym	Pao_d	Bu_1	Na_1	$OSPf_d$	$OfG_d(F'o_n$ and $F'f_m)$
Ju_1	Zy_d	Te	Pao_1	Bu_w	Pm_h	SPf_1	$OPAf_d(F'o_n$ and $F'f_m)$
Ju_t	Mx_d	$Di_{(2)}$	Bs_w	Sq_1	Sk_1	Of_1	$Nc_1(F'g_{sk}$ and $F'f_m)$
M^*_w	Mn_w	Inm_1	Pao_w	Pa_1	Cr_w	Of_w	
M_1	$Ms_{(1)}$	In_w		Sqm_1		MOf_d	

the anatomy and life-processes. Even with these for reference, however, the formation of formal F-groups from purely biological information proved unsatisfactory, for we ourselves could not agree upon a plan of grouping that was satisfactory. This was true in spite of the fact that the pentremites provide anatomical detail to a degree rarely met among the fossil invertebrates. With all the advantages that they have, interpretation of some areas was uncertain. The situation for most fossil groups, particularly among the invertebrates, is much less satisfactory. Nevertheless, biological interpretation inevitably must be made, even though it is not formalized.

The first stage of the work at which biology becomes a consideration is in the decision on the measurements to be used. The objective of measuring the pentremites was to express as completely as possible the details revealed by the calyx, which contains a large portion of the functional systems of the organisms. The selection of measures involved an understanding of the systems so that they could be represented properly and economically. This step was carried out by the junior author and Dr. Keith Chave, who studied the materials in great detail by serial sections and by "dissection" after preparation by a softening agent. The notes compiled during this work depict various systems, their interrelationships, and their functions. It is quite clear that a conceptual grouping of the parts of the animals was involved in the study carried out prior to measuring. An array of measures was then laid out to express what were considered to be the fundamental aspects of the systems, to include what appeared to be differences between the two stocks—pyriform and godoniform —and to represent major features related to the over-all size of the individuals and the maximum dimensions of some of the hard parts related to particular systems.

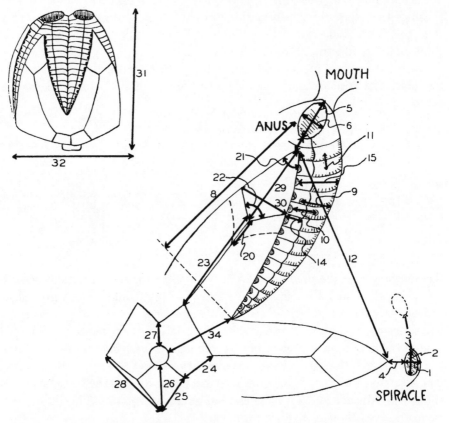

Fig. 31.—Diagrammatic, plan view, of *Pentremites* showing measurements used in various studies. (For explanation of measurements see Table 15.)

A total of thirty-five measures was finally decided upon as the minimum for representation of the fundamental aspects of the organism. Several of these were eliminated as measurement proved them to be not repeatable within allowable error. The residue was then taken and used for calculations. These measures are illustrated in Figure 31 and briefly described in Table 20.

Linear, angular, and count measures were used in this study, in contrast to the use of only linear measures in others. Various arrangements into F-groups are suggested by the measurements. One that seems suitable for the purposes of study of the evolution of groups is as follows: (1) general size and shape measures, (2) basal architecture, (3) respiratory group, (4) digestive group, and (5) deltoid group.

TABLE 20

MEASURES OF *Pentremites*

Measure	Description
1	Outer edge of spiracle to margin of oral opening
2	Width of spiracle
3	Distance between two spiracles
4	Exposed tip of deltoid plate to oral opening along line bisecting spiracle
5	Outer edge of anal opening to margin of oral opening
6	Width of anal opening
7	Exposed tip of deltoid plate to outer edge of anal opening
8	Length of ambulacral groove
9	Length of food groove
10	Length of respiratory groove
11	Distance between adjacent ambulacral ridges
12	Width of ambulacrum measured between tips of adjacent deltoid plates
14	Width of side plate*
15	Width of lancet plates
20	Length of deltoid plate covered by radial plates
21	Angle of deltoid plate made by two sides at apex
22	Angle between sides of radial plates as base of (exposed) deltoid
23	Length of interradial suture
24	Length of radial-basal suture adjacent to azygous plate
25	Length of radial-azygous plate suture
26	Length of azygous plate from center of base to distal apex of azygous
27	Length of interbasal suture
28	Width between proximal corners of radials on azygous basal side
29	Length of exposed portion of deltoid
30	Width of deltoid plate from tips of lateral margins of paired radials
31	Total length of calyx
32	Maximum width of calyx
34	Base of ambulacrum to base of calyx
35	Number of side plates

* Measures in sequence not listed were dropped as inaccurate. Numbers are retained as taken to correspond to diagrams and tables.

Two of these groups are specifically functional because they contribute to the life-processes—groups 3 and 4. These might be expected to show important aspects of change as adaptations to new circumstances were accomplished. The deltoid group (5) is presumably functional, since it is intimately related to various essential elements of the anatomy, but it is difficult to state a single, clear-

cut function. The group representing basal architecture is thought to be important in view of the difference between the two lines, pyriform and godoniform,
in this region. The general size and shape group is important in this respect
as well and also provides major measures for partial correlations in which the
over-all effect of size may be held constant. Part of the measures can be readily
assigned to one or another group, if it is desired to form quantitative F-groups,
but others overlap two or more groups, for example: measure 4—deltoid and
respiratory group; measure 8—size and digestive groups; measure 14—size,
digestive, and respiratory groups. Some measures, such as 34, are difficult to
assign to any of the groups. It is evident, in addition to these difficulties, that
other groupings with about equal merit can be made. Thus, while the general
outlines are clear, detailed assignment and precise formulation of groups cannot
be made with confidence.

Biological information also becomes a primary concern at a later stage, once
quantitative groups have been formed and their meanings sought. At this
stage the F-groups have been derived from ρ-groups, a practice that we considered legitimate only after a basis for assessing relationships had been established, as will be discussed in chapter vi. Thus it is premature to discuss most
aspects of F-group formation at present, but one very important item is well
illustrated by the data on *Pentremites*, that is, the part the original measures
play in the interpretation of the quantitatively formed groups. The effectiveness of the ρ-groups is limited by the measurements and the integrative patterns determined. Interpretations depend basically upon the initial judgments.
As the number of measures over the total animal is increased, the scope of interpretation is enlarged, but in all cases there is a practical limit to the size of
the matrix to be used, both because of the time consumed in measurement and
because of the rapid increase in calculations that results from the addition of
each measure. Initial decisions on measurement, in short, are critical and must
be considered both in the formulation of the problem and in the assessment
of the results.

Relationships of ρ and F

The first section of this chapter is devoted to tests of the capacity of mathematically derived groups of measures (ρ-groups, in which relationships are expressed as bonds formed by correlation) to predict biologically associated groups of measures (F-groups). In later sections attention is directed to the quantitative detection of F-groups by various techniques that have proved valuable in one or another situation. The final part includes discussion and illustration of an index of morphological integration, designed to provide a single number for expressing the total integration of an organism or of one or more systems within the organism. In these sections the theoretical and empirical materials necessary to research into the problems of morphological integration are gathered together. Repetition of some aspects of the earlier chapters is made in various places in order to aid in completeness and full understanding of particular problems. Most of the data used also provide the basis for the work of the final chapters and are presented in this chapter no more fully than is required for clarity. The details of the samples are presented in the Appendixes which may be used as supplements should the reader wish more specific information.

Some of the operations performed are inevitably complex, since large numbers of variables are treated simultaneously. To make sure that we do not lose sight of the basic biological concepts back of the manipulations, we repeat briefly the basis for the various operations taken up in this chapter. All the points considered stem from (1) the evident fact of the morphological stability of the component parts of members of a species, (2) the proposition that the components of systems in a species form a morphological community in which dimensions vary together to a degree that is positively greater than that between components of different systems, and (3) the conclusion that there is thus provided a pattern of morphological integration that can be expressed and studied by means of the coefficient of correlation.

The hierarchy of procedures used later form a unified approach to the problems of morphological integration and are, we believe, based on sound biological theory. All depend ultimately upon the validity of the assumed relationships of groups of measures associated by correlation to groups of measures that have a biological basis for association. The following section is devoted to studies of this relationship.

EMPIRICAL STUDIES OF THE RELATIONSHIPS OF ρ AND F

Rana pipiens and *Sciurus niger* were selected for study of the relationships of ρ and F because it was possible to obtain samples of known composition and to make studies of their anatomy and behavior without serious difficulties. F-groups were formed qualitatively to provide arrays of measures as outlined in chapter v. Simple correlation coefficients were calculated for all measures of the two networks, and ρ-groups were formed at various levels of ρ for each sample. These groups were then compared with the previously formed F-groups by study of the intersects, with the results detailed below. As will be shown, clear-cut results were forthcoming, but it must be noted that in these two cases, as in all that have been studied, solutions have never been simple.

Certain procedures aid in a reduction of the exploratory process. A first problem concerns the level or levels of ρ to be used and the confidence level at which determination is to be made. A solution provides the answer to the problem of what we have termed "the best level of ρ." There is, however, no single answer, for the best level must apply to the specific question whose answer is sought. For example, if we are interested in total integration in the species, it is desirable to set the level as low as possible, for instance, where ρ includes all values of r significantly greater than zero. This varies, of course, with the size of the sample and the confidence level. It may be necessary, if the sample is very small, to take a large risk, say, $P = .1$. With high correlation and/or a large sample, $P = .01$, or even $P = .001$, may be preferable. On the other hand, if interest is centered in ρ-groups that do not intersect, the level selected should be that at which intersects of groups are reduced but large (relatively) groups still exist. In some cases, particularly where partial correlation is involved, this may be the value of ρ (at some confidence level) that includes all values of r significantly greater than zero. Commonly, however, it is at some higher level of ρ. To obtain a first approximation of the appropriate level of ρ, it is useful to scan the array of values of r, to determine whether or not "clusters of measures" exist and to determine the approximate levels at which "clusters" occur. A preliminary ranking by number of bonds at different levels of ρ provides an important basis for estimations of the most effective levels for various purposes, particularly when combined with information about the levels of "clusters." This is accomplished, after scanning, by spacing levels of ρ, regularly or irregularly as seems best, between the highest and lowest ($r \overset{\text{sig}}{>} 0$) levels, and tabulating the number of bonds present between each of the set levels. A small amount of time devoted to this operation immensely shortens the tedious process of forming groups, for it avoids the necessity of search for the best level after groups have been formed and the usual need for repeating group formation at this level.

Once groups have been formed, the problems of how to treat them arise. Seldom have these been solved in our work without considerable fumbling. That this has been steadily reduced as experience has broadened suggests that a formal program of treatment will eventually arise, but this has not been fully accomplished at the present time. When a problem is specific, the way may be fairly clear. If, for example, as in the study of *Rana*, the aim is the definition of ρ-groups with a specific meaning (function), the immediate use of the basic-pair technique is indicated. This practice tends to establish groups that are discrete, if the levels of ρ are appropriate. The basis for a second step—the study of intersects—is thereby established. Without this step the array of groups in *Rana* is so confusing that only the vague outlines of a pattern of functional groups is evident. In other cases, partial correlation may be indicated, especially where there is reason to believe that some overriding factor, such as size, plays a dominant role in the formation of groups. In the study of the dentition of *Hyopsodus* (chaps. vii and viii) partial correlation proved indispensable to the recognition of significant groups of measures. In still another case, that of *Pentremites*, a crude layout of measures and bonds at a series of decreasing levels of ρ provided the most useful means of entering the problem of group relationships.

Once problems of individual samples have been considered, those of comparisons between samples arise if the study is one that involves intersample relationships, as in ontogenetic or evolutionary series, or comparisons of temporally equivalent populations. Again, a trial-and-error method has been necessary in exploratory studies. This type of problem is more pertinent to the studies reported in the following chapters, and consideration is deferred for the time being.

The results obtained for *R. pipiens* and *S. niger* are presented below in brief, without discussion of the preliminary difficulties that were encountered. Relevant data on the samples are kept to a minimum in the text but will be found in greater detail in Appendixes A and B, respectively. Every effort has been made to highlight the important aspects of these studies at the expense of detail, since they are crucial to the concept of the relationship of ρ and F that is used throughout the rest of the book.

Rana pipiens

As noted in chapter v, twenty adult and subadult individuals from a breeding population form the sample of *R. pipiens*. Fifty measurements were taken on each individual to cover the skeleton as fully as possible. The calculations of simple coefficients of correlation resulted in 1,225 values of *r;* and F-groups were formed as shown in Table 21.

These, as shown in the symbols and explanation, are all considered to be functional groups.

From the 1,225 r values, ρ-groups were formed at the following levels: $\rho \geq .97$; $\rho \geq .95$; $\rho \geq .93$; $\rho \geq .91$; and $\rho \geq .86$. The probability level $P = .01$ in a two-tailed distribution was used. Below $\rho \geq .86$, r is not significantly greater than zero at this level with $N = 20$. Basic pairs were determined, and clusters around these pairs were derived. The original ρ-groups and those derived by the use of basic pairs were studied to determine the relationships between ρ- and F'f-groups.

TABLE 21

F-Groups, *Rana pipiens*

F'f$_j$.........Jaw movement group
F'f$_{ho}$........Head orientation and movement group
F'f$_a$.........Axial group
F'f$_{fl}$........Forelimb locomotor group
F'f$_{hl}$........Hind-limb locomotor group

TABLE 22*

Basic-Pair Analysis of *Rana pipiens*

F'f$_{fl}$........ Hu$_l$–Hu$_w$–An$_l$–An$_w$–Epic$_p$–St$_p$–Epic$_{de}$–Hu$_{cr}$
F'f$_{hl}$........ Cr$_l$–Ta$_l$–Fe$_l$–Mt$_l$
F'f$_a$......... $\begin{cases} \text{Pz3}_w\text{–Pz2}_w\text{–Pz7}_w \\ \text{Pz4}_w\text{–Pz3}_w\text{–Pz2}_w \\ \text{TVP3–TVP2–TVP4} \end{cases}$
F'f$_{ho}$........ $\begin{cases} \text{FP}_{ld}\text{–Pas}_{lc}\text{–Par}_c \\ \text{Pas}_{lc}\text{–Par}_c\text{–Sca}_c \end{cases}$
F'f$_j$......... None

* $\rho \geq .91$.

The primary objective of this study required the detection of ρ-groups that were discrete, in order that direct comparison with the single F-group might be made. This somewhat limited aim, of course, resulted in loss of information that is extremely interesting from the broader point of view, which involves the interrelationship of functional systems. Much of the information relevant to interrelationships has been compiled in Tables 22, 23, and 24, and these are the basis for a short commentary later in this section. For the present we shall be concerned with the smaller and less complex groups determined by a basic-pair analysis.

BASIC-PAIR ANALYSIS

The basic pairs, determined as described in chapter iii and used to determine ρ-groups, were as follows:

Ur_l–Il_l (urostyle length–ilial length)

Cr_l–Fe_l (crus length–femoral length)

Hu_l–Hu_w (humerus length–humerus width)

Pas_{lc}–Par_e (parasphenoid wing–insertion *M. cucullaris*)

TVP4–TVP2 (transverse process 4–transverse process 2)

$Pz3_w$–$Pz2_w$ (postzygapophysis 3–postzygapophysis 2)

The study was made at the level $\rho \geq .91$ to assure inclusion of a sufficient number of measures and to avoid increasing the difficulty of analysis by the complexities introduced at the $\rho \geq .86$ level. At this level ($\rho \geq .91$), 133 ρ-groups are formed in the initial analysis before the application of basic pairs. These were distributed with respect to size of the ρ-group as shown in Table 25.

The ρ-groups in Table 25 had many intersects (see Tables 22, 23, and 24), and, although some of the smaller groups contained only measures of a single F'f-group, many included measures of two or more. Intersects were reduced by the use of the basic pairs. Comparisons of the resulting ρ-groups and F'f-groups (chap. v) reveal several items of interest. The most important single fact from the comparisons is that, after basic-pair analysis, no derived ρ-group includes measures that occur in more than a single F'f-group. On the other hand, at the $\rho \geq .91$ level no one ρ-group includes *all* the measures of any single F'f-group. Each of the ρ-groups shows individual features of importance as follows:

1. $F'f_{fl}$ (*forelimb locomotor group*).—Two measures in $F'f_{fl}$ are absent from the single large ρ-group that includes measures of the forelimb locomotor group. This extent of coincidence is the greatest encountered between a ρ-group and an F'f-group in this study. The coracoid length and the length of the second metacarpal are absent. Inasmuch as our studies are dealing with samples of measures, not the total possible suite, it is somewhat meaningless to speculate on reasons for the absence of particular measures at a particular level of integration. The ρ-groups serve to associate some of the measures in F'f. Quite different associations might be encountered if the sample of measures were different. The importance lies not in the specific measures included or excluded but in the fact that only measures in one and only one particular F'f-group are related by correlation.

2. $F'f_{hl}$ (*hind-limb locomotor group*).—A single ρ-group includes four measures that are in $F'f_{hl}$. No measures from other F'f-groups are present. This ρ-group fails to include as high a percentage of measures as was the case for $F'f_{fl}$, and all measures relate to length of elements. It may be that this difference is related to function, but the present state of knowledge renders this highly speculative. At a lower level of ρ ($\rho \geq .86$), some width measures are present, but at this level the forelimb and hind-limb measures merge into a single group, since the basic pairs about which they form are totally bonded.

It will be noted that few pelvic measures were used and that those that were thought to be most intimately involved in locomotor activity do not appear in

TABLE 23*

ρ-Groups, *Rana pipiens*, Intersecting Two F'f-Groups at Successively Lower Levels of ρ

$\rho \geq .97$, no intersections

$\rho \geq .95$

	$F'f_{ho}$	$F'f_{hl}$	$F'f_a$	$F'f_j$
$F'f_{fl}$	None	$\underline{Ta_l}-Hu_l-Hu_w-St_{ps}$	None	None
$F'f_{ho}$		None	None	None
$F'f_{hl}$			None	None
$F'f_u$				None

$\rho \geq .93$

	$F'f_{ho}$	$F'f_{hl}$	$F'f_a$	$F'f_j$
$F'f_{fl}$	$An_l-An_w-\underline{Pas_{lo}}-Epic_{pe}$	$Ta_l-Fe_l-Hu_l-Hu_w-An_w-Epic_{pe}$ $Ta_l-Hu_l-Hu_w-An_l-Epic_{pe}-St_{pc}$	Pairs only	None
$F'f_{ho}$		None	None	None
$F'f_{hl}$			None	None
$F'f_a$				None

* Underscored measures belong to the F'f-group which heads the column in which the measures so designated occur.

TABLE 23 (Continued)

ρ-GROUPS, *Rana pipiens*, INTERSECTING TWO F'f-GROUPS AT SUCCESSIVELY LOWER LEVELS OF ρ

$\rho \geq .92$

	$F'f_{ho}$	$F'f_{hl}$	$F'f_a$	$F'f_j$
$F'f_{fl}$	6 groups of 4 meas. each including: Hu_l, An_l, An_w, $Epic_{pe}$, Pas_{lc}, Sca_c, Par_c	Ta_l–Fe_l–Hu_l–Hu_w–An_l–An_w–$Epic_{pe}$–St_{ps}–$Epic_{de}$	An_l–$Epic_{pe}$–$Pz3_w$–$Pz7_w$ / Hu_l–Hu_w–An_l–An_w–$Epic_{pe}$–St_{ps}–$Epic_{de}$–Hu_{cr}–$Pz7_w$	Pair
$F'f_{ho}$		None	Par_c–Pas_{lc}–Sca_e–$Pz4_w$	None
$F'f_{hl}$			Pairs only	None
$F'f_a$				None

$\rho \geq .91$

	$F'f_{ho}$	$F'f_{hl}$	$F'f_a$	$F'f_j$
$F'f_{fl}$	Hu_l–An_l–$Epic_{pe}$–Pas_{lc}–Par_c and other similar groups	Ta_l–Fe_l–Fe_w–Hu_l–Hu_w–An_w–$Epic_{pe}$–St_{ps}–$Epic_{de}$–Hu_{cr}	Hu_l–Hu_w–An_l–An_w–$Epic_{pe}$–St_{ps}–$Epic_{de}$–Hu_{cr}–$Pz7_w$ / And other smaller groups with $Pz2_w$, $Pz4_w$, Tvp3	2 pairs
$F'f_{ho}$		None	Pas_{lc}–Par_c–Sca_c–$Pz4_w$ / Par_c–Oc_w–$Pz2_w$–$Pz3_w$ / Pas_{lc}–Par_c–Fp_{ld}–$Pz2_w$	2 pairs
$F'f_{hl}$			Pairs only	1 pair
$F'f_a$				None

the ρ-group. The reason that few were used is simply that very few definitive measures are possible on the pelvis of *Rana*. This is a difficulty that is inherent in studies that involve the relationships of measurement and function. At times, as in this case, it is not possible to establish measures that express certain aspects of function. The particular difficulty in the pelvis of *R. pipiens* stems from the fact that the origins of most of the muscles to the hind limb

TABLE 24*

ρ-Groups, *Rana pipiens*, Intersecting Three or More F'f-Groups

F'f-Groups	$\rho \geq .92$	$\rho \geq .91$
F'f$_{fl}$–F'f$_{ho}$–F'f$_{hl}$	None	An$_w$–Pas$_{lc}$–Pas$_c$–Pp$_l$
		An$_w$–Pas$_{lc}$–Epic$_{de}$–Pp$_l$
		Hu$_{cr}$–Sca$_c$–Ta$_l$
F'f$_{fl}$–F'f$_{ho}$–F'f$_a$	An$_l$–Pas$_{lc}$–Pz3$_w$–Pz4$_w$	Hu$_l$–An$_l$–An$_w$–Epic$_{pe}$–Oc$_w$–Pas$_{lc}$–Par$_c$–Pz3$_w$–Pz4$_w$
F'f$_{fl}$–F'f$_{hl}$–F'f$_a$	An$_w$–Pas$_{lc}$–Par$_c$–Pz4$_w$	Epic$_{pe}$–Pas$_{lc}$–Par$_c$–Pz7$_w$
	None	Hu$_l$–Me$_l$–Epic$_{pe}$–Epic$_{de}$–Ta$_l$–TVP3 (plus similar intersects with one measures different)
		Hu$_l$–Ta$_l$–TVP2
F'f$_{ho}$–F'f$_{hl}$–F'f$_a$	None	Oc$_w$–Pas$_{lc}$–Pp$_l$–Pz2$_w$
F'f$_{fl}$–F'f$_{ho}$–F'f$_{hl}$–F'f$_a$	None	An$_w$–Par$_c$–Pp$_l$–Pz4$_w$

* Underscoring (with one, two, or three lines) of measures in the ρ-groups of the second and third columns assigns the measures to the F'f-groups in the first column which are similarly underscored.
No ρ-groups occur at $\rho \geq .93$. None include F'f$_j$.

TABLE 25

Distribution of p-Groups in *Rana pipiens*

No. of Meas. in ρ-Group	No. of ρ-Groups*
2	38
3	35
4	29
5	8
6	14
7	5
8	2
9	1
10	1

* Technically, these are all "non-contained" ρ-groups as defined on p. 53.

are on the pelvic plate, around the acetabulum, and the muscles are so small, so closely spaced, and so largely fibrous in their origins that there is no visible expression of the origin on the bone.

3. *F'f$_a$ (axial group)*.—Four small ρ-groups, three of which include the basic pair Pz2$_w$–Pz3$_w$ (zygapophyses) in the intersects, are formed.. All measures in F'f$_a$ are included in one or another group except TVP7. It is of interest to note that the measure TVP7 (the width of the transverse process in the 7th vertebra) is related to the others, so far as soft tissue is concerned, only by insertion of one of the intramuscular septa. The combination of three of the four ρ-groups into a single, incomplete ρ-group is shown in Figure 32. In this instance, as well as others, a more complete representation of F'f can be obtained by combinations of small ρ-groups with a common basic pair into a large but incomplete

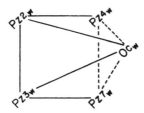

FIG. 32.—Small ρ-groups of *Rana pipiens*, axial ρ F-groups, arrayed for an incomplete ρ-group. Dashed lines indicate bonds below level ρ ≥ .91, but above level ρ ≥ .86.

group. As shown in the figure, the missing bonds are just below the selected level of ρ. This group is complete at the lowest level used, ρ ≥ .86.

The five measures in the incomplete ρ-group include the width of the occipital condyle and the widths of the postzygapophyses of vertebrae 2, 3, 4, and 7. The other ρ-group, a triplet, is made up of measures of the transverse processes of vertebrae 2, 3, and 4. *Musculus intertransversarius* attaches to these three, whereas this is not the case for the process of 7, which is not included. It seems noteworthy that the widths of the zygapophyses and of the transverse processes of the same vertebrae are not intercorrelated. They are sites of origins and insertions of different sets of muscles and appear, probably for this reason, to be somewhat independent in functional relationship.

4. *F'f$_{ho}$ (head orientation group)*.—Four of the six measures in F'f$_{ho}$ occur in two small ρ-groups. Eo$_{int}$, the width of the scar lateral to the exoccipital, which marks the site of the insertion of *M. intertransversarius capitis superior*, and Pro$_{lc}$, the site of insertion of *M. levator scapulae superior* on the pro-otic, are not in the groups. The intimate relationship of the origin and insertion of *M. cucullaris* on the scapula and skull, respectively, is particularly interesting, for it shows a relationship in function that is revealed in the correlations in spite of the fact that the two measures occur on separate parts of the skeleton, each of which is intimately related topographically to other functional groups. The

association of the four measures in the ρ-groups to form a single large, but incomplete, group is shown in Figure 33.

5. *F'f$_j$ (jaw movement group).*—No groups which include measures in F'f$_j$ form at the level of $\rho \geq .91$. This appears to be in accord with the relatively low functional integration originally recognized in this group from observations of behavior. This would not have been apparent from dissection alone, for the majority of measures used appeared to form a nicely integrated series related to a single function. At a lower level of ρ, $\rho \geq .72$ ($r > 0$, $p \overset{sig}{=} 0.1$), triplets, including measures of the F'f$_j$-group, are formed, but no more complete association occurs.

The analysis by basic pairs offers considerable support to the theoretical proposition that the pairs and the measures that cluster around them imply only a single F-group. In this instance measures all fall in functional groups

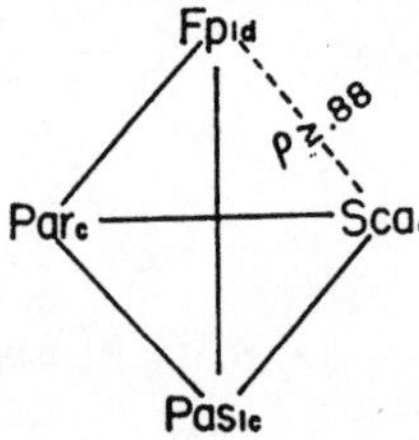

Fig. 33.—Single ρ-group of four measures of head orientation F-group of *Rana pipiens*, at level $\rho \geq .91$. One band present at $\rho \geq .88$, as shown by dashed line.

(Ff) with the exception of the three mixed measures of the axial and locomotor complexes noted earlier. In some instances the ρ-groups that were formed are smaller than desirable for evolutionary studies involving functional groups. This can be remedied in part either by forming incomplete groups, in which measures attached to a basic pair that occur in the intersect of two ρ-groups are considered to be associated, or by lowering the level of ρ. Other aspects of the method and the problems that it poses are considered under the analysis of *Sciurus*.

INTERSECTIONS OF ρ-GROUPS

Information not available in the basic-pair analyses comes from the "raw" ρ-groups. Interpretation of these groups is generally difficult because of their number and the complexities of intersects. When this is the case, the best procedure is to make an initial analysis by the use of basic pairs and subsequently a study of the "raw" ρ-groups in the light of the information so gained. Such studies are usually most instructive when the groups are formed at a series of levels of ρ. Tables 22 and 23 show such series for *R. pipiens*, in which the intersections, with relationship to the various F'f groups, are graphically illustrated.

Here the F'f groups formed by qualitative study are used as the basis for analysis.

The highest level at which an intersection of two F'f groups by a single ρ-group occurs is $\rho \geq .95$. This intersect involves measures in F'f$_{fl}$ and F'f$_h$ (forelimb and hind-limb locomotion). The intersection is increased at the level $\rho \geq .93$, and, at this level, a single ρ-group also intersects F'f$_{fl}$ and F'f$_{ho}$ (forelimb locomotion and head orientation). At the $\rho \geq .92$ level, a very complex intersection occurs (F'f$_{fl}$ and F'f$_{hl}$, F'f$_{fl}$ and F'f$_a$, and F'f$_{ho}$ and Ff$_a$, and two ρ-groups that intersect Ff$_{fl}$, Ff$_{ho}$, and Ff$_a$ [Table 24]). At the $\rho \geq .91$ level, all F'f groups are involved in triple intersections, except for Ff$_j$ (jaw movement group), and two ρ-groups intersect all four.

The rather bewildering array of ρ-groups that is reduced to a relatively simple situation by use of the basic pairs is indicated by this complex of intersects. Also, important biological information appears to be available from a careful study of the intersections. The pattern of increased size of ρ-groups and the increase in the intersections of more than one F'f-group by single ρ-groups are to be expected. It is, however, important to note that, in *R. pipiens* at least, the increase of intersection occurs in a way that is highly consistent with what would be expected from the biological relationships of the systems described earlier.

The highest intersection, in terms of the level of ρ, involves the locomotor groups F'f$_{fl}$ and Ff$_{hl}$. These two systems are, of course, the ones most intimately involved in the highly adaptive function of locomotion. Forelimb locomotion and head orientation (F'f$_{fl}$ and F'f$_{ho}$) are next related, and, as pointed out earlier, these two systems act in accord during feeding and in head orientation while the individual is at rest. Forelimb and axial groups (F'f$_{fl}$ and F'f$_a$) are related only at a lower level, $\rho \geq .92$. At this level the head orientation (F'f$_{ho}$) and axial groups (F'f$_a$) are also intersected by single ρ-groups. The discreteness of the hind-limb group (F'f$_{hl}$) from all other systems except that of the forelimb to the $\rho \geq .91$ level is in excellent accord with the fact that this system has a highly specific function, not intimately related to joint actions of the other systems.

Finally, it is noteworthy that measures in the jaw movement group (F'f$_j$), which seem to require only a low-order biological association for the functions performed, not only fail to occur together in any ρ-groups above the $\rho \geq .86$ level but also are not in intersects of ρ-groups with any other F'f-groups above or at this level.

MEASURES NOT IN A SINGLE F'f-GROUP

It seems inevitable that in most organisms there occur structures that cannot be allocated, even through the most searching study, to any single Ff-group. Measures of urostyle length, iliac length, and sacral rib length (Ur$_l$, Il$_l$, and

Sr₁), were deliberately selected to represent this category in the study of *R. pipiens*. Each of the three appears to relate more or less equally to the hind-limb and axial systems. These measures were included purely for exploratory purposes, for there was little basis for an a priori judgment of how they might be associated by correlation with measures of the two Ff-groups or with other measures of the skeleton. It would appear that the structures expressed by the three measures would have been acted upon more or less equally by various factors that were contributing to the high integration of structures in each of the Ff-systems to which they were related. Insofar as these factors acted upon structures in the two Ff-groups in a similar way, their action on the intermediate structures might be thought to be the same. But the discreteness of Ff_{hl} (hind-limb locomotion) and Ff_a (axial movement) in function as well as in

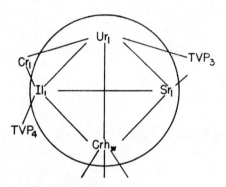

Fɪɢ. 34.—Bonding, to form ρ-group, of "mixed" measures at $\rho \geq .86$, in *Rana pipiens*. Crh_w, width of head of crus, measure of hind-limb locomotor group. All bonds for Ur_1, Sr_1, and Il_1 shown, but not all for Crh_w are entered.

development, inferred from biological studies and brought out in the mathematical analysis, suggests that this would not be the case. Thus a high degree of mathematical association with measures in the two Ff-groups seemed unlikely. If there was a "resultant" of the actions of the various factors of the two Ff-groups, which was reflected in these intermediate structures, a new grouping involving only these measures seemed a possibility.

In Figure 34 the total bonding of the "mixed" measures (Ur_1, Il_1, and Sr_1) at the $\rho \geq .86$ level is shown. With the width of the head of the crus (Crh_w), a measure in the hind-limb complex, they form a complete ρ-group. The full extent of bonding of Crh_w is not shown, but it is not extensive. Shown in its entirety, the bonding of the three measures to others taken on the skeleton is low, especially in view of the fact that there are 138 bonds possible, in addition to those in the figure. The few bonds that do exist are to measures in the hind-limb and axial groups. In general, this supports the theoretical considerations presented in the paragraph above.

From these various analyses a partial picture of *R. pipiens* itself emerges. The integration of the organism, to the extent that it is expressed in the measurements, is weak, for only a small percentage of the total possible bonds between measures are realized, even when ρ is reduced to the low level of $\rho \geq .86$. Furthermore, ρ-groups tend to be small and numerous rather than large and few, as is true when total integration is strong. In contrast to the general low level of integration, however, is the strong association of measures in some of the functional systems of the body. From these there is apparent a hierarchy of systems that can be readily arrayed upon a basis of the extent and intensity of integration. This array presents the general order of adaptive specificity of the systems. Heading the list are the systems specifically adapted to locomotion, the forelimb and hind-limb complexes. They form separate integrated complexes at the very highest levels of ρ that can be used, suggesting the existence of partly independent roles, but they merge into overlapping systems as the level of intensity of the relationships of parts is lowered. Structures involved in trunk movement and head orientation appear in separate integrated systems at levels somewhat below those that exist for the locomotor systems. Their discreteness is lost just below the levels at which their measures first show the existence of separate groups. Finally, at a very low level, measures involved in jaw movement begin to exhibit relationship. Large groups are not, however, formed within this complex by the values of r that are significantly different from zero. In this ordered array is revealed the picture of the organization of several functional systems of *R. pipiens*.

The intersections of the various ρ-groups provides a general picture of the way that the different functional systems come into co-operative relationships in the development of over-all body integration. The following stages are revealed at successively lower levels of ρ.

1. Forelimb and hind-limb locomotor systems show partial intersection ($\rho \geq .95$).

2. Forelimb and hind-limb intersection markedly increased. Forelimb and head orientation systems undergo partial intersection ($\rho \geq .93$).

3. Forelimb system strongly related to hind limb, head orientation, and axial systems. Head orientation shows first relationship to axial structures. Axial structures show first intersections with hind-limb measures, but the relationship is weak. Incomplete merging of three systems, forelimb, head orientation, and axial, appears ($\rho \geq .92$).

4. All systems show some relationships, with forelimb, head orientation, and axial systems exhibiting a high degree of intersection. Relationship of head orientation and hind-limb measures still weak. Jaw movement weakly related to forelimb, hind-limb, and head orientation but not to axial structures ($\rho \geq .91$).

At level $\rho \geq .91$, a beginning of full group association is first in evidence.

Continuation of increased complexity of intersection of groups occurs at lower levels of ρ, but even at the very lowest levels, i.e., where all values of $r > 0$ are included with $P = 0.1$, intersection is far from complete. The frog, *R. pipiens*, viewed in the light of this analysis, appears as an animal with highly differentiated functional systems and rather low-level total organization as expressed by the interrelationships of the systems.

Sciurus niger ruviventer

Measurements taken on a sample of skulls of *Sciurus niger ruviventer*, from the collections of the Chicago Natural History Museum, are used as a second source of data. The measures used were selected from a series of 235 made by Robert L. Miller on a sample of fifty individuals (Miller, 1950). These specimens were drawn from a local population, which assures a reasonably high degree of genetic homogeneity. Forty-five measures, from the total array, were selected as representative, and N was reduced to thirty by utilization of only those individuals for which the suite of forty-five measures was available. These measures are shown in Figures 29 and 30 and described in Table 18 (chap. v).

All measurements were taken on skulls and jaws, so that the situation is very different from that in *R. pipiens*. Detailed dissections of the head preceded measurement and served as the basis for selection and interpretation of the measures. The formation of qualitative groups on the basis of biological associations was considered in chapter v, where the following F'-groups were described (see Table 19): $F'f_m$—masticatory function group; $F'f_{ho}$—head orientation functional group; $F'g_{sk}$—skull group based on growth of measures in topographic proximity; $F'o_n$—"ontogenetic" group, pertaining to brain and nervous system, and "mixed"—measures of uncertain affinities that overlap two or more other groups.

THE ρ-GROUPS

The analytical procedure used for *S. niger* is the same as that for *R. pipiens*, but ρ-groups were formed at the following levels: $\rho \geq .91$; $\rho \geq .89$; $\rho \geq .87$; $\rho \geq .85$; $\rho \geq .82$; $\rho \geq .80$; and $\rho \geq .78$. The confidence level was $P = .01$, in a two-tailed distribution. At this level with $N = 30$, where $\rho \geq .71$, r is not significantly different from zero. Few bonds exist at the $\rho \geq .91$ level, but there is a rapid increase in the number of bonds at successively lower levels. The fact that levels of ρ lower than those used for *R. pipiens* were suitable may give the false impression that association is relatively less intense in *S. niger*. A proper comparison, involving only measures of the skull and jaws, shows a generally higher level of association in *S. niger* than in *R. pipiens* in this region.

BASIC-PAIR ANALYSIS

Six basic pairs form, as listed in Table 26. Of the six basic pairs, two, Mn_w–$Ms_{(2)}$ and Pm_h–Sk_1, form "cores" of clusters of measures (see Table 27). This is not the case for the others and, although there is room for interesting speculation in their meanings, they are not of primary importance in the present study. Pair $Di_{(2)}$–Of_w indicates an association of the width of the optic foramen and the diastema of the lower jaw. There is no evident biological explanation for this association in the information available. Two measures, as noted later in tables, are associated with this pair by negative bonds; hence it seems probable that there is some underlying, but obscure, basis for the existence of the basic pair. Pair M_w^3–OPA_d involves molar width and a distance between two foramina—the foramen ovale and palatine foramen—and So_w and Cr_w are measures of the cranial vault. They were considered as members of $F'g_{sk}$, but they do not show association with other measures, either in or out of this group. Items In_1 and Inm_1 are two measures of incisor length. It seems probable that

TABLE 26

BASIC PAIRS IN SKULL OF *Sciurus niger*

Mn_w–$Ms_{(2)}$. . . . . . $\rho \geq$ .91 (mandibular width–masseteric length)
Pm_h–Sk_1. . . . . . . $\rho \geq$ .89 (premaxillary width–skull length)
$Di_{(2)}$–Of_w. . . . . . . $\rho \geq$ .87 (lower diastema length–width optic foramen)
M_w^3–OPA_d. . . . . . $\rho \geq$ .82 (width third upper molar–foramen ovale and
                                        palatine foramen)
So_w–Cr_w. . . . . . . $\rho \geq$ .82 (supraoccipital width–width cranial vault)
In_1–Inm_1. . . . . . . $\rho \geq$ .78 (incisor lengths)

at least some of these pairs are parts of groups for which other measures were not taken. The fact that only a sample of all possible measures is used is likely to produce such a result, but, of course, the conclusions can be confirmed only by an increase in the number and coverage of the measures.

Basic pairs that are not group-formers are occasionally encountered in analyses. When measurements have been taken over the whole animal, the occurrence of such pairs is usually less extensive than that found in *S. niger*, in which only skull measurements were taken. In some instances such pairs suggest a meaningful relationship, as in the case of incisor lengths (In_1–In_{n1}) in *Sciurus*. Sometimes, however, they involve measures for which there is no plausible morphological or functional relationship. Such correlations may have a causal relationship in factors not recognized, or the correlation may be spurious. Where they do not relate to any aspect of a particular study, they are dropped from consideration as indeterminate. No case in which such a basic pair has acted as a group-former has been encountered in our work, although this is not out of the question.

No basic pair in *S. niger* includes measures in the head orientation ($F'f_{ho}$) or ontogenetic groups ($F'o_n$). Equally significant is the fact that the measures assigned to these two groups, except for measures of the optic foramen (Of_1 and Of_w), do not occur in any of the ρ-groups formed under the basic-pair analysis. In the case of the head orientation group, there is every reason to

TABLE 27*

Basic-Pair ρ-Groups at Various ρ Levels, *Sciurus niger*

Level of ρ	$F'f_m$	$F'f_{ho}$	$F'g_{ak}$	$F'o_n$	Joint Groups
$\rho \geq .91$	$Mn_w–Ms_{(2)}$	None	None	None	None
$\rho \geq .89$	$Mn_w–Ms_{(2)}–In_w$	None	$Pm_h–Sk_1$	None	None
$\rho \geq .87$	$Mn_w–Ms_{(2)}–In_w–Zy_d$ Bonded to Mn_w and $Ms_{(2)}$: Or_n; $Ms_{(1)}$	None	$Pm_h–Sk_1$	None	$Di_{(2)}–Of_w$
$\rho \geq .85$	$Mn_w–Ms_{(2)}–In_w–Zy_d–Ms_{(1)}–Or_n$	None	$Pm_h–Sk_1$	None	$Di_{(2)}–Of_w$ Zym ....¹
$\rho \geq .82$	$Mn_w–Ms_{(2)}–In_w–Zy_d–Ms_{(1)}–Or_n$ [Bonded to most members of group: $Di_{(1)}$] $M^3_w–OPA_{fd}$†	None	$Pm_h–Sk_1$ (Bonded to both: Bu_1, Bu_w, Sq_1) $So_w–Cr_w$	None	$Di_{(2)}–Of_w$ Zym
$\rho \geq .80$	$Mn_w–Ms_{(2)}–Zy_d–Ju_t–Di_{(1)}$ [Nc_1–Bonded to $Di_{(1)}$; Mn_w; $Ms_{(2)}$] $M^3_w–OPA_{fd}$† (Note: Or_n and In_w now in joint group, not bonded to Ju_t)	None	$Pm_h–Sk_1–Bu_1–Bu_w$ (Bonded to Pm_h, Sk_1: Sq_1) $So_w–Cr_w$	None	$Di_{(2)}–Of_w$ Zym $Zy_1–Or_n–Ju_t–Pm_h–Sk_1–Zy_d–Mn_w–Ms_{(2)}–Ms_{(1)}–In_w$ Bonded measures: Di_1 (8 bonds); OPA_{fd} (6 bonds)†
$\rho \geq .78$	$In_1–Inm_1$; $M^3_w–OPA_{fd}$†; $Mn_w–Ms_{(2)}–In_w–Or_n–Ju_t$ (Note: Measures not in $F'f_m$ at this level but present at higher level of ρ, now in joint group)	None	$Pm_h–Sk_1–Bu_1–Bu_w–Sq_1$ $So_w–Cr_w$	None	$Di_{(2)}–Op_f$ Zym $Di_{(2)}–Of_w–Of_1$ $Zy_1–Or_n–Ju_t–Pm_h–Sk_1–Di_{(1)}–Zy_d–Mn_w–Ms_{(1)}–Ms_{(2)}–In_w$

* Joint groups include (1) those in which basic pair crosses F'f group (or F'f and "mixed" measure) and (2) groups in which there are two basic pairs whose clusters of measures are in a single F'f group at some higher level of ρ.

† Mixed measure.

¹ Negative bond.

believe that the measures do have biological relationships, as was the case for comparable measures in *R. pipiens*. Only one part of the total group is represented in the measures, however, for there are no measures on the postcranium. It appears likely that failure to form a basic pair, or pairs, is due to this sample bias. The association of measures in the ontogenetic group $(F'o_n)$ was questionable, as indicated in chapter v. Whatever correlation may have existed at some stage in ontogeny is evidently absent in the stages studied.

As shown in Table 27, a number of measures assigned to the masticatory $(F'f_m)$ and growth groups $(F'g_{sk})$ are bonded into ρ-groups, groups that increase in size as the level of ρ is lowered. Only the basic pair that involves mandibular width and masseteric length $[Mn_w-Ms_{(2)}]$ is formed at $\rho \geq .91$. The group formed around this pair includes six measures in the masticatory group at $\rho \geq .85$, with one other, the upper diastema $[Di_{(1)}]$ bonded to the group. At this level the pair Pm_h-Sk_1 is discrete, but certain measures in $F'g_{sk}$ (growth group) are bonded to it $(Bu_1, Bu_w, \text{ and } Sq_1)$. Also at this level, a second basic pair with measures in the growth group $(F'g_{sk})$ occurs, So_w-Cr_w. At $\rho \geq .80$, four measures, including premaxillary width and skull length (Pm_h-Sk_1), from $F'g_{sk}$, form a ρ-group. A large joint group including measures from both the masticatory and the growth groups $(F'f_m \text{ and } F'g_{sk})$ with, of course, the two basic pairs is formed at this level. It should be noted that the method of formation of ρ-groups from the array of intercorrelations provides only the largest discrete groups and, as a result, certain measures in ρ that intersect $F'f_m$ at $\rho \geq .82$ are not in ρ that intersects $F'f_m$ at levels $\rho \geq .80$ and $\rho \geq .78$ but are in the joint group. The group formed around Pm_h-Sk_1 is not similarly affected.

The order of formation of ρ-groups by the basic-pair analysis supports the theoretical contention that the highest levels of association should occur between measures of dynamically functional groups. For this to occur, however, it is necessary that some of the most intensely associated measures of the Ff-group be present in the sample of the total array, for otherwise a basic pair may fail to form. Correlation of measures related by topographic proximity during growth, in this case, first became evident at a lower level, $\rho \geq .89$, and widespread association is found only at $\rho \geq .80$. At this level, this factor also is apparently involved in the relationship of measures in $F'f_m$ to other measures of the skull.

The absence of certain gross measures of bone dimensions in the growth group $(F'g_{sk})$ from the ρ-groups that include some measures of this F-group is worthy of note. None of the longitudinal measures of roofing bones of the skull—Pa_1, Fr_1, or Na_1—is in ρ, nor is squamosal length (Sq_1). All others in $F'g_{sk}$ are included in the Pm_h-Sk_1, or So_w-Cr_w, groups. These measures, except for Sq_1, were taken along the mid-line, and the length for each measurement was determined by the intersection of the highly irregular transverse suture and the median suture between the paired bones. This presumably accounts for the lack of correlation with skull length (Sk_1), which certainly might be expected, since these measures contribute to this measure. So_w and Cr_w, width of the supraoccipital and of the cranial vault, are related topographically to measures in the premaxillary width and skull length (Pm_h–Sk_1) group. In spite of this, they do not appear to have been subject to the same effects in development as the other measures. It is assumed that they form the "core" of another Fg-group for which no other measures were included.

Finally, the negative correlation of the zygomatic measure, Zym, with $Di_{(2)}$–Of_w (lower diastema and width of the optic foramen) must be noted. The meaning of this relationship, while it may be a basis for interesting speculation, is not evident from the measures taken or from their associations. Except in a rather ideal case, such as *R. pipiens*, associations that defy sound interpretation are probably to be expected.

INTERSECTIONS OF ρ-GROUPS

As for *Rana*, the principal interest in the "raw" ρ-groups, in the phase of study concerned with relationships of the correlation and the various aspects of F, lies in the intersections of two or more F-groups by a single ρ-group. Table 28 shows the majority of intersections of two F-groups by one ρ-group, with some of the smallest intersections omitted, and a representative group listed when there are several that differ by a single measure. Only pairs are formed by intersects of two groups at $\rho \geq .91$ and $\rho \geq .89$. The inclusion of measures in basic pairs of the masticatory and growth groups, $F'fm$ and $F'g_{sk}$, at these levels is of interest in showing the delicate relationship between the dominance of dynamic function and the growth relationship. The slight dominance of the former, evident in the basic-pair analysis, is increasingly masked at successively lower levels of correlation, with the basic pairs of $F'f_m$ and $F'g_{sk}$ present in an intersect at $\rho \geq .80$.

A relationship between the masticatory and ontogenetic groups, $F'f_m$ and $F'o_n$, occurs at $\rho \geq .87$. The relationship between measures in the growth group, $F'g_{sk}$, and $F'o_n$, appears at $\rho \geq .85$. The number of measures of these two intersected by a single ρ-group increases rapidly at lower levels. Two measures—length and width of the optic foramen, Of_1 and Of_w, the ontogenetic group ($F'o_n$)—occur with the highest frequency in these groups, and two others —$OSPf_d$ (optic foramen to sphenopalatine foramen) and Fi_1 (length of anterior intraorbital foramen)—are present. The width of the foramen ovale (Orfw), of $F'o_n$, also occurs in intersects with "mixed" measures. The measures in $F'o_n$ do not, however, form intercorrelated groups with each other, and whatever dominance their early association may have had has been lost in later stages of ontogeny. In large part, their association with other skull measures seems to be related merely to factors of growth.

Measures of the head orientation group, $F'f_{ho}$, which did not form ρ-groups which could be detected by the use of basic pairs, also fail to occur in intersects with measures of other F-groups, except in the case of the distance between paroccipital tips (Pao_w), which is encountered at the $\rho \geq .82$ level.

Table 29 shows the principal intersects of a ρ-group with three F-groups. Only at the level $\rho \geq .80$ is an appreciable number of intersects present. The situation is highly complex at $\rho \geq .78$. The effects of function and growth are indistinguishable, there is no evidence of a segregating influence of early formation of the central nervous system, and mixed measures show little tendency to be discrete. At this level, in short, the general integration of the skull appears to play the dominant role.

In this case, as in the less difficult case of *R. pipiens*, independent qualitative and quantitative studies and the synthesis of the results by a study of intersects, after basic-pair analysis, show that biologically meaningful groups of measures can be derived through analysis of quantitative association alone. Such groups do not necessarily include all measures that can be associated by qualitative means. The quantitative groups, however, do not include measures of more than one F-group, except in cases in which two or more basic pairs occur in a single intersecting ρ-group. This situation, of course, is easily recognized from the formation of basic pairs. No measures of the head orientation group, $F'f_{ho}$, are in the intersects, except in the case of mixed measures. It is evident, assuming the qualitative assignments to be correct, that some measures of an F-group, such as $F'f_{ho}$, may be present but not revealed in association. This presumably arises when none of the most highly correlated measures of such an F-group are included in the total array of measures.

The relatively slight dominance of the factor of dynamic function, Ff, in the association of complexes of measures, over the factor of growth in topographic proximity is evident in the tendency for formation of large ρ-groups that intersect both masticatory and growth, $F'f_m$ and $F'g_{sk}$, in both the basic-pair analysis and the treatment of "raw" ρ-groups.

The skull of *S. niger* is rather completely expressed by the measures that have been used, for they were selected from a much larger array to insure representation of all major systems subject to measurement. The strong tendency for the measures related to mastication and to topographic proximity to dominate the integration is clearly shown by the analyses. These two are separated, that is, they do not intersect, only at the highest level of integration found in the squirrel skull. Below this level they rapidly become indistinguishable and also tend to "take on" measures of other groups. The over-all integration of the skull is developed under the influence of the masticatory and topographic growth factors. Although total integration is far from realized, the fact that the skull and jaws tend to form an integrated entity, with discreteness of groups distinctly secondary, is clearly revealed.

A marked contrast between the skulls and jaws of *S. niger* and *R. pipiens* is

TABLE 28*

INTERSECTIONS OF TWO F'f-GROUPS BY ρ-GROUPS, *Sciurus niger*

$\rho \geq .87$

	$F'g_{sk}$	$F'f_{ho}$	$F'o_n$	M
$F'f_m$	Zy_1–$\underline{Sk_1}$–$Ms_{(2)}$ $\underline{Pm_h}$–Sk_1–$Ms_{(2)}$–Or_n $\underline{Sk_1}$–$Ms_{(2)}$–Zy_d–Or_n $\underline{Sk_1}$–$Ms_{(2)}$–Zy_d–$Di_{(1)}$	None	Zym–$\underline{Of_w}$ $Di_{(2)}$–$\underline{Of_w}$	None
$F'g_{sk}$		None	None	None
$F'f_{ho}$			None	None
$F'o_n$				None

$\rho \geq .85$

	$F'g_{sk}$	$F'f_{ho}$	$F'o_n$	M
$F'f_m$	$\underline{Pm_h}$–Sk_1–Zy_d–$Ms_{(1)}$–$Ms_{(2)}$–Or_n [Plus $\underline{Zy_1}$, $Di_{(1)}$, In_w: bonded] $Di_{(1)}$–$Di_{(2)}$–Sk_1 $\underline{Bu_w}$–Zy_1; $\underline{Bu_w}$–$Di_{(2)}$	None	Zy_1–Of_w Zym–$\underline{Of_w}$ $Di_{(2)}$–$\underline{Of_w}$	None
$F'g_{sk}$		None	Sq_1–Of_w Bu_1–Sk_1–$\underline{Of_w}$	None
$F'f_{ho}$			None	None
$F'o_n$				Bu_1–Ovf_w

* At $\rho \geq .91$, 1 intersection, Sk_1–$Ms_{(2)}$; at $\rho \geq .89$, 2 intersections, Sk_1–$Ms_{(2)}$, Pm_h–$Ms_{(0)}$. Underscored measures belong to the F'f-group which heads the column in which the measures so designated occur.

126

INTERSECTIONS OF TWO F'f-GROUPS BY ρ-GROUPS, *Sciurus niger*

$\rho \geq .82$

	$F'g_{sk}$	$F'f_{ho}$	$F'o_n$	M
$F'f_m$	$\underline{Pm_h-Sk_1-Ms_{(2)}-Zy_1-Ju_1-Di_{(1)}}$ $\underline{Pm_h-Sk_1-Ms_{(1)}-Ms_{(2)}-Zy_d-Or_n-In_w}$ [Plus smaller groups with meas.: Bu_1, Bu_w, Sq_1, $\underline{Pa_1}$, Te, $Di_{(2)}$]	None	$In_w-\underline{Mof_a}$ $Zy_1-\underline{Of_w}$ $\underline{Zy_m-Di_{(2)}-Of_w}$	$M_w^3-\underline{OPA_{fd}}$
$F'g_{sk}$		Bu_w-Pao_w	$Bu_1-\underline{OSP_{fd}}; Bu_w-\underline{Fi_1}$ $Sq_1-\underline{Fi_1}$ $Sq_1-Sk_1-\underline{Of_w}$ $Fr_1-Sk_1-\underline{Of_w}$ $Bu_1-Pa_1-Sk_1-\underline{Of_w}$	None
$F'f_{ho}$			None	None
$F'o_n$				$Bu_1-\underline{Ovf_w}$

$\rho \geq .80$

	$F'g_{sk}$	$F'f_{ho}$	$F'o_n$	$\dot{M}$
$F'f_m$	$\underline{Pm_h-Sk_1-Zy_d-Ms_{(2)}-Or_n-Ju_1-In_w}$ [Bonded with most of above: Zy_1; $Di_{(2)}$, Ju_1] [In smaller groups with some of above: Bu_1, Bu_w, Sq_1, $\underline{Pa_1}$, $\underline{Sqm_1}$, Te, $Di_{(2)}$]	None	$Zy_m-Di_{(2)}-\underline{Of_w}$ $In_w-\underline{Mof_a}$ $Zy_1-\underline{Of_w}$	$M_w^3-\underline{OPA_{fd}}$ $Ms_{(1)}-\underline{Te-In_w-Nc_1}$
$F'g_{sk}$		Bu_w-Pao_w	$Bu_w-\underline{Fi_1}; Bu_1-\underline{OSP_{fn}}$ $Sq_1-\underline{Fi_1}$ $Bu_w-Fr_1-Sk_1-\underline{Of_w}$ $Bu_w-Sq_1-Sk_1-\underline{Of_w}$ $Sq_1-Zy_1-Sk_1-\underline{Of_w}$ $Bu_1-Pa_1-Sk_1-\underline{Of_w}$	None
$F'f_{ho}$			None	None
$F'o_n$				$Bu_1-\underline{Ovf_w}$

TABLE 28 (*Continued*)

INTERSECTIONS OF TWO F'f-GROUPS BY ρ-GROUPS, *Sciurus niger*

$\rho \geq .78$

	F'g$_{sk}$	F'f$_{ho}$	F'o$_n$	M
F'f$_m$	Pm$_h$–Sk$_l$–Zy$_d$–Ms$_{(1)}$–Ms$_{(2)}$–Or$_n$–Ju$_l$–In$_w$–Zy$_l$–Ju$_t$–Mn$_w$–Zy$_d$ (Bonded to most of above: Bu$_w$, Sq$_l$) [Smaller groups with above meas.: So$_w$, Bu$_l$, Pa$_l$, Sqm$_l$, In$_l$, Di$_{(2)}$, Inm$_l$]	Pao$_l$–Ms$_{(2)}$	Zy$_l$–Of$_w$ In$_w$–Mof$_a$ Zy$_m$–Di$_{(2)}$–Of$_w$	M^3_w–OPA$_{fd}$ Ms$_{(1)}$–Ms$_{(2)}$–Te–In$_w$–Nc$_l$
F'g$_{sk}$		Bu$_w$–Pao$_w$	Bu$_w$–Fi$_l$; Bu$_l$–OSP$_{fd}$ Sq$_l$–Fi$_l$ Bu$_w$–Fr$_l$–Sk$_l$–Of$_w$ Bu$_w$–Sq$_l$–Sk$_l$–Of$_w$ Sq$_l$–Zy$_l$–Sk$_l$–Of$_w$ Bu$_l$–Pa$_l$–Sk$_l$–Of$_w$ So$_w$–Bu$_w$–OSP$_{fd}$	None
F'f$_{ho}$			None	None
F'o$_n$				None

128

Table 29

INTERSECTIONS OF THREE F'f-GROUPS BY ρ-GROUPS, *Sciurus niger*

None at $\rho \geq .87$

	$\rho \geq .85$	$\rho \geq .82$	$\rho \geq .80$	$\rho \geq .78$
$F'f_m$–$F'g_{sk}$–$F'o_n$	Sk_1–$Si_{(2)}$–Of_w	Zy_1–Sk_1–Of_w Bu_w–Sk_1–$Di_{(2)}$–Of_w	Bu_w–Zy_1–Fi_1 (plus Sq_1 for Bu_w) Bu_w–Sk_1–$Di_{(2)}$–Of_w (plus Sq_1 for Bu_1) Sq_1–Sk_1–$Ms_{(2)}$–Of_1 Bu_1–Pa_1–Sk_1–Zy_a–Orf_w Sk_1–Zy_1–$Ms_{(2)}$–Orf_w	Zy_1–Sk_1–$Ms_{(1)}$–OPA_{fd}–Of_w (meas. correl. with most of above: Bu_w, Bu_1, Sq_1, Fi_1, Of_1) Or_n–Pm_h–Te–In_w–Mo_{fd} Bu_1–Sq_1–Sk_1–Zy_a–Orf_w [Meas. correl. with most of above: Pa_1, $Ms_{(2)}$] Bu_w–Zy_m–$Di_{(2)}$–Inm_1
$F'f_m$–$F'g_{sk}$–M	None	Bu_1–Pa_1–Zy_a–Or_f	Pm_h–Sk_1–Mn_w–$Ms_{(1)}$–$Ms_{(2)}$–In_w–OPA_{fd} [Plus meas. correl. with most of above: Ju_1, $Di_{(1)}$, Ncl]	Zy_1–Or_n–Pm_h–Sk_1–Mn_w–$Ms_{(2)}$–In_w–OPA_{fd} [Plus meas. correl. with most of above: Ju_1, $Di_{(1)}$, Te, Ncl]
$F'f_m$–$F'o_n$–M	None	None	None	Mn_w–Ncl–SPf_1

found in the analyses of the two species. First, a higher level of integration of skull measures is found in *Sciurus*. Presumably this is related to the higher specificity of the function of mastication and perhaps to the more intimate relationships of the bonds of the cranium and the cranial parts of the nervous system. Second, the ρ-groups in *S. niger* are very closely interrelated, whereas in *R. pipiens* the measures of the skull tend to be associated only in small groups, which are discrete one from the other. In general, only measures of single small structures, such as the measures related to a single muscle complex, tend to be related in *R. pipiens*.

It is unsafe to generalize from these two cases, one a mammal and the other an amphibian, to characteristics of the major taxonomic categories that they represent. The skulls of two amphibians, *Diplocaulus* and *Trimerorhachis* (Olson 1951, 1953), for example, proved to be very highly integrated with respect to measures of the dermal surface and openings for the sense organs of the head. This integration, however, was in large part related to the size and shape factors of the skulls and not to a specific functional aspect. For the present, each case must be treated separately to serve as a source of evidence from which it may some day be possible to arrive at more general conclusions.

THE QUANTITATIVE DETECTION OF F-GROUPS

The two experiments described in the preceding section were considered by the writers to be sufficient in themselves to validate the assumption that F-groups could be inferred from properly constituted ρ-groups. Further work, on other samples of various populations, has lent additional support to this conclusion, but by the reasonableness of results rather than by the independent establishment of F- and ρ-groups. These studies have involved the following:

Five ontogenetic stages of *Rattus norvegicus*
Five samples of domestic pigeons drawn from a single breeding population
One sample of the extinct amphibian *Trimerorhachis insignis*
Two samples of the extinct amphibian *Diplocaulus* (*D. magnicornis* and *D. recurvatus*)
Eight samples of lower molar dentitions from different species of the extinct Eocene mammal
   Hyopsodus
One sample of upper and lower molar dentitions of the monkey *Aotus trivirgatus*
Six samples from different populations of the extinct blastoid *Pentremites*
Four samples of the Eocene teleost *Knightia*

For all these thirty-two samples, the ρ-groups, when analyzed by one or another technique, provided arrays of measures so grouped that they were subject to reasonable and consistent biological interpretation. Correlation as treated in the model and its extensions does, without any question, provide a reliable and powerful means of arriving at important information about morphological integration.

The Formation of ρ-Groups

CRUDE METHODS

There are two general types of approach to the formation of ρ-groups, once the correlation coefficients have been calculated. One, discussed in detail later, is that which has been followed in the text to this point. This involves formal establishment of ρ-groups by procedures that assure, through sequential steps, that all existing groups have been recognized. The method technically involves determination of all pairs of measures bonded at a selected level and the subsequent formation from the pairs, of triplets, quadruplets, and so forth. Only non-contained ρ-groups are retained. This is an extremely tedious task and becomes virtually impossible for large series of variables in which the majority of measures are bonded at the level being used. It is possible to shorten the procedure drastically by a scanning process in which the larger groups are carried through intermediate stages by grouping mentally. This short method is the one that has been used in all our later studies, but it is one that requires considerable experience and is prone to error in the absence of such experience. For studies in which detailed analyses are to be carried out, either the formal method or the short-cut modification should be used, for only in this way are the ρ-groups to be studied obtained in their entirety.

A second approach involves the use of less formal, graphic methods. If there are only a few measures, say ten variables, or if correlations are few even though the number of measures is large, it is an easy matter to list the measures and to plot the bonds between them, as illustrated in Figure 2, chapter iii. As the number of bonds increases, however, this approach soon reaches a state of confusion and is impractical. When only an approximation is required and interest lies in general tendencies of group formation or in rough comparisons of bonding at different levels of ρ in or between populations, a listing of measures and plotting of the bonds between them provides a quick and constructive approach. The data from *Pentremites* are used in illustration of this process in Figures 35, 36, and 37. It is convenient to arrange the variables on the sheet in some way that appears to approximate the way in which they may group, either from knowledge gained in scanning the matrix or from some estimate of probable biological relationship.

The measures of the godoniform *Pentremites* from the Paint Creek formation of the Chester series in Illinois (Mississippian), as shown in Figure 31, were arrayed to bring into proximity (1) *general size measures* (8, 12, 15, 23, 31, 32, 34, and 35); (2) a *basal structural group* (24, 25, 26, 27, 28); (3) a *deltoid group*, proximity and structure (20, 21, 22, 29, 30); and (4) two "functional" groups, *respiratory* (1, 2, 3, 4, 7) and *digestive* (5, 6, 9, 11, 14). It turned out that this was only a rather poor approximation of the integrative grouping as a whole, but, so long as the arrangement was the same for all plots, this was not serious for a first approximation of grouping, or for comparisons.

In Figure 35, the Paint Creek godoniform pentremite is shown at the level $\rho \geq .97$, with all bonds entered, and in Figure 36 it is shown at $\rho \geq .95$. From Figure 35 it is evident that "size" measures tend to be closely grouped and that measures 29 (length of the deltoid) and 9 (length of the food groove) are closely related to size. The bond of 26 and 27 gives slight indication of a basal group, and that between 4 and 7 shows a relationship of the distance of the mouth and of the anus from the deltoids. *It is at once evident that over-all size is a controlling factor. Partial correlation in which at least major size elements are held constant is indicated.*

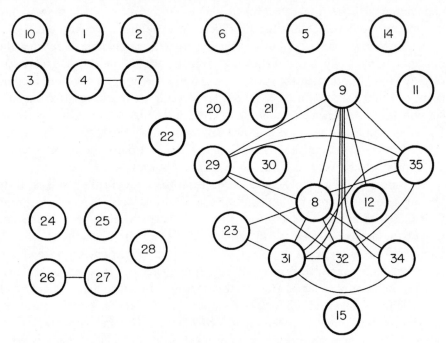

FIG. 35.—Illustrations of crude method of analysis of bonding. Based on godoniform *Pentremites* from the Paint Creek formation, $\rho \geq .97$. (See also Figs. 26 and 27.)

For the same sample $\rho \geq .95$ (see Fig. 36) gives immediate evidence of a very rapid increase in bonds but shows little evidence of a marked change in patterns. The dominance of size is still evident, and the basal group is strengthened. As the level of ρ is carried lower, the complexity greatly increases, as shown in Figure 36, where $\rho \geq .93$. Some relationship between measures of arrays plotted in proximity emerges, but in general it is clear that the arrangement does not express the underlying organization except for a *basal group* and a *size group*. Even at lower levels, not figured, the isolation of a basal group is striking, but bonding between other measures becomes so complex that little can be made out of it.

One of the objectives of this study was to compare the godoniform and pyriform lines of *Pentremites*. In Figures 35 and 38 the patterns of the Paint Creek godoniform and pyriform samples are shown and can be compared on the basis of crude plots. Higher integration at the same level of ρ is evident in the pyriform example. The size group and basal group are both apparent at the high level, more fully expressed than in the godoniform plot. Also, however, there is strong bonding of the digestive measures and bonding between these measures and the measures of the size group. At a lower level of ρ, ρ ≥ .97, in the pyriform type (not figured) intercorrelation nears completeness except that the deltoid measures are little bonded. From such a partial survey, a tabular summary may be made (Table 30).

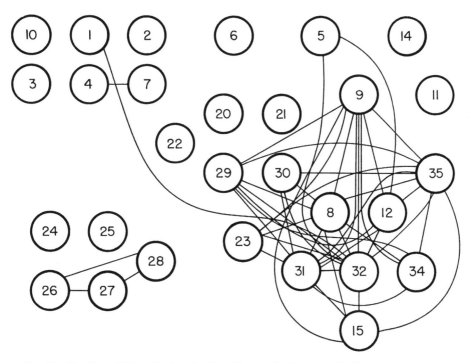

FIG. 36.—Bonding of Paint Creek godoniform *Pentremites* formed as in Figure 35, but ρ ≥ .95

The information in Table 30, while of a very general nature, gives a reasonably good concept of the broad outlines of integration in these two forms. A further objective of work on *Pentremites* was the study of temporal change in a single line. Figure 39 shows three successive stages in the godoniform line from the Paint Creek, Golconda, and Glen Dean, successively higher formations of the Chester in this order. These stages are shown at the high level of ρ ≥ .97. A sharp drop in integration is evident to the extent that there is almost no pattern in the Golconda stage. There is a minor retention of the size group in the Golconda but almost none in the Glen Dean. What bonding does exist,

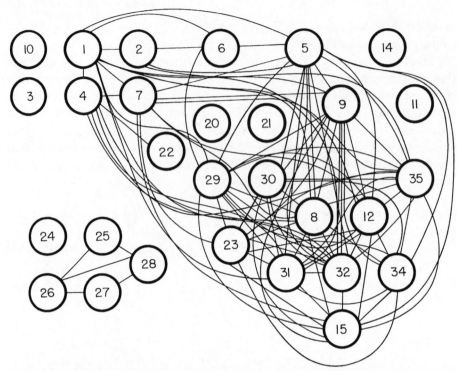

FIG. 37.—Bonding of Paint Creek godoniform *Pentremites* formed as in Figure 35, $\rho \geq .93$

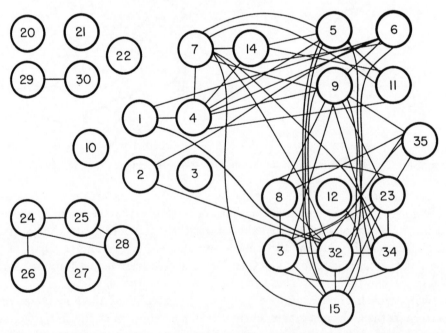

FIG. 38.—Bonding formed by crude method for pyriform *Pentremites* from Paint Creek formation, $\rho \geq .98$. Compare with bonding for godoniform *Pentremites* from same formation (Fig. 35).

however, is related to size measures. At lower levels of ρ, of course, groupings do occur, but this single level is sufficient for illustration.

This crude method, which may be carried out very rapidly, yields considerable information. This may be enough for some purposes, but, if the full potential of morphological integration is to be used, the initial information serves more as a guide to subsequent operations than as an end in itself. The illustrated examples show the general outlines of groups and indicate that details of discrete groupings are obscured by the predominant size-shape effect. This suggests that partial correlation rather than basic-pair analysis will be necessary, for it is probable that the basic pairs would include predominantly size-shape measures. The results indicate as well that different levels of ρ will be

TABLE 30

Summary of ρ-Groups in Paint Creek *Pentremites* (Based on Crude Analysis)

	Godoniform	Pyriform
$p \geq .98$......	Size and shape measures consolidated; deltoid size and food groove brought in	Size-shape and digestive areas highly consolidated; basal independent integration well along
$p \geq .97$......	Some additional consolidation; no new groups	Nearly total consolidation of all but deltoid area
$p \geq .95$......	Further consolidation of above; beginnings of independent formation of basal group	Little additional change at this or lower levels; deltoid area gradually brought into total association with others
$p \geq .93$......	Crown and anus brought in to previous major group; basals further independently consolidated	
$p \geq .90$......	High integration size-shape, crown, and digestive areas; basals independent group	

needed to arrive at the optimum grouping for different systems. Also, they show that changes of integration are important in evolution and that a study of the index of morphological integration should be carried out. Comparisons of the two lines, godoniform and pyriform, indicate that different scales will be needed to obtain comparable intensity of grouping in the lines. Such preliminary information is a very important guide, for the sake of economy, in the work of formal analysis.

The Levels of ρ

Use has been made at various places in the preceding pages of different levels of ρ to express different levels of integration. We shall now examine this practice more thoroughly with examples that bring out details of the effects as they relate to the detection of meaningful relationships between ρ and F and the quantitative detection of F-groups. A special case that arises from this practice is the segregation of basic pairs. This will be discussed separately in the following section.

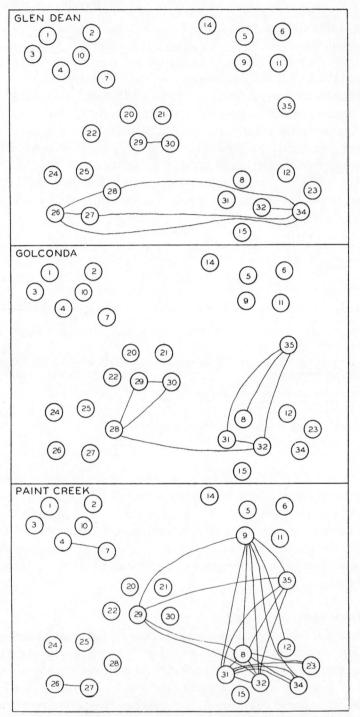

Fig. 39.—Three temporal stages of godoniform *Pentremites*, Paint Creek, Golconda, and Glen Dean, showing bonds at $\rho \geq .97$. In this instance the bonds are drawn in the informal style used in carrying out quick analyses. It is the general clustering of bonds rather than the precise composition that is important in such an analysis.

In the presentation and subsequent discussion of the working model in chapter iii, ρ-groups were defined as depending on intercorrelations greater than or equal to an arbitrarily selected ρ level. We are interested in estimating ρ for each pair of measurements, and much of the complexity that arises in our analysis results from the fact that an *interval* is used rather than a point estimate of the true correlation coefficient ρ. The nature of the material with which we work, however, is such that single samples of a species are all that are available. Since comparison of compatible animal populations is one of the purposes of the study, we must deal in terms of ρ rather than *r*. Some discussion of the arbitrary level of ρ selected for analysis in a given case seems required. The following discussion outlines the way in which this problem has been handled. Meaningful ρ levels for analysis and biological interpretation fall into two classes in our treatment.

1. "INDEPENDENCE VERSUS DEPENDENCE"

We have in some cases formed ρ-groups at a level such that all correlations significantly greater than zero are used. The writers realize that a correlation coefficient not significantly greater than zero for a given sample size does not necessarily signify statistical independence (see Feller, 1950, p. 186). However, we deal with the converse, namely, that the level just significantly greater than zero is the weakest and most inclusive "bonding" analysis that we can use. At this level, dependence is established. It is clear from the nature of the ρ intervals used that the greatest numbers of intercorrelations occur at the lowest level of significant correlation. Thus, as the level of ρ is successively raised, fewer and fewer pairs of measurements are correlated at a sufficiently high level; inversely, the maximum number of bonds (intercorrelated pairs of measurements) occurs at the lowest possible level of correlation, namely, just significantly greater than zero.

Our interpretation of ρ-groups formed at this lowest possible level is based in part on the intuitive feeling that in an organism we do, in fact, deal with a situation where *correlation, at least in part, implies causation!* This statement is discussed in some detail, in the final chapter of the book. If it is carried one step further, we may say that the *degree of correlation is a function of the degree of causation.* This statement is intended to apply to patterns arising from treatment of a representative number of measurements taken on the organ or on the whole animal. From the above we specifically exclude consideration of a single correlation and stress that we are dealing here with interacting parts of a biological system.

2. THE SPECTRUM OF CHANGE IN ρ-GROUPS AS THE ARBITRARY LEVEL IS RAISED

Interest may lie in the pattern of ρ-groups in which bonds are very strong, for example, $\rho \geq .98$. In some cases, however, all bonds may have vanished as the level was raised to $\rho \geq .98$. It is thus not possible to choose, for the general case, a specific high ρ level for examination of exceptionally strongly bonded ρ-groups. Instead the matter must be treated on a *relative* basis. For a given animal population the level of ρ is successively raised from the lowest level, just significantly greater than zero. Change in the patterns and membership of ρ-groups is noted. In one population, for example, all groups may vanish by the time the level $\rho \geq .75$ is reached, whereas in another population, large ρ-groups may still persist at the level $\rho \geq .99$.

As mentioned above, successive rises in the ρ level result in a regularly decreasing *number of bonds*, and, as the ρ level approaches 1, the bonds tend to disappear totally, although animal populations vary greatly in this respect. The pattern of ρ-groups, however, does not change in a simple way as the ρ level is raised. The tendency is for the number of ρ-groups to be small at the lowest ρ level, to increase (in some cases very rapidly) as the ρ level is raised, and then decrease as a result of the loss of bonds.

A simplified illustration will demonstrate this phenomenon. Following this, an actual case is presented in some detail. Consider the case, illustrated in Figure 40, in which four different anatomical measurements are bonded in all possible ways at ρ just significantly greater than zero. The successive rise of the ρ level is simulated by deletion of one bond at a time. At each successive stage the maximum and minimum possible number of ρ groups is given, an illustration of several of the possible arrangements accompanying each stage. The minimum possible number of ρ-groups in this very simple example fluctuates from 1 to zero with zero at the highest level. The maximum increases from the level significantly greater than zero and then decreases to zero at the highest ρ level. Enumeration of possible numbers of ρ-groups and of the number of ways in which bonds may be arrayed to form the ρ-groups at the various levels of ρ presents an interesting problem which does not have an immediately obvious solution. The writers have attempted a detailed counting of possible ρ-groups and of ways of arranging these groups. By utilizing the symmetry properties, complete enumerations up to six measurements have been made. Even with this low number of measurements the enumeration becomes extremely laborious. Unfortunately, we have not been able to detect an over-all trend in the number of ρ-groups at various ρ levels.

An example, from forty-day-old individuals of the ontogenetic series of rats, described in detail in chapter vii, will be used in illustration of the effects of shifting levels of ρ. This particular example was selected because it is neither too simple nor too complex to obscure the essential features. A total of thirty-four measurements was taken on each individual of a sample of twenty. Measurements were spread over the entire skeleton. Since it is necessary to use only measurements which could be taken on the one-day-old individuals in order to maintain continuity throughout the ontogenetic series, there are limitations

STAGES	NUMBER OF BONDS	NUMBER OF P-GRP'S, POSSIBLE	P-GROUPS
1 P JUST SIGNI-FICANTLY > 0	6	1	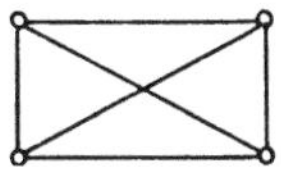
2	5	2	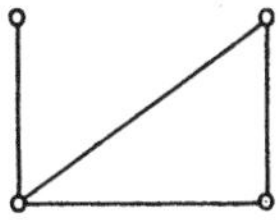
3	4	MINIMUM − 2 MAXIMUM− 4	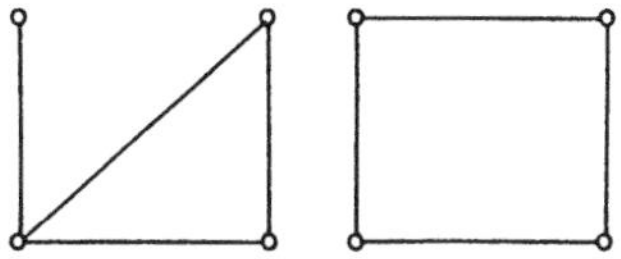
4	3	MINIMUM − 1 MAXIMUM − 3	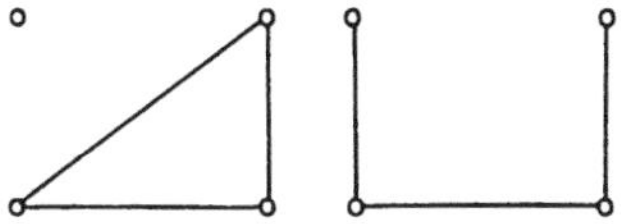
5	2	2	
6	1	1	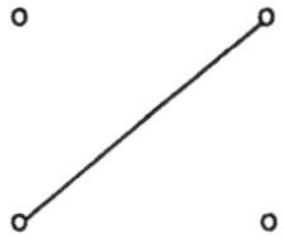
7	0	0	

FIG. 40.—Simple illustration of increase and decrease of number of ρ-groups as ρ is raised from a level just significantly greater than zero ($\overset{\text{sig}}{>} 0$) to the level at which all bonds vanish.

that required the use of predominantly gross dimensions. The measurements used are given in Table 31 (see also Fig. 41).

The effects of shifts in the level of ρ are shown graphically in Figure 42 and in tabulations for $\rho \geq .90$ and $\rho \geq .86$ (Table 32), since the latter arrays are too complex to be meaningful in graphic form. Numbers are used for designations rather than abbreviations, since they are more readily distinguished and

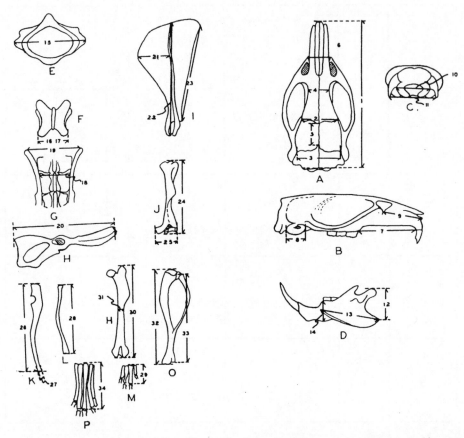

Fig. 41.—Measurements used in the study of the albino rat. All measurements shown used in one-, ten-, twenty-, and forty-day stages. Measure 19 not used in adults. (See Table 22 for explanation of measurements.)

our purposes are general rather than specific. The abbreviations are inserted in the text, where they are particularly important.

At the extremely high level of ρ, $\rho \geq .995$, there is but a single pair, automatically a basic pair, 20–33 (Pe_1–Ti_1). These are measures of the hind-limb complex. At $\rho \geq .990$, two bonds are added, giving pairs as follows: 20–33 (Pe_1–Ti_1), 22–23 (Sc_{sp}–Sc_{pl}), and 20–23 (Pe_1–Sc_{pl}). These are measures of the forelimb and hind-limb complexes, and three of the four dimensions refer to the girdles. The situation at $\rho \geq .980$ remains simple. The pelvic-scapular measures form a triplet, 20–22–23. Then 24 (Hu$_l$) is added by a bond to 33. Continued increase in bonds occurs at $\rho \geq .97$, with consolidation of 20–22–23–33 and 24–32–33, and a similar trend is followed to the level $\rho \geq .96$. At this level, two measures not in the locomotor complex, but bonded to it, appear: 17 (Zy22) of the axis and 13 (Ms$_l$) of the lower jaw. Below this level, complexity sets in, for, although the locomotor complex remains central (20–22–23–24–26–32–33–34), measures from other systems are added. The number of non-

TABLE 31

MEASURES OF *Rattus norvegicus*

No.	Abbr.	Description
1	Sk$_l$	Length of the skull from the occipital crest to the tip of the nasals, along the mid-line
2	Fr–Pa$_s$	Width of the fronto-parietal suture across the vault of the skull, measured as segment of arc, from ends of sutures as seen in dorsal aspect
3	Pa–Ip$_s$	Width of the parietal-interparietal suture measure as a segment of arc, from tips of interparietal
4	Io$_w$	Minimum interorbital width, measured normal to skull length
5	Ip–Fr$_d$	Distance along the mid-line from the junction of the interparietal and parietal to the junction of the frontal and parietal
6	Zyg$_w$	Width of the zygomatic arch, measured between the anterior tips of the two arches
7	Dia$_l$	Length of the dental diastema on the upper jaw, from base of first cheek tooth (or its position in young specimens) to base of incisor
8	Bu$_l$	Length of bulla (auditory) from junction with paroccipital process to junction with basicranial axis
9	Zy–Pa$_d$	Distance from the anterior process of zygomatic arch to most anterior extension of premaxillary
10	Occ$_w$	Maximum distance between outer margins of occipital condyles
11	Par$_w$	Distance between basal tips of paroccipital processes
12	Jp$_h$	Height of jaw from base of angular process to top of coronoid condyle
13	Mas$_l$	Distance from posterior termination of angular processes to most anterior point of insertion to masseteric
14	Jd$_m$	Depth of jaw at level of anterior margin of M_1
15	At$_w$	Maximum width of atlas
16	Zy9	Maximum width of postzygapophysis of 9th vertebra
17	Zy22	Maximum width of postzygapophysis of 22d vertebra
18	Sr$_w$	Maximum width of sacral rib
19	Pe$_w$	Maximum distance between anterior tips of two halves of ilium
20	Pe$_l$	Maximum length of pelvis, from anterior tip of ilium to posterior tip of ischium
21	Sc$_{ps}$	Maximum width of prespinose part of scapula, measured from crest of spine to anterior edge, normal to plane of spine
22	Sc$_{sp}$	Maximum length of spine of scapula
23	Sc$_{pl}$	Posterior length of scapula, from crest at angle with dorsal margin, to base at glenoid fossa
24	Hu$_l$	Maximum length of the humerus
25	Hu$_{dw}$	Maximum distal width of humerus
26	Ul$_l$	Maximum length of ulna
27	Ul$_{bw}$	Maximum basal width of ulna
28	Ra$_l$	Maximum length of radius
29	Mc$_l$	Maximum length of 3d metacarpal
30	Fe$_l$	Maximum length of femur
31	Fe$_w$	Minimum width of femoral shaft
32	Ti$_l$	Maximum length of tibia
33	Fi$_l$	Maximum length of fibula
34	Mt$_l$	Maximum length of 3d metatarsal

contained ρ-groups at $\rho \geq .90$ and $\rho \geq .86$ is high—16 and 28, respectively—and at first glance these seem to present an almost insoluble tangle.

This example of the forty-day-old rat is neither the simplest nor the most complex that can be expected, but it is reasonably representative. It clearly illustrates the reason for earlier statements that procedures, after grouping has been done, are not necessarily simple or straightforward. The need for simplification for interpretation of groups at the lower levels of ρ is amply shown.

As a rule, analysis by the use of a series of levels of ρ is effective in the detection of the most highly integrated groups of measures but fails to give clear

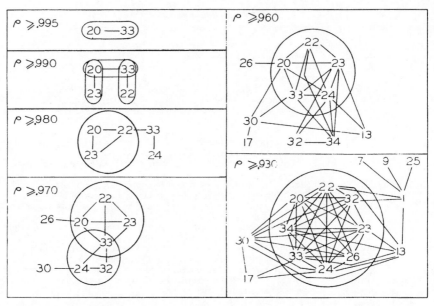

Fig. 42.—Bonding in the forty-day sample of the albino rat at successively lower levels of ρ, as indicated. The finest gradation feasible used for illustrative purposes. Such fine levels are usually not drawn in practical problems. Principal ρ-groups circled, bonds indicated by solid lines.

definition to groups that may exist at lower levels of intensity. These tend to be masked by complex series of intersections of ρ-groups, which include measures from more than one F-group. This linkage, of course, has very important biological connotations, for we see in this intersect the interrelationship of systems and through these gain information leading to the concept of the total organism. But this often cannot be detected in the absence of clearly defined "core" ρ-groups. Procedure that is most meaningful requires, as a rule, some process by which the groups can be recognized and limited, prior to the study of group intersects. Two methods of formal analysis, as noted before, are important in group determination—the basic-pair approach and the use of partial correlation. The pattern of grouping in the rat sample and in the other samples of the ontogenetic series strongly indicates the use of basic pairs, for

TABLE 32*

ρ-GROUPS OF THE FORTY-DAY-OLD RAT AT $\rho \geq .90$, $\rho \geq .86$

$\rho \geq .90$
15–25
28–34
15–17–24–30
15–18–24–30
17–21–24–30–33
17–24–30–32–33–34
18–20–24–30
1–7–9–13–24
1–7–13–20–23–24–33–34
1–9–13–14
1–13–14–33–34
1–13–20–22–23–24–33–34
13–20–21–23–24–30–33
13–20–22–23–24–26–30–32–33–34
13–22–25
1–9–13–25
1–13–22–25

$\rho \geq .86$
4–17
8–12
11–19
16–34
28–29
10–15–17–24–33–34
1–7–13–18–20–22–23–24–30–32–33
1–9–13–15–20–22–23–24–25–30–33–34
7–13–14–20–22–23–24–31–32–33–34
1–13–14–20–22–23–24–31–32–33–34
9–13–15–17–20–22–23–24–30–33
13–15–17–18–20–22–23–24–30–33
13–17–18–20–22–23–24–30–32–33
13–17–20–21–22–23–24–30–32–33–34
14–20–24–26–28–32–33–34
1–7–10–15–24–33–34
1–7–11–18–20–24–33
1–7–9–13–14–20–22–23–24–26–32–33–34
1–7–9–13–15–18–20–22–23–24–30–33–34
1–7–9–13–20–22–23–24–26–30–32–33–34
1–7–13–15–18–20–22–23–24–30–33
1–9–13–14–20–22–23–24–25–26–33–34
1–9–13–14–20–22–23–24–26–32–33–34
1–9–13–20–22–23–24–25–26–30–33–34
9–13–17–20–22–23–24–25–26–30–33–34
9–13–17–20–22–23–24–26–30–32–33–34

* For identity of measures, see Fig. 41 and Table 31.

the dominant group is functional and there are vague suggestions of other functional groups under the masking effect of the locomotor group. In contrast, the *Pentremites* samples, as noted on pages 132–35, suggest the use of partial correlation to remove the factor of size and shape, which has a masking influence. This can be followed, if needed, by basic-pair analysis. A rat sample and one sample of *Pentremites* will be carried through these steps to bring out the various aspects of the operations and interpretations.

BASIC-PAIR ANALYSIS

The theoretical and technical aspects of basic-pair analysis have been taken up at various places in the preceding pages (pp. 46–48, 54, 110–11). The effectiveness of the technique in establishing groups in *R. pipiens* is appropriate to the present consideration, but the aim of that study was validation, not detection, of ρ-groups. In the ontogenetic series of rats, basic pairs provide the key to the study of changes of integration with ontogenetic time as groups formed on the one-day-old sample are carried through the series (chap. vii). Any one of the stages in this series would serve for illustration of the use of basic pairs

TABLE 33

BASIC PAIRS IN SAMPLE OF TWENTY-DAY-OLD RATS

20–24 . . . . (Pe$_1$–Hu$_1$) (pelvic length–humeral length) at $\rho \geq .96$
28–32 . . . . (Ra$_1$–Ti$_1$) (radial length–tibial length) at $\rho \geq .96$
16–17 . . . . (Zy9–Zy22) (zygopophysial width, vertebrae 9 and 22) at $\rho \geq .93$
6–7 . . . . (Zyg$_w$–Dia$_1$) (zygomatic width and diastema length) at $\rho \geq .90$

in the detection of ρ-groups and their relationships to F-groups. We shall use the twenty-day stage, since correlations are relatively high and there is a very complex array of intersecting ρ-groups at the lower levels of ρ.

Four basic pairs are present in the twenty-day rat sample as shown in Table 33. The different levels of ρ of the basic pairs in Table 33 should be noted. As the level of ρ is raised, various pairs may be isolated. If neither measure of a pair so isolated occurs at this level or at any higher level in any other group, then the measures of the isolated pair have the characteristic that each is more highly correlated with the other than with any other measure, that is, they form a basic pair. Such pairs represent the highest integration for each group that can form from the *sample* but do not necessarily express the highest actual relationship for the group inasmuch as appropriate measures may not have been taken.

The pair 20–24 (pelvic length and humeral length) may be a case in point. A measure of the posterior locomotor complex and one of the anterior locomotor complex, Pe$_1$ and Hu$_1$, respectively, are involved. It may well be that some higher integration exists between each of these measures and another possible measure in its own complex; it is not illogical to suppose that there is some tendency toward higher integration *within* the two locomotor complexes than *between* them. There is no way of checking this without recourse to additional measures, calculations, and formation of new groups. Except at extremely high levels of integration, the net effect is relatively unimportant, for it is the composition not of the pair but of the group formed by the pair that is critical. Pair 28–32 (radial and tibia lengths) shows a strong relationship between the length of radius and length of tibia. This certainly is not an unexpected association, although it may not actually be the highest one for either of the components. It is to be expected that 20–24 and 28–32 will merge at a relatively high level of ρ. The problems of an incomplete array of measures and the possibility that the highest integration has not been represented are always present in basic-pair analyses and cannot be ignored, either in the interpretation of meaningful relationships or in the case of basic pairs in which the basis for association is not evident.

If we look for a moment at the basic pairs of the five samples of the rat series, Table 34, and at the groups they form, Table 35, the relatively minor

TABLE 34

BASIC PAIRS FOR FIVE SAMPLES OF RAT SERIES

Age	Basic Pairs and Level of ρ
One day.........	22–24 (Sc_{sp}–Hu_1) $\rho \geq .90$; 25–29 (Hu_w–Mc_l) $\rho \geq .95$
	28–32 (Ra_l–Ti_l) $\rho \geq .95$
Ten days........	22–23 (Sc_{sp}–Sc_{pl}) $\rho \geq .93$
Twenty days.....	20–24 (Pe_l–Hu_l) $\rho \geq .96$; 28–32 (Ra_l–Ti_l) $\rho \geq .96$
Forty days......	20–23 (Pe_l–Sc_{pl}) $\rho \geq .96$
Adult..........	26–28 (Ra_l–U_l) $\rho \geq .96$; 32–33 (Ti_l–Fi_l) $\rho \geq .96$

effects of differences in composition of pairs will become apparent. There are, of course, differences in the complexes at the successive stages, but, so far as the locomotor complex is concerned, there is reasonable stability. We shall restrict ourselves to the groups formed by pairs that pertain to this series of measures. The basic pairs differ in specific composition, which in part reflects real differences between the samples. For the most part, however, they act much the same as group-formers. The important differences are considered in the detailed discussions of chapter vii. The similarities are shown in the examples in Table 35 of groups drawn from the ρ-levels that show the closest resemblances.

The resemblances of the groups, even where basic pairs are different, are so striking that they need little comment. It will be noted that in the twenty-day-old and adult samples two basic pairs are present in the groups, that is, they have merged at the level of ρ indicated. They are, of course, separate at higher levels and act as independent group-formers.

On occasion, as noted particularly in *S. niger* (p. 121), basic pairs are formed between measures that have no evident relationship. To the present time no such pair has been found to have any group-forming capacities, and, in the absence of this, they have been noted and dropped from further consideration. There are many possible explanations—deficiencies of the sample of measures, definition of a real group whose meaning is not recognized, spurious correlation, and so forth. No such pairs appeared in the rat series. The closest approach is found in 25–29 in the one-day-old rat, for, although both measures are in a single complex, this pair does not form groups independently of another basic pair. It also does not appear in any other samples in the series.

We shall now return to the study of the basic pairs at the twenty-day stage in the rat series. Four pairs were noted, two from the locomotor measures, one from the axial skeleton, and one from the measures of the skull and jaw. These pairs act as group-formers as the level of ρ is shifted downward, as illustrated in Table 36.

TABLE 35

BASIC-PAIR GROUPS OF LOCOMOTOR COMPLEX RAT SERIES*

Age	Level of ρ	Basic-Pair Groups
One day...........	$\begin{cases} \rho \geq .86 \\ \rho \geq .86 \end{cases}$	22–24–<u>25</u>–<u>29</u> <u>20</u>–<u>23</u>–28–30–<u>32</u>–33–34
Ten days..........	$\rho \geq .86$	22–23–24–26–30†
Twenty days.......	$\rho \geq .93$	<u>20</u>–<u>22</u>–23–24–26–<u>28</u>–30–<u>32</u>–33–34
Forty days........	$\rho \geq .90$	<u>20</u>–22–23–<u>24</u>–26–<u>28</u>–30–<u>32</u>–33–34
Adult.............	$\rho \geq .93$	<u>20</u>–22–<u>23</u>–24–<u>26</u>–<u>28</u>–30–<u>32</u>–<u>33</u>

* Underscored measures represent basic pairs at the level of ρ indicated. Numbers only, without reference to the nature of the measure, are used here, since the importance is in the identity of numbers. The same number refers to identical measures in each of the samples. All measures are in the locomotor complex.

† Integration is very poor at this stage (see pp. 169–70), and, even at the lowest levels of ρ, groups are small.

The flow of group formation as the level of ρ is lowered is clear from Table 36. It should be noted that the groups formed at the lowest level of ρ used pertain to locomotion, to axial structures, and to the skull. The relationships of ρ to F are clearly portrayed. The different levels at which the groups appear is an excellent illustration of the different degrees of intensity of integration in the various systems. At the lowest level, the locomotor complex includes almost all the measures taken on the system, only 19 (pelvis width), 21 (posterior length of scapula), and 27 (ulna, basal width) being absent. The axial group is complete. It should be noted for this group at $\rho \geq .90$ the anterior measure At_w (atlas width) and the posterior measure Sr_w (width sacral rib) are bonded to the basic pair but not to each other. The least satisfactory group is 1–6–7–8, of the skull and jaws. The restrictive aspect of the basic pair in this instance is such that other integrative patterns are excluded at this level. At the lowest significant level with $P = .01$, $\rho \geq .84$, the group 1–6–7–8–9–10–11–12–13 (9, zygomatic-nasal distance; 10, occipital condyle width; 11, parietal width; 12, jaw height; 13, length masseteric ridge on lower jaw) occurs. At no level are well-differentiated functional groups of skull measures defined. It appears that the necessity for gross measurements, imposed by use of the one- and ten-day samples, has eliminated the selection of measures to portray any finely integrated functional complexes that may exist. Over-all size is the dominant factor. Partial correlation might possibly aid in the detection of masked systems.

TABLE 36*

BASIC-PAIR GROUPS FORMED IN TWENTY-DAY-OLD RATS
AS LEVEL OF ρ IS LOWERED

Leve of ρ	Basic-Pair Groups Formed
$\rho \geq .995$	28–32 (radial length–tibial length)
$\rho \geq .990$	28–30–32 (30, femoral length)
$\rho \geq .985$	28–30–32
	20–24 (pelvis length–humeral length)
$\rho \geq .960$	20–24–26–28–30–32–33 (26, ulnar length; 33, tibular length)
$\rho \geq .950$	20–24–26–28–30–32–33–34† (34, metacarpal length)
	20–22–23–24–26–28–30–32† (23, scapula, posterior length)
	20–22–24–26–28–30–32–33†
	15–16–17 (At_w–Pz9–Pz22) (atlas width, postzygapophysis width, vertebrae 9 and 22)
$\rho \geq .930$	20–22–23–24–26–28–30–32–33–34
	15–16–17
	6–7 (zygomatic width, diastema length)
$\rho \geq .900$	20–22–23–24–26–28–30–32–33–44
	15–16–17, 16–17–18 (18, width of sacral rib)
	6–7
$\rho \geq .860$	20–22–23–24–25–26–28–29–30–31–32–33–34 (25, distal width humerus; 29, metacarpal length; 31, femoral width)
	15–16–17–18
	1–6–7–8 (1, skull width; 8, bulla width)

* Underscored measures represent basic pairs at the level of ρ indicated.

† Very similar groups, with one or two different measures, can appear only when basic pairs have merged, here 20–24 and 28-32. Otherwise, measures will fall into a single group.

Results similar to those in the illustration are encountered wherever basic-pair analyses are made upon appropriate materials. The cases of *R. pipiens* and *S. niger,* taken up earlier in this chapter, and other examples used later in the text provide additional illustrations. Two things in particular stand out. First, the basic pair provides a powerful tool for the simplification of complex arrays of intersecting ρ-groups, and the simplified ρ-groups can generally be interpreted without difficulty. Second, the action of basic pairs masks the true extent of integration and, in so doing, results in a departure from the ideal of consideration of the organism as a unit. With regard to the second point, if the results of basic-pair analyses are viewed as a means to clarification and as the basis for interpretation of the more complex pattern, the loss of totality is minimized. This, in general, is the procedure that we have found most profitable.

PARTIAL CORRELATION

The theoretical and technical aspects of partial correlation as they relate to problems of morphological integration have been presented in some detail in chapter iv. Our purpose at this time is to explore the use of partial correlation as a means of detection of ρ-groups and the relationships of these groups to F-groups. We shall consider examples taken from *Pentremites*, since preliminary studies suggested that this series of samples would yield results under this approach. The factor of size and shape appeared to be a predominant influence in formation of ρ-groups. It is well expressed by measures 31 (length [height] of the calyx) and 32 (width of the calyx). Consequently, these measures are held constant to arrive at $r_{xy.31,\ 32}$.

The godoniform sample of the Paint Creek, which proved too complex for graphic illustration or clear-cut analysis at the level $\rho \geq .90$, yields the patterns shown in Figure 43 under partial correlation ($r_{xy.31,\ 32}$), where $\rho \geq .74$, which includes all values of r significantly greater than zero, $P = .01$. The pattern is simple and lends itself to logical interpretation. The pyriform *Pentremites* from the Paint Creek formation, however, presents a more complex array at this level, one that does not lend itself to clear graphic representation. This pattern, in terms of groups, is shown in Table 37 (for meanings of measurements, see Table 20).

The presence of negative bonds, in considerable number, is in striking contrast to the situation in the godoniform pentremite from the same horizon. It will be noted, however, that every negatively bonded pair includes one of five measures, as follows: 8 (length of ambularcal groove), 23 (length of radial plate), 29 (length of deltoid plate), 30 (width of deltoid plate), and 34 (distance from base of ambulacrum to base of calyx). All these measures are expressions of size, either of some large portion of the organism or of a major element in one of the systems. They all show a strong integration under simple correlation with the measures that were held constant and tend to be less strongly bonded to most of the measures where a negative bond was developed in partial correlation.

Figure 43 presents a graphic portrayal of the Paint Creek pyriform pentremite at $\rho \geq .84$, at which level integration is about as intense as for the godoniform sample at $\rho \geq .74$. Here, three groups stand out. One, 24–25–27–28, is identical with a group in the godoniform pentremite; 4–7–11–14, of the pyriform pentremite, includes measures related to digestion, as does 2–5–6–12 of the godoniform sample. Measure 5 is strongly bonded to this group in the

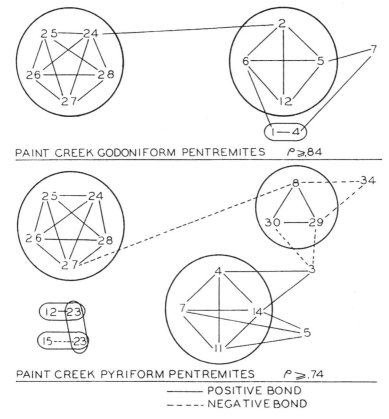

PAINT CREEK GODONIFORM PENTREMITES $\rho \geqslant .84$

PAINT CREEK PYRIFORM PENTREMITES $\rho \geqslant .74$

——————— POSITIVE BOND
– – – – – NEGATIVE BOND

FIG. 43.—Godoniform and pyriform *Pentremites* ρ-groups, formed after partial correlation, based on Paint Creek samples. (For godoniform, $\rho \geq .84$; for pyriform, $\rho \geq .74$.)

TABLE 37

PYRIFORM PAINT CREEK *Pentremites;* GROUPS FORMED AT $\rho \geq .74$, $r_{xy\cdot 31,\ 32}$

1. *ρ-groups, including both positive and negative bonds:*

1–6	1–4–8	4–7–11–14–25
2–6	8–27–34	4–7–14–25–27
2–34	8–30–35	7–14–25–27–28
3–22	3–5–7–15	14–25–26–27–28
4–9	3–12–15–20	24–25–26–27–28
9–12	3–12–20–29	3–4–5–7–11–14
10–12	2–8–26–27	3–4–7–11–14–29
15–32	4–8–26–30	3–4–8–11–21–29
23–26	8–20–29–34	3–4–8–11–29–30

2. *Negative bonds:*

1–8	3–30	7–29	8–26	11–30	21–29
2–34	4–8	8–11	8–27	12–29	23–26
3–8	4–29	8–20	8–34	14–29	26–30
3–29	4–30	8–21	11–29	20–29	29–34

pyriform pentremite, and 7 and 4 show some relationship to the group in the godoniform pentremite. Thus, although the composition is different, which may be considered significant, there is a strong implication of resemblance in the integration of these systems in the two lines. Group 8–29–30 (basal size and two deltoid size measures) of the pyriform pentremite is not, however, present in the other line. The negative correlations of this group to the other two groups are evident at the $\rho \geq .84$ but much less fully developed than at $\rho \geq .74$. The persistence of a size group with measures 31 and 32 held constant and its negative relationships to other groups in the pyriform pentremite marks a sharp contrast between pyriform and godoniform phases in the Paint Creek samples.

We shall not include efforts toward interpretation of the groups at this time, for the illustration is designed to show the effects of partial correlations in increasing clarity of groups by the removal of masking features. It is evident, however, without special interpretation, that such a ρ-group as 24–25–26–27–28 (all basal measures) does imply an F-group with a readily apparent meaning, and that the same is true, although somewhat less immediately obvious, for the other groups, except for pairs, that have been differentiated in the two lines. The effectiveness of such groups for comparison of two forms, such as the pyriform and godoniform pentremites, seems evident without further comment.

Basic-pair analysis can be applied after partial correlation in precisely the same way as has been done for groups formed by simple correlation. This can be used as an alternative to shifting levels of ρ, used to obtain comparable levels of integration for the two samples of *Pentremites*. With the removal of size factors, basic pairs tend to sort out groups that have formed as a result of some other factor. In the case of *Pentremites*, these groups appear to be predominantly related to function and to proximity. For illustration of the results of this process, the six samples of *Pentremites* that have been studied are shown, arranged stratigraphically (see Fig. 67). Basic-pair analysis was performed after partial correlation, $r_{xy.31, 32}$, with ρ at a level for each sample such that all values of r greater than zero are included, where $P = .01$. The interpretation of this series will be taken up in chapter viii. Its importance in the present discussion is found in the clarity and simplicity of the patterns and in the opportunities for comparative studies in and between the two lines.

INDEX OF MORPHOLOGICAL INTEGRATION

The various types of study designed for the detection of ρ-groups and an understanding of their relationships to F-groups provide a number of ways of looking at the problems of morphological integration. Partial correlation and the use of basic pairs are both directed toward reduction in complexities to aid in interpretation of particular systems and the interrelationships of systems.

The shift of levels of ρ permits investigation of integration in its entirety to the extent that this is represented in the sample, or at some level of intensity of integration at which less than the totality is expressed. From such studies it is evident that in some instances (see Fig. 44) the morphology is highly integrated, that is, most proportions are highly intercorrelated. In other cases (Fig. 45) certain parts are largely intercorrelated, but, as a whole, the morphology does not exhibit a high degree of integration. In still other instances widely

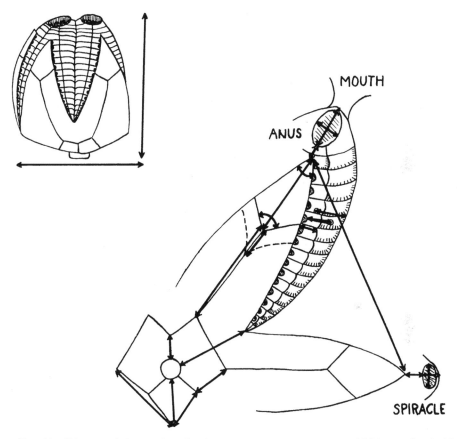

Fig. 44.—Diagram of *Pentremites*, showing a large number of measures highly correlated. All measures indicated by arrows form a single ρ-group.

separated dimensions are highly integrated, apparently independent of topological affinity (Fig. 46).

In the formation of ρ-groups, as discussed earlier in this chapter, these various characteristics become evident as the original groups are analyzed and subjected to the techniques designed to bring out their identities and relationships. Over and above this concern with ρ-groups as such, however, there exists the need for expression of a unifying measurement of morphological integra-

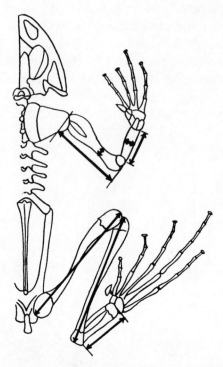

FIG. 45.—*Rana pipiens*, showing two highly integrated areas, correlated measures indicated by arrows, with total morphology not highly integrated.

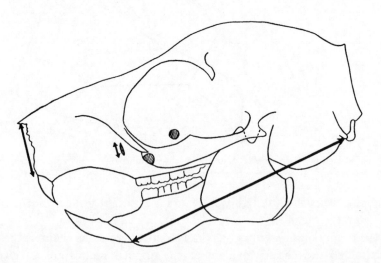

FIG. 46.—*Sciurus niger* skull and jaw, showing a few widely separated measures (indicated by arrows) correlated, with intervening measures not correlated.

tion. It is essential for our aims, in which the totality is important, to have available a simple numerical statement of the extent of total integration, in which all measures and all bonds play some role. The need for such an expression has arisen repeatedly in studies that involved comparisons of over-all integrative patterns, in contradistinction to studies in which the composition of groups and group relationships are of primary concern. The expression, which is considered in detail below, we defined as the *index of morphological integration*.

The required index must simultaneously take into account the "strength" of intercorrelation of the total array of morphological measurements and the way in which the degree of correlation of the various pairs of measurements is distributed, for these two aspects of measure-relationships are basic to the expression of integration.

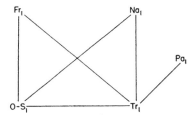

Fig. 47.—Illustration of bonds and non-contained ρ-groups. Based on data on *Captorhinus aguti*, $\rho \geq .95$.

The index of morphological integration is developed formally as follows:

Bo; ρ = The number of observed bonds, evaluated for selected lower limit of ρ for a suite of measurements under analysis. Thus Bo;.95 is the number of bonds evaluated at the level $\rho \geq .95$, when all the measurements have been correlated in all possible ways with each other.

$K\rho$ = The number of non-contained ρ-groups (Fig. 47) evaluated for a selected level of ρ. Note that Bo; ρ is always greater than or equal to $K\rho$. The ratio of the number of observed bonds to the number of non-contained ρ-groups is used to estimate the distribution of the bonds over the total suite of measurements.

$$\frac{\text{Bo}; \rho}{K\rho}. \tag{1}$$

The ratio of the number of observed bonds to the total number of possible bonds estimates the strength of intercorrelation. The greatest possible number of bonds is the value obtained when all measurements in the total suite under study are correlated with all other measurements at a level exceeding the arbitrary level of ρ. This value is simply $(n^2 - n)/2$, where n is the number of measurements in the total suite available for analysis. We thus have

$$\frac{\text{Bo}; \rho}{(n^2 - n)/2}. \tag{2}$$

The least possible integration is found when no bonds exist. The greatest possible integration is achieved when all measurements are bonded in all possible ways (see Fig. 48). In Figure 48, *a*, Bo; $\rho = 0$ and $K\rho = 0$. In Figure 48, *b*, Bo; $\rho = 10$ and $K\rho = 1$. In both cases, $n = 10$, so that the greatest possible number of bonds is $(n^2 - n)/2 = 45$.

When integration is at a maximum, the product of the two ratios described above is as follows:

$$\frac{\text{Bo};\rho}{K\rho} \frac{\text{Bo};\rho}{(n^2 - n)/2} = \frac{(n^2 - n)/2}{1} \frac{(n^2 - n)/2}{(n^2 - n)/2} = \frac{n^2 - n}{2}. \tag{3}$$

For convenience, the writers prefer that the index range from 0 to 1. The product of two ratios is multiplied by the factor

$$\frac{1}{(n^2 - n)/2},$$

so that the maximum integration becomes 1. When integration is at a mini-

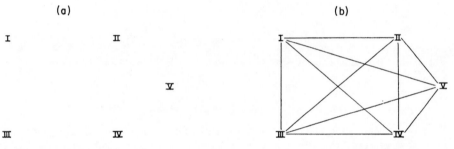

(a)　　　　　　　　　　　　　　(b)

FIG. 48.—Hypothetical bonding, showing (*a*) measures with no bonds and (*b*) measures completely bonded.

mum, for example, when Bo; $\rho = 0$, the product of the two ratios multiplied by the correction factor

$$\frac{1}{(n^2 - n)/2}$$

is

$$\frac{\text{Bo};\rho}{K\rho} \frac{\text{Bo};\rho}{(n^2 - n)/2} \frac{1}{(n^2 - n)/2} = \frac{0}{0} \frac{0}{(n^2 - n)/2} \frac{1}{(n^2 - n)/2}, \tag{4}$$

which is indeterminate. We shall therefore define the function $I\rho$ as is shown in equations (5) below, for all Bo; $\rho \geq 1$. When Bo; $\rho = 0$, we arbitrarily define $I\rho$ to be 0, as we require for a consistent index. We now have the full index of morphological integration, which will be denoted by $I\rho$, where

$$I\rho = \frac{\text{Bo};\rho}{K\rho} \frac{\text{Bo};\rho}{(n^2 - n)/2} \frac{1}{(n^2 - n)/2},$$

which reduces to

$$I\rho = \frac{4\,(Bo;\rho)^2}{K\rho\,(n^2-n)^2}\,, \qquad\qquad \text{when } Bo;\rho \gtrless 1,\quad (5)$$

$$I\rho = 0\,, \qquad\qquad \text{when } Bo;\rho = 0\,.$$

For illustration of the behavior of the index $I\rho$, we shall use a simple hypothetical case, in which a suite of six measures is involved and the increase

ρ Level	$Bo_;$	$K\rho$	$I\rho$	
$\geq .99$	$Bo_;\,.99 = 0$	0	$\dfrac{4(0)^2}{0(30)^2}=0$	
$\geq .95$	$Bo_,\,.95 = 3$	3 (I,II) (II,V) (III,IV)	$\dfrac{4(3)^2}{3(30)^2}=.0122$	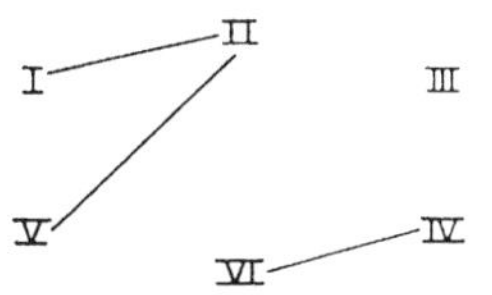
≥ 85	$Bo_,\,85 = 7$	3 (I,II,V,VI) (II,III,VI) (III,IV)	$\dfrac{4(9)^2}{3(30)^2}=.1200$	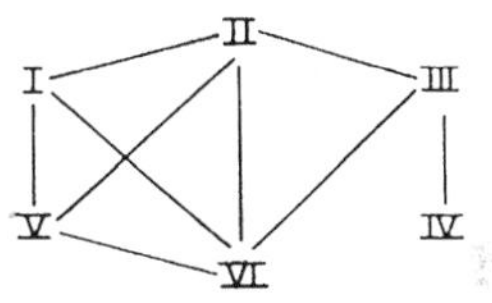
> 0	$Bo_;\,>0 = 11$	2 (I,II,V,VI) (II,III,IV,VI)	$\dfrac{4(11)^2}{2(30)}=.2688$	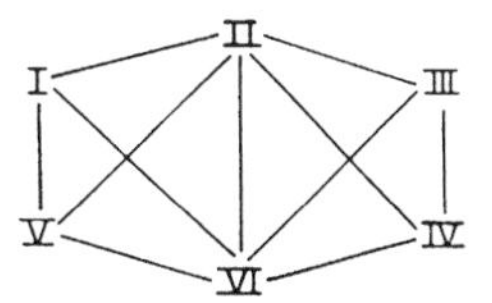

FIG. 49.—Diagram of behavior of $I\rho$ as level of ρ is lowered, as explained in text

in the number of bonds as the level of ρ is dropped is illustrated. At each level the index is computed. Figure 49 contains the relevant data and a diagrammatic representation of the arrangement and number of bonds at each level. Since the number of measurements is six, the maximum number of possible bonds is $(6^2 - 6)/2 = 15$. As the level of ρ drops, the number of bonds observed $(Bo;\rho)$ increases, and, in this simplified example, the value of the index

Animal	$I\rho$ is read as Index (I) evaluated $\geq \rho$			
<u>Sciurus niger</u> (extant squirrel)	$I_{.91} = 3 \times 10^{-7}$	$I_{.89} = .000011$	$I_{.87} = .000062$	$I_0 = .00014$
Sample size, N=30 No. of measurements, n=45	$I_{.82} = .00029$		$I_{.80} = .00040$	$I_0 = .00066$
<u>Diplocaulus recurv.</u> (fossil amphibian of Permian age) N=8, n=14	$I_0 = .00731$			
<u>Diplocaulus magnicornis</u> N=32, n=14	$I_{.98} = .00031$		$I_0 = .11425$	
<u>Trimerorachis insignis</u> (fossil amphibian of Permian age) N=20, n=14	$I_{.95} = .01195$		$I_0 = 1.0000$	
<u>Hyopsodus</u> (fossil mammal of Eocene time)	<u>H. miticulus</u> $I_0 = .0017$		<u>H. despiciens</u> $I_0 = .0014$	
	<u>H. lysitensis</u> $I_0 = .0004$		<u>H. mentalis</u> $I_0 = .0016$	
	<u>H. paulus</u> $I_0 = .0002$		<u>H. sp.</u> $I_0 = .0008$	

FIG. 50.—Examples of $I\rho$ in various organisms. Read as indicated in directions in the chart

156

Animal	$I\rho$ is read as Index (I) evaluated $\geq \rho$		
Knightia #1 (fossil fish of Eocene age) Sample size, N=15 No. of measurements, n=18	$I_{.95} = .00615$	$I_{.90} = .02939$	$I_0 = .15385$
Knightia #2 N=12, n=18	$I_{.95} = 0$	$I_{.90} = 0$	$I_0 = .00103$
Knightia #4 N=11, n=18	$I_{.95} = 0$	$I_{.90} = .000539$	$I_0 = .004709$
Rana Pipiens (extant frog) N=30, n=50	$I_{.97} = .0000004$	$I_{.95} = .000642$	$I_{.91} = .000202$
Pentremites sp. (extinct Blastoids of Mississippian age)	"Godoniform" $I_0 = .0379$ (Renault) "Godoniform" $I_0 = .0553$ (Paint Creek) accept $\|r > .781\|$ "Godoniform" $I_0 = .0031$ (Golconda) "Godoniform" $I_0 = .0037$ (Glen Dean)	Pyriform $I_0 = .4951$ (Paint Creek) Pyriform $I_0 = .0177$ (Golconda)	

FIG. 51.—Additional examples of $I\rho$ in various organisms (see also Fig. 50). Read as indicated in the chart.

of morphological integration $(I\rho)$ increases. It is thus clear that comparisons of one population with another, in terms of the index $I\rho$, must be made at approximately the same levels.

A survey of various analyses of different animal groups, both extant and fossil, is shown in Figures 50 and 51 to illustrate the behavior of the index of morphological integration as ρ is lowered within a single population and also to allow comparison of one animal population with another at commensurate levels of ρ. The levels used are those that have been found practical for work upon the various groups of organisms listed. The discussions of various studies of morphological integration that follow in the succeeding chapters make various uses of this index, particularly in ontogenetic and evolutionary series. Interpretations of the meanings of many of the values of $I\rho$ in the tables will be in these later sections of the book.

CHAPTER VII

Morphological Integration in Ontogeny
and in Non-growing Systems

Irrespective of the stage or stages in the life-history of organisms at which integration is studied, the patterns that are revealed must be considered the result of the totality of developmental processes active prior to the beginning of the ontogenetic periods studied. Clues to the underlying genetic and developmental factors should thus be found in the patterns of integration, just as they are in particular single characters. The pattern constitutes a more sensitive gauge, however, since even slight changes are often detectable in modifications of the relationships between measures. A broad area of investigation of the beginnings, development, and modifications of morphological integration lies open for study, and it is to be hoped that eventually research may be specifically oriented in this direction.

Severe limitations in the field of ontogenetic study are imposed upon our work by the use of only skeletal elements, since these do not come into existence until rather late ontogenetic stages in most organisms. Our preoccupation with studies directed toward interpretation of fossil organisms and our lack of competence in the fields of experimental embryology pose this limitation; investigators with other orientations could probably conduct successful studies of integration on early embryological materials and upon later stages as well by the use of characteristics of the soft anatomy. Even within the limited range available to us, however, some interesting insight into the ontogeny of morphological integration is possible, although a probe of fundamental causal factors is largely denied. This chapter is devoted in part to a consideration of certain empirical studies and will, we hope, serve as an invitation to studies at other levels of organization.

There are, in general, two ways in which the problems of morphological integration in ontogeny may be viewed. One involves consideration of the members of an ontogenetic series as a sample of a population composed of the individuals from all ontogenetic stages at which measurements can be made or, because of practical limitations, of a considerable portion of this ontogenetic

159

span. The other considers ontogenetic stages defined separately, either temporally or physiologically. In this situation, the story of ontogeny is revealed
through analyses and comparisons of discrete, successive samples of the series
of populations at the selected stages. Any single sample provides knowledge of
the patterns that result from the events of its development prior to the stage
sampled. Adults, provided that there is terminal growth, may be considered as
one stage from this point of view, just as may any earlier assemblage that has
been properly defined and limited.

A special case arises in structures that do not grow after histogenesis has been
completed, a case exemplified by the dentitions of many mammals. Since non-
growing structures also reflect the ontogenetic processes that have contributed
to their formation, "non-growing systems" will be considered in the later parts
of this chapter.

The general mathematical basis for studies of morphological integration in
samples of populations, those with a considerable span of ontogenetic time as
well as those with a limited span, was developed in chapter iv. We shall now
be concerned more specifically with the biological aspects of these types of
samples with special reference to studies that involve each type. We shall term
the type of sample that has been drawn from a population that covers a major
spread of ontogenetic time a *sample of an ontogenetic series*. Samples drawn
from individuals of a single ontogenetic time, we shall term a *sample of an ontogenetic stage*.

COMPARISONS OF SAMPLES OF ONTOGENETIC SERIES
AND ONTOGENETIC STAGES

The best examples of ontogenetic series that have been analyzed for patterns
of integration in our work are those of *Pentremites*, briefly discussed in chapters
v and vi. In the six samples the size range over which measurements could be
made was widely represented. We may infer with confidence that a considerable span of ontogenetic time was involved. The five samples of the albino rat,
two of which were used for illustration in the last chapter, represent samples of
ontogenetic stages. The results of the analyses of *Pentremites* are presented in
various places in the book, as they are pertinent to one or another phase of the
discussions and do not need to be treated specifically from the standpoint of
ontogeny in the present chapter. The contrary is true for the rats, however,
for these samples were obtained specifically for investigations of ontogenetic
stages and have not been treated in their entirety elsewhere in the book. These
samples are considered in *detail* later in the present chapter. At present we shall
be concerned more with the general theoretical aspects of the characteristics
of samples from ontogenetic series and stages.

The samples of *Pentremites* demonstrate clearly that tightly knit integrative
groups emerge, as samples of ontogenetic series are analyzed for morphological

integration. This is, of course, confirmed by other studies, such as those of *Diplocaulus* and *Trimerorhachis* (Olson, 1953) and *Captorhinus* (Olson and Miller, 1951*b*). Likewise, highly integrated patterns of measures emerge when ontogenetic stages are analyzed, as exemplified by the rat series, the dentitions in *Aotus* and *Hyopsodus* (see pp. 178–210), and the skeletal elements in the samples of pigeons (Appendix C).

We should not expect patterns at a particular ontogenetic stage to show more than an approximate resemblance to patterns of the same species derived from a single sample ranging over an extended ontogenetic series. An ontogenetic series, in the sense used, designates a group of individuals whose sizes, which imply ages, range over most of the growth stages of the hard parts. Both the mathematical considerations in chapter iv and biological considerations support our contention. At present, this problem must be considered only from a theoretical point of view, since our materials have not permitted an empirical test for comparisons in an actual case.

Integrated groups of measures in ontogenetic series in general represent biological systems that persist with a high level of integration, resulting from growth relationships of the parts over a significant part of the ontogeny of a species and are revealed when the individuals that exist over such a range are analyzed. Such systems are presumably little affected by fluctuations in environment during their existence and are adapted to carry out functions of life under a wide variety of internal and external circumstances. Selection in such systems would be in the direction of optimal fitness to the gross conditions of the existence of the organism. Such systems may be considered to provide insight into heredity as it contributes to major, adaptive, stable patterns.

Although the systems revealed in ontogenetic series commonly provide a part of the total integrative pattern, by no means all the groups of associated measures that occur at stages in ontogeny are necessarily revealed in such series. The fact that different systems and components of the same systems grow at different rates during ontogeny and that rates of different parts do not always change together renders this result inevitable. It is clear, for example, that the complete organization of the adult is not expressed *in toto* at some much earlier stage. Differences tend to increase with the degree of developmental disparity of the stages considered, but no direct relationship of a temporal nature occurs. This is most evident in the major changes that occur over very short periods of time as developmental thresholds are reached and passed, as in metamorphism, rapid changes at the time of onset of sexual maturity, and so forth. Comparison of the constitution of integrative patterns requires that the same measures be made on all stages involved, thus comparisons such as a pre-metamorphic and a post-metamorphic stage cannot be made from this point of view. It is possible, however, to compare *levels* of integration between such stages by application of the index of morphological integration.

The phenomena represented by the disparity of integration between onto-genetic stages are most meaningfully conceived for the present discussion as ex-pressions of ontogenetic adjustments to the altering circumstances of develop-ment that maintain functional integrity as organization continuously alters toward the establishment of the final adult pattern. Integration at a particular ontogenetic stage may thus be considered (1) *a product of the ontogeny to that stage* and (2) *related to adaptation within the broader ontogenetic pattern manda-tory to the development of viable adults*. Patterns of integration that are present but do not persist to other stages or are not expressions of patterns of the on-togenetic series may be presumed to relate specifically to a phase of develop-ment critical to that stage in the dual, but not independent, roles of assuring temporary adaptation and of producing an organizational basis for succeeding stages.

Selection, of course, is operative at all stages and must be considered effective in the establishment and preservation of integration at particular stages and in the maintenance or modification of patterns that persist in ontogenetic series. Morphological integration at a given stage can provide important infor-mation on ontogeny from this point of view. If it is followed through a series of integrative stages, there emerges a quantitative statement of the flow of organization toward the production of the adult pattern or any other terminal stage that may be sought.

Before we turn to the study of the rat, we shall note briefly the special case on non-growing structures, exemplified in our studies by the dentitions in *Aotus* and *Hyopsodus*, a primate and condylarth, respectively. From one point of view, such structures are no different from others, for they develop through a series of ontogenetic steps, genetically determined, and progress in harmony with the other structures of the organism. Such structures, however, are fully developed at the time that they first assume their special functional role, and they do not participate in the functional activities of the organism at stages prior to the time that they assume this role. They are not then subject to the restrictions that result from the need for participation in activities of the organism during their formation, nor are they subject to such modifications of form as usage, prior to full formation, may directly impose upon structures that are operative during their growth.

It seems logical to suppose that structures such as the teeth in question would present a more explicit insight into the effects of genetic constitution upon the phenotype than would those that assume their basic form early and pass through growth stages that are necessarily functional prior to the attain-ment of adult form. They pose a particular series of problems from the stand-point of integration, however, since it is impossible to obtain anything but the grossest measures prior to complete development. Ontogeny cannot be studied from the basis of integration in the same way that it can be in structures in which comparable measures may be taken on individuals at many stages of ontogeny. Integration is obtained only in the final product. The situation is thus comparable to that which obtains when a study of integration is carried out on a sample in which elements have reached the point of terminal growth.

Studies of dentitions are of primary importance among fossil vertebrates, especially in mammals, because of good preservation and because teeth are excellent indicators of genetic affinities and differences. Studies of ontogenetic *series* are likewise important in the analysis of extinct groups and the study of evolution through geological time. In general, this is not the case for studies that involve ontogenetic *stages* prior to the final, adult, stage. Rarely is it possible to determine with the degree of accuracy necessary the ontogenetic age of fossil individuals. There are, of course, exceptions where there is morphological evidence of age in growth lines, molt stages, and so forth or where preservation has sorted out age groups. Such cases are rare, and the results must always be open to some question. Morphological integration in ontogenetic *stages* is best studied among living organisms and, thus limited, contributes to our knowledge primarily as it is used for the determination and establishment of principles of development. In this role, however, it can contribute in an extremely important way to our knowledge of evolution, providing not only an insight into the patterns of development among living organisms but a basis for interpretation of phenomena witnessed over longer periods of time in fossil organisms. A healthy balance between studies of modern and extinct populations is necessary to derive the maximum information on the many facts of evolution to which morphological integration can be related.

ONTOGENETIC STAGES OF THE ALBINO RAT

The samples used in the study reported in the following pages consist of individuals drawn from five temporal, ontogenetic stages of the albino rat. The stages are as follows: one-day, ten-day, twenty-day, forty-day, and adult (250-day), based on time after birth. Details of the constitution of the samples are included in Appendix G. Thirty-four measurements were taken on each individual of the first four samples and were, of course, the same for each individual. Only thirty-three were taken on the adults, since disarticulation precluded the possibility of measuring pelvic width in a number of specimens. The measurements used are listed in Table 31 and shown graphically in Figure 41.

The objectives of the study, from a limited point of view, involved evaluation of the phenomena of morphological integration in ontogeny of the rat, as follows: (1) the course of $I\rho$, the index of morphological integration, in ontogenetic time; (2) the origin and development of groups of measures (ρ-groups) in ontogeny and the biological meanings of groups and their changes; (3) an interpretation of the causal factors of the phenomena revealed in the attack on items 1 and 2 above.

In a broader sense, this experiment was designed to provide information necessary for additional and more definitive work. This second role will presumably be the most important eventually, but for the present we shall be concerned only with the results as they pertain to the group studied. These results are of considerable interest in themselves and suggest various lines of profitable work for further study. They do not, however, provide a sound basis for generalizing about the roles of integration in ontogeny, and we have attempted no extrapolations from the specific case treated in the report.

The procedures of analysis in general follow the patterns that have been developed earlier in this book. Each sample was treated separately. Simple intercorrelation was carried out, and ρ-groups for each sample were formed at the following levels, where $P = .01$: $\rho \geq .96$, $\rho \geq .95$, $\rho \geq .93$, $\rho \geq .90$, and $\rho \geq .86$.[1]

A complex series of intersecting ρ-groups was found to exist below the very highest levels ($\rho \geq .96$, $\rho \geq .95$), and it was necessary to use a simplifying procedure in analysis. The basic-pair technique was employed, since our interest was primarily in the meanings of the groups in terms of composition rather than in the general factors revealed by partial correlation. Basic pairs, at the level $\rho \geq .86$, were determined for each sample separately. In the analysis the basic pairs of the one-day sample formed the basis of grouping throughout, except as new groups appeared, as taken up in detail in a later section.

The matter of rates of growth clearly is important in the analysis of change of $I\rho$. We shall now assess the variability of growth rates of the several measurements within the span of a defined ontogenetic range. This will be done by devising an index of the variability of increment-increase for the various measurements. This index will be referred to as V_{pi}. We seek a single number which represents all the measures over a particular ontogenetic range and will be available for comparison with $I\rho$. Note that $I\rho$ and V_{pi} represent two quite different analyses of the same data; $I\rho$ is based on the covariance of pairs of measurements, while V_{pi} is based on an averaging over the increment-increase in size of the various measurements. We may thus treat the information given below as derived from two complementary sources. The method of assessment of the average increase in size is now described. We shall consider morphological measurements taken at various time intervals and use the following rotation:

X_{ij} = Morphological measurement i, which takes values
$$X_{i_1}, X_{i_2}, \ldots, X_{i_n} \text{ for } n \text{ individuals}.$$

Measurement X_i is taken on the various individuals at times $t_1, t_2, \ldots, t_k$, where t_k represents the termination of the kth ontogenetic stage. The "ontoge-

---

[1] For the one-, ten-, and forty-day rats only. The high order of integration at the twenty-day and adult stages rendered grouping at this level unnecessary, except for the calculation on $I\rho$. No evaluation below $\rho \geq .86$ was used. Comparisons between samples were of little value in view of the high integration at the twenty-day and adult stages.

netic stages" are arbitrary unequal time intervals spanning the life-period of the animal.

We thus denote $X_{i,\,t_k}$ as the morphological measurement X_i taken at time t_k. And now we define the average value of measurement X_i for n individuals measured at the termination of the kth ontogenetic stage at time t_k as

$$\bar{X}_{i,\,t_k} = \sum_{j=1}^{n} \left(\frac{X_{ij,\,t_k}}{n} \right) \qquad j = 1,\ldots,n; \qquad i,k \text{ fixed}.$$

The percentage increase for measurement i from time t_{k-1} to t_k (implying percentage increase accured during ontogenetic stage k) is computed as follows:

$$\left(\frac{\bar{X}_{i,\,t_k} - \bar{X}_{i,\,t_{k-1}}}{\bar{X}_{i,\,t_{k-1}}} \right) = P_i \Big]_{t_{k-1}}^{t_k},$$

where $P_i \Big]_{t_{k-1}}^{t_k}$ represents a percentage increase for $\bar{X}_i$ evaluated over the interval $t_k - t_{k-1}$. We shall consider the frequency distribution of all P_i's ($i = 1, 2$, etc.; N for the number of different anatomical measurements). For a fixed time interval, the distribution of P_i's may be characterized by a mean and a variance, where

$$\bar{P}_i = \sum_{i=1}^{N} \frac{P_i}{N}$$

and

$$\text{Var}\,(P_i) = \sum_{i=1}^{N} \frac{(\bar{P}_i - P_i)^2}{N}.$$

A final term is now defined. It is the coefficient of variability, V, for those P_i's which contribute to the frequency distribution just described. Note that a separate V may be computed for each time interval of the ontogenetic range:

$$V_{P_i} = \frac{100\,\sqrt{\text{Var}\,(P_i)}}{\bar{P}_i}.$$

Ip and Growth

For all measures in each sample at $\rho \geq .90$, $\geq .93$, $\geq .96$, calculated according to the formula

$$I\rho = \frac{4\,(Bo;\rho)^2}{K\rho\,(n^2 - n)^2}$$

(see p. 155), $I\rho$ is shown in Table 38. At all recorded levels of ρ, $I\rho$ shows a pattern of decreasing from one to ten days of life (except, of course, at $I_{.96}$, where $I\rho = .00000$ at one day and ten days), a sharp rise from ten days to twenty days, a moderate drop at forty days, from the twenty-day stage, and a rise at the adult stage from the forty-day stage. There appear to be phases of "high" and "low" integration. The fact that there are only a few selected stages does not permit a specific statement of the precise time at which shifts are initiated, and selection of other stages might present a somewhat different picture. The general phase relationships, however, almost surely express a real phenomenon in the ontogeny of the rat.

The use of $\bar{X}_i$ (see p. 165) provides the results expressed in Table 39, where V of $\bar{X}_i$ between the stages is entered. The comparison of $I\rho$ and V_{P_i} for the one- to ten-day sample gives a biased result with respect to other stages. With $I\rho$ very low, as values of $I\rho$ are in the second stage, the shift in its value depends

TABLE 38

$I\rho$ for Totality of Measures of the Albino Rat

Age of Sample	$I_{.90}$	$I_{.93}$	$I_{.96}$
One day.............	.00059	.00019	.00000
Ten days............	.00028	.00001	.00000
Twenty days........	.01029	.00317	.00076
Forty days.........	.00432	.00145	.00031
Adult..............	.00804	.00414	.00079

TABLE 39

V_{P_i} at Ontogenetic Stages of Albino Rat

Stages	V_{P_i}
One-day to ten-day....................	40.42
Ten-day to twenty-day................	44.90
Twenty-day to forty-day..............	85.30
Forty-day to adult...................	47.80

on relatively few measures. We know nothing of the other measures, but V_{P_i} takes into account all measures. At the other stages, essentially all measures are involved in $I\rho$. The limiting case for the one–ten stage, of course, is seen at the level of $\rho \geq .96$, where $I\rho$ is zero for both and *no* relationship is expressed.

The three later stages, in which comparison is based on roughly comparable suites of measures in $I\rho$, show a striking relationship between the two. The strong rise of $I\rho$ from ten to twenty days is accompanied by a value of V_{P_i} of 44.90, with the ten-day stage initial and the twenty-day stage terminal. The drop in $I\rho$ from twenty to forty days is accompanied by a V_{P_i} of 85.30, and the rise in $I\rho$ from forty days to the adult stage, by a V_{P_i} of 47.80. The relatively high value of V_{P_i} from the twenty- to the forty-day stage marks a relatively wide divergence of rates of growth of the variables relative to their means at the twenty-day stage. During this time the *number* of bonds and the *intensity* of their formation into ρ-groups are reduced, as witnessed by the change in $I\rho$.

Between the forty-day and adult stages, on the contrary, $I\rho$ increases, and V_{P_i} is relatively low.

The over-all trend in ontogeny, as seen in the series of stages, is toward an increase in morphological integration. This is clear in Table 38 for $\rho \geq .93$, .96, but the adult stage has a lower $I\rho$ for $\rho \geq .90$ than does the twenty-day stage. This reflects a differentiation in the locomotor groups, as is brought out in the next section. Beneath the general trend for increase of $I\rho$, there appear to be phases of increase and decrease. These are related to modification of the relative growth rates, expressed in V_{P_i} of the structures involved. Where wide divergence of rates is the rule, as shown by a high value of V_{P_i}, integration tends to drop, and increase in integration is accompanied by less divergence in rates. Whether this pattern of phases in $I\rho$ is general or not we cannot say from the present evidence. Similar studies must be performed on other groups, and more detailed studies, with respect to both characters and stages, are required to relate the observed patterns to the various physiological aspects of growth that are presumably basic to them. Some additional light, however, is shed upon development by a study of the formation of ρ-groups in ontogeny, as presented in the section that follows.

The Formation of ρ-Groups and the F-Groups

In the following analysis, F-groups are derived in a strictly quantitative way through the formation of ρ-groups and the use of basic pairs to reduce intersects between them. Initially, basic pairs, where $\rho \geq .86$, $P = .01$, were formed for each of the five samples separately. Tests showed that similar large ρ-groups, ones that differed by only a few bonds, were formed at each stage, even though the basic pairs were somewhat different. This was less true for small ρ-groups, but most of the basic pairs that failed to form large ρ-groups did not themselves form discrete groups, that is, they were in large ρ-groups only when merged with another basic pair and could thus be dropped from consideration. The exceptions, where the basic pairs not found at the one-day stage did form important groups, were considered to represent new significant groups that were developed during ontogeny. In view of these considerations the method of attack was as follows: Basic pairs from the one-day sample were carried through the ontogenetic series, that is, applied to the one, ten, twenty, forty and adult samples. As new basic-pair ρ-groups appeared, based on pairs not present in the one-day sample, these were carried on through the remaining stages as group-formers. All basic pairs were dropped from consideration which were (1) represented by a similar[2] pair in the one-day sample, (2) failed to form ρ-groups, or (3) formed groups only with each other.

[2] Similar pairs are those which are different in separate samples but have the same group-forming capacities. Thus in two samples of the series, 27–28 and 26–28 may both form the group 24–25–26–27–28. In this event one, say 27–28, may be used in both samples without altering the results that would be obtained, were 27–28 used for one and 26–28 for the other.

The basic pairs in the one-day stages, $\rho \geq .86$, $P = .01$, are as follows:

28–32............	Ra_l–Ti_l[3]		1–13.........	Sk_l–Mas_d
25–29............	Hu_w–Mc_l		3–15.........	Pa–Ips–At_w
22–24............	Sc_l–Hu_l			

Three of these basic pairs are related to limb measures, 28–32, 25–29, and 22–24. One relates to the skull, 1–13; and one includes a skull and vertebral measure, 3–15. Of the five basic pairs, 28–32 and 1–13 are the principal group-formers; pair 3–15 does not form an independent group; and 22–24 and 25–29 act discretely, as they coalesce below the level of ρ of independence to form the ρ-group 22–24–25–29, a group found only in the one-day stage. Hence 28–32 and 1–13 are followed through the ontogenetic series, with attention to 3–15, 22–24, and 25–29 only in order to determine their fate in more advanced ontogenetic stages.

The group formed by 28–32 will be called the *locomotor group*, and that formed by 1–13 the *skull group*. The former is clearly a functional group, and the F-group that it represents will be designated $F'f_{lo}$. The latter involves both functional aspects and others that relate to proximity. It will be designated $F'g_{sk}$.[4]

The Groups at the Ontogenetic Stages

ONE-DAY STAGE

Group composition at the one-day stage is shown in Table 40. Both $F'f_{lo}$ and $F'g_{sk}$ are present at $\rho \geq .93$. The former becomes relatively large at $\rho \geq .86$,

TABLE 40

ρ-GROUPS OF THE ONE-DAY SAMPLE OF THE ALBINO RAT

Level of ρ	$F'f_{lo}$	$F'g_{sk}$	Others
$\geq .93$......	28–30–32	1–13	25–29
$\geq .90$......	20–23–28–30–32–33	1–13–14	25–29, 13–15
$\geq .86$......	20–23–28–30–32–33–34	1–13–14	22–24–25–29–13–15

but the latter includes only one measure in addition to the basic pair. Locomotor function is well expressed in the integration, but integration of the skull is poorly developed. Group 22–24–25–29 is an independent forelimb group at this stage, including Hu_l, Hu_{dw}, Scs_l, and Mc_l. This differentiation does not

[3] Numbers will be used throughout for reference for the sake of simplicity, but abbreviations are added when critical. For meanings of numbers see Table 31 and Fig. 41. It will be noted that the numbers are arranged in series that more or less correspond to the various functional groups.

[4] Because of the relatively poor development of many of the definitive measures of the head in the one- and ten-day samples, it was necessary to use only gross measures, to assure that the same measures could be obtained on each sample. Strictly functional groups are thus not well represented, although a detailed breakdown will give some evidence of their existence.

occur in later ontogenetic stages. Here the relationship of the humerus and measures of the spine of the scapula presumably reflects some sort of functional differentiation, but the studies necessary for confirmation have not been made.

TEN-DAY STAGE

Low integration at the ten-day stage, as expressed by $I\rho$ (Table 38), is found in the discrete groups as well as in the over-all integration. This low integration is thus a general condition and not the result of the absence of some groups and the "normal" development of others. No groups occur at $\rho \geq .93$; only the pair 22–23 is present. The locomotor group, $F'f_{lo}$, is present at $\rho \geq .90$ but is only partially developed (see Fig. 52). It will be noted in this figure that there

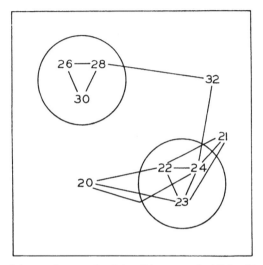

FIG. 52.—ρ-groups of the ten-day sample of the albino rat and additional bonds at level $\rho \geq .90$

are two "centers," 22–23–24 (Scs_1, Scp_1, Hu_1) and 26–28–30 (U_1, Ra_1, Fe_1). Numbers 20 (Pe_1) and 21 (Sc_w) are closely bonded to the former. Except for the length of the humerus, this is a "girdle group," in contrast to the "limb group" of 26–28–30. There is no evidence of the forelimb group, 22–24–25–29, at the one-day stage. No skull group appears at $\rho \geq .90$. At $\rho \geq .86$, not figured, the only modification is a slight increase in the number of bonds of $F'f_{lo}$. To bring out the most complete grouping possible, with r significantly greater than zero at $P = .01$, groups were formed at $\rho \geq .86$. These are shown in Table 41.

The condition of the locomotor group is of particular interest. The group 22–24–25–29 does occur as a discrete group, as at the one-day stage, *but* 22–24 and 25–29 also occur with 28–32 in large locomotor complexes. The "forelimb complex," 22–24–25–29, which was discrete and unmerged at the one-day stage, here is showing a tendency to merge with the general locomotor group. The coalescence, however, is not complete, and this fact has a disruptive effect, producing small incomplete ρ-groups, which in summation result in a low level of integration. The ten-day stage appears to have "caught" ontogeny in a period of reorganization, the results of which are evident at the twenty-day stage.

One other item of interest is seen in the ten-day stage. At $\rho \geq .86$, measures 16 (Pz9$_w$) and 17 (Pz22$_w$) appear as shown in Figure 53. They are strongly bonded at this time of first appearance to 24 and 32, measures of F$'f_{lo}$. The emergence does not appear to have any striking significance when viewed at

TABLE 41*

ρ-GROUPS OF THE TEN-DAY SAMPLE OF THE ALBINO RAT
($\rho \geq .86, P = .01$)

$$F'f_{lo}\ldots \begin{cases} 20\text{--}23\text{--}\underline{25}\text{--}\underline{28}\text{--}\underline{29}\text{--}30\text{--}\underline{32}\text{--}33 \text{ (largest group} \\ \quad \text{present)} \\ 22\text{--}23\text{--}24\text{--}26\text{--}\underline{28}\text{--}32 \\ \underline{22}\text{--}24\text{--}\underline{25}\text{--}29 \end{cases}$$

$F'g_{sk}\ldots$ $1\text{--}3\text{--}\underline{13}\text{--}14$

* Basic pairs belonging to the F-group indicated are underscored.

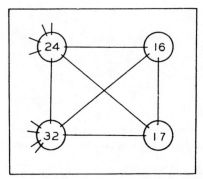

Fig. 53.—Bonding of measures 16 and 17, axial group, at ten-day stage in albino rat, at level $\rho \geq .86$.

this stage alone, for occasional pairs such as this are often encountered. It is, however, of particular interest in view of the development of the group including these measures at later stages. In this guise, it presents further evidence of the highly formative character of the ten-day stage.

TWENTY-DAY STAGE

The twenty-day stage represents a time of very high integration. At $\rho \geq .95$, the F$'f_{lo}$ group is virtually complete,[5] as follows: 20–22–24–26–28–30–32–33,

[5] "Completeness" in the sense used here refers to inclusion of the maximum number of measures included in the largest ρ-group at the maximum integration (given the lowest value of ρ, where r is greater than zero at $P = .01$) and at any stage studied. Not all measures of the limbs or girdles are necessarily included. This definition is a direct consequence of the purely quantitative approach to F. The complete F$'f_{lo}$ group is thus 20–22–23–24–26–28–30–32–33–34.

with 23 and 34 missing only one bond each. Measures 25 and 29 are not present, but 22 and 24 are an integral part of the group. Measures 25 and 29 appear occasionally here and later as single entries in small ρ-groups, and group 22–24–25–29 is completely gone, with 22–24 incorporated into $F'f_{lo}$. The disruptive tendency of these measures, seen in the ten-day stage, is now obliterated, and, in the process, 25 and 29 have been eliminated from the locomotor complex as important elements in integration.

Measures 16 and 17 form a basic pair at this stage, which was not the case at earlier stages. This pair, with measure 15 (At_w), forms a new F-group, an axial group which we shall designate "$F'f_{ax}$." It is present at $\rho \geq .95$, in great contrast to the level of $\rho \geq .86$, at which 16 and 17 occurred at the ten-day stage, bonded to $F'f_{lo}$.

TABLE 42

ρ-GROUPS OF TWENTY-DAY SAMPLE OF RATS

($\rho \geq .90$)

3–11, 21–33, 4–18–31, 5–17–34, 6–7–22–23–33
6–15–22–23–26–33–34, 7–20–22–23–24–25–28–30–32–33
1–8–10–13–15–16–17–20–22–23–24–25–26–28–30–32–33–34
1–9–15–20–22–24–26–28–30–31–32–33–34
1–10–12–20–22–23–24–26–28–30–32–33–34
1–10–13–15–16–17–20–22–23–24–26–28–30–31–32–33–34
1–13–17–20–22–26–28–29–30–31–32–33–34
1–8–13–17–20–22–26–28–29–30–32–33–34
1–9–20–22–26–28–29–30–31–32–33–34
1–10–11–12–20–22–23–24–26–28–30–32
1–10–11–13–20–22–23–24–26–28–30–31–32
13–14–20–24–28–30–32
13–14–20–28–29–30–32
16–17–18–19–22–26–31
16–17–18–22–26–31–32
13–15–16–17–19–20–22–23–24–25–26–30
13–15–16–17–19–20–22–23–24–26–30–31

Here $F'g_{sk}$ is represented only by pairs at $\rho \geq .95$, but at $\rho \geq .90$ the ρ-group 1–8–10–13 is developed, and measures 9, 11, 12, and 14 are strongly bonded to it. Although this group forms at a lower level of integration than $F'f_{lo}$ and $F'f_{ax}$ at the twenty-day stage, it has a distinctly higher integration than at either the one- or the ten-day stage.

The evidence of the various groups formed at the twenty-day stage argues strongly that this stage represents a culmination, or near-culmination, of the organization witnessed first at the one-day stage. Three main groups are highly integrated and in the intersects of ρ-groups before basic-pair analysis (see Table 42) are shown to be strongly related to each other. *The rat has reached a high state of morphological integration. It might be expected that this would carry on, perhaps increasing, to the adult stage, but this turns out not to be the case!*

FORTY-DAY STAGE

General integration at the forty-day stage is lower than that at twenty days (see Table 38). Elements of $F'f_{lo}$, however, appear at the high level of $\rho \geq .96$: 20–23, 22–23, 24–33. A suggestion of separation of girdle measures (20, 22, 23) and limb measures (24–33) is present, and, even at this very high level, relationships between the forelimb and hind-limb and girdle measures exist, whereas strictly "within-limb" integration is not expressed. $F'f_{lo}$ approaches completeness at $\rho \geq .93$, but the total absence of measure 28 is striking because this has been a central measure in the three earlier stages. The significance of this loss becomes evident in the adult stage, in which 26–28 are discrete and the core of a second locomotor group. The forty-day stage shows evidence of a second phase of reorganization but one less drastic than that between the one- and twenty-day stages. Details will be seen more clearly in the following section, in which intensities of integration of the F-groups are considered.

While $F'g_{sk}$ is represented only by pairs 1–7 and 1–9 at $\rho \geq .95$, at $\rho \geq .93$ the following group occurs: 1–9–13–14. Measure 7 is bonded to 1, 9, and 13, of this group. At $\rho \geq .93$, $F'f_{ax}$ is represented by only 15–17.

ADULT STAGE: 250 DAYS

The most significant feature at this stage, at which integration is high, is the presence of two locomotor groups. At an extremely high level of ρ, $\rho \geq .995$, two groups are present: $F'f_{lo}$, 24–30–32–33, with 20 bonded to 24 and 32; and $F'f_{dl}$, 26–28–34. Now $F'f_{dl}$ is a new F-group, which we will call the "distal-limb group." It includes only distal elements, 26 (Ul_1), 28 (Ra_1), and 34 (Mc_1). This has been differentiated from the earlier, more general, locomotor complex. The beginning of differentiation was noted in the forty-day stage. The two F-groups of locomotion remain discrete to the $\rho \geq .93$ level in the adult, but at this level they merge into a single group.

At a lower level, $F'f_{ax}$ appears, $\rho \geq .93$, where 15–17 occurs. At $\rho \geq .90$, 15–17–18 is present. At $\rho \geq .96$, $F'g_{sk}$ is developed as the pair 1–13. At $\rho \geq .90$, 1–7–9–11–12–13 occurs, and *all* measures of the skull and jaws are bonded to it. There is undoubtedly an internal differentiation of the skull and jaws, as occurred in *Sciurus* (see pp. 121–23), but the grossness of the structures that could be measured in all stages presumably serves to mask the details.

The history of the development and modifications of F-groups through the ontogenetic stages may be presented in tabular form as follows:

One-day stage:
  Three F-groups
    $F'f_{lo}$ —hind-limb and some forelimb measures
    $F'f_{fl}$ —forelimb locomotor group
    $F'g_{sk}$ —proximity, growth group of the skull and jaws

Ten-day stage:
  Two distinct F-groups
    $F'f_{lo}$
    $F'g_{sk}$
  One partially distinct, partially merged (with $F'f_{lo}$) group
    $F'f_{fl}$
  One emerging group (bonded with $F'f_{lo}$)
    $F'f_{ax}$—axial group
Twenty-day stage:
  Three distinct groups[6]
    $F'f_{lo}$ (with 25–29 absent)
    $F'g_{sk}$
    $F'f_{ax}$
Forty-day stage:
  Three distinct groups
    $F'f_{lo}$—26–28 being lost and 34 not fully bonded; beginning of differentiation of $F'f_{dl}$
    $F'g_{sk}$
    $F'f_{ax}$
Adult stage:
  Four distinct groups
    $F'f_{lo}$
    $F'f_{dl}$
    $F'g_{sk}$
    $F'f_{ax}$

Levels of Integration and V_{P_i}

DISTRIBUTION OF GROWTH INCREMENTS BETWEEN STAGES

The quantitative data that are presented in tabular form to accompany this discussion are intended to provide supplemental details to the more general account of the analyses of groups and group comparisons just presented. In addition, they serve for analysis from a different point of view, which relates variability of the measures and the factor of growth to the patterns of integration given above and the intensity of integration shown in the tables. Data are given only for $F'f_{lo}$, $F'f_{dl}$, and $F'g_{sk}$, since the small number of measures in $F'f_{ax}$ detracts from the meaning of this type of collation.

Tables 43–48 show that the various F-groups conform roughly in their general behavior during ontogeny with respect to $I\rho$ with the behavior of the indexes of the total suite of measures (Table 38). The modifications of $I\rho$ that we have interpreted as indications of organization adjustment show that these changes are general, insofar as the parts measured are concerned, and are not confined to one particular system or another over a given period of growth. The consistency in changes of V_{P_i} with respect to $I\rho$ suggests that growth, in

[6] Note that $F'f_{fl}$ is lost and does not reappear in later stages. This appears to be a group organized only in "early" ontogeny. We may speculate, in view of its anatomical position and time of existence, that it may be related to suckling behavior in feeding.

terms of differential relative rates, is a primary factor in the adjustments. Where the various dimensions increase with divergences in the relative rates, integration tends to be reduced. Tightening of integration follows the development of higher conformity of relative rates. Through these changes, old integrative systems may be lost or altered, and new systems may appear. So far as

TABLE 43*

$I\rho$ FOR LOCOMOTOR GROUP ($F'f_{lo}$), ALBINO RAT

Stage	$\rho \geq .86$	$\rho \geq .90$	$\rho \geq .93$	$\rho \geq .95$	$\rho \geq .96$
One-day..............	.08889	.02209	.00494	.00000	.00000
Ten-day..............	.05144	.02039	.00049	.00000	.00000
Twenty-day	1.00000	1.00000	1.00000	.27037	.11864
Forty-day............	1.00000	.43556†	.33783	.30247	.03984
Adult................	1.00000	.64000‡	.64000	.15819	.10283

* $F'f_{dl}$, 26–28–32; included at all stages. Measures considered in $F'f_{lo}$: 20, 22, 23, 24, 26, 28, 30, 32, 33, 34.
† Beginning of isolation of $F'f_{dl}$.
‡ $F'f_{dl}$ completely isolated at high levels of ρ.

TABLE 44

$I\rho$ FOR TWO LOCOMOTOR GROUPS IN FORTY-DAY AND ADULT RAT

Stage	Group	$\rho \geq .86$	$\rho \geq .90$	$\rho \geq .93$	$\rho \geq .95$	$\rho \geq .96$
Forty-day....	$F'f_{lo}$	1.00000	1.00000	1.00000	1.00000	.17007
	$F'f_{dl}$	1.00000	1.00000	.22222	.11111	.00000
Adult........	$F'f_{lo}$	1.00000	1.00000	1.00000	1.00000	1.00000
	$F'f_{dl}$	1.00000	1.00000	.22222	.11111	.00000

TABLE 45

$I\rho$ FOR SKULL GROUP ($F'f_{sk}$), ALBINO RAT

Stage	$\rho \geq .86$	$\rho \geq .90$	$\rho \geq .93$	$\rho \geq .95$	$\rho \geq .96$
One-day.....	.00541	.00101	.00033	.00000	.00000
Ten-day.....	.00082	.00000	.00000	.00000	.00000
Twenty-day..	.02765	.00679	.00099	.00000	.00000
Forty-day....	.00728	.00592	.00351	.00005	.00000
Adult........	.03241	.02237	.00403	.00049	.00000

TABLE 46

CHANGES OF V_{P_i} AND $I\rho$ (LOCOMOTOR GROUP)
ALBINO RAT

Stage Interval	V_{P_i}	$I\rho$
One- to ten-day............	29.6	Decreases
Ten- to twenty-day.........	17.5	Increases
Twenty- to forty-day.......	48.1	Decreases
Forty-day to adult.........	16.7	Increases

the limited suite of stages reveals, modifications of the various systems appear to take place more or less simultaneously.

It was noted earlier that there are complex intersections between the ρ-groups formed by groupings based on simple correlation without the use of basic pairs. A meaningful and simple estimate of degree of group integration can be made by an analysis of the percentage of bonds between various F-groups, as established by the use of basic-pair analysis. In this instance this method presents a clearer portrayal of the stages and the levels of ρ than a study of the intersects as

TABLE 47

CHANGES OF V_{P_i} AND $I\rho$ FOR SKULL GROUP (F$'$g$_{sk}$)
ALBINO RAT

Stage Interval	V_{P_i}	$I\rho$
One- to ten-day............	35.6	Decreases
Ten- to twenty-day.........	15.7	Increases
Twenty- to forty-day........	95.0	Decreases
Forty-day to adult..........	53.0	Increases

TABLE 48

$I\rho$ FOR AXIAL GROUP (F$'$f$_{ax}$), ALBINO RAT
(Measures in F: 15–16–17)

Stage	$\rho \geq .86$	$\rho \geq .90$	$\rho \geq .93$	$\rho \geq .95$	$\rho \geq .96$
One-day............	.00000	.00000	.00000	.00000	.00000
Ten-day............	.02778	.00000	.00000	.00000	.00000
Twenty-day.........	1.00000	1.00000	1.00000	1.00000	.00000
Forty-day..........	1.00000	1.00000	.25000	.00000	.00000
Adult.............	1.00000	1.00000	.55556	.02778	.00000

used for *Rana pipiens* and *Sciurus ruviventer* (see pp. 116, 124) since the existence of five stages in the ontogenetic series introduces a complexity that obscures meaning in simple group diagrams. The method used to compute the percentage of possible intergroup bonds, presented in Tables 49, 50, 51, and 52, is as follows:

An intergroup bond represents correlation between a pair of measures where the members of the pair are not in the same F-group. The following describes a ratio which we shall call the *index of intergroup integration*, G_I. Let $n_i =$ the number of measurements in the ith F-group. For two F-groups, denoted by F-group 1 and F-group 2, and with n_1 measures in the first F-group and n_2 measures in the second F-group, the maximum number of intergroup bonds is

$n_1 n_2$. In general, for any number of F-groups, the maximum number of inter-group bonds will be denoted by B_p, where B_p is defined as follows:

$$Bp \stackrel{\text{def}}{=} \sum_{i=1}^{n} \sum_{j=1}^{n} (n_i n_j) \qquad i \neq j,$$

where n_i = number of measures in the ith group, n_j = number of measures in the jth group, and for k groups, $i, j = 1, 2, \ldots, k$. We shall denote the number of *observed* intergroup bonds as B_o. Then the index of intergroup integration, G_I, is defined as

$$\frac{B_o}{B_p} = G_I,$$

which ranges from 0 to 1.

TABLE 49

VALUES OF G_I FOR LOCOMOTOR (F′f$_{\text{lo}}$) AND AXIAL (F′f$_{\text{ax}}$) GROUPS

Stage	$\rho \geq .86$	$\rho \geq .90$	$\rho \geq .93$	$\rho \geq .95$	$\rho \geq .96$
One-day..............	.20	.08	.00	.00	.00
Ten-day..............	.20	.13	.00	.00	.00
Twenty-day..........	1.00	.83	.65	.38	.10
Forty-day*...........	.93	.60	.25	.10	.03
Adult*...............	.88	.73	.48	.25	.15

* For total F′f$_{\text{lo}}$ (F′f$_{\text{lo}}$ plus F′f$_{\text{ld}}$). As 28–28–34 (F′f$_{\text{ld}}$) is differentiated, the drop from the forty-day to adult results. This reflects differentiation rather than a real difference in integration.

TABLE 50

VALUES OF G_I FOR SKULL (F′g$_{\text{sk}}$) AND LOCOMOTOR (F′f$_{\text{lo}}$)* GROUPS

Stage	$\rho \geq .86$	$\rho \geq .90$	$\rho \geq .93$	$\rho \geq .95$	$\rho \geq .96$
One-day..............	.25	.08	.00	.00	.00
Ten-day..............	.13	.05	.01	.00	.00
Twenty-day..........	.69	.53	.21	.16	.05
Forty-day............	.54	.29	.16	.07	.01
Adult................	.67	.55	.33	.15	.05

* F′f$_{\text{lo}}$ includes measures of both this group and F′f$_{\text{dl}}$.

TABLE 51

VALUES OF G_I FOR AXIAL (F′f$_{\text{ax}}$) AND SKULL (F′g$_{\text{sk}}$) GROUPS

Stage	$\rho \geq .86$	$\rho \geq .90$	$\rho \geq .93$	$\rho \geq .95$	$\rho \geq .96$
One-day..............	.17	.08	.00	.00	.00
Ten-day..............	.10	.02	.00	.00	.00
Twenty-day..........	.52	.31	.04	.02	.00
Forty-day............	.46	.23	.00	.00	.00
Adult................	.65	.35	.19	.08	.00

The values that show group relationships in Tables 49–52 are in themselves independent of the existence of ρ-groups and are based upon associations in F-groups revealed upon basic-pair analysis. The intensities of bondings are thus independent of the intensity of $I\rho$. Limiting cases of G_I (0.00 or 1.00) and I (.0000 and 1.00) can exist simultaneously, with $I\rho$ for the groups considered and G_I, the relationships between the groups considered. In the cases studied, however, the patterns of increase and decrease of G_I in the successive stages, one-day through adult with increasing values of ρ, show strong resemblances to the patterns of $I\rho$ as viewed both for the totality of measures (see Table 38) and for the measures of the various F-groups. There is, of course, no basis for estimation of quantitative equivalence of the two indexes, but the similarities of trends must be considered highly significant, as they show the pervasiveness of changes of integrative levels not only within groups but also between groups.

TABLE 52

VALUES OF G_I FOR AXIAL (F'f_{ax}), SKULL (F'g_{sk}), AND LOCO-
MOTOR (F'f_{lo}) GROUPS, INCLUDING 26–28–34

Stage	$\rho \geq .86$	$\rho \geq .90$	$\rho \geq .93$	$\rho \geq .95$	$\rho \geq .96$
One-day............	.23	.07	.00	.00	.00
Ten-day............	.14	.06	.00	.00	.00
Twenty-day.........	.71	.53	.25	.17	.05
Forty-day..........	.59	.33	.14	.06	.01
Adult..............	.71	.53	.32	.15	.05

The lack of intersects between ρ-groups that represent F-groups at high levels of ρ, of course, gives clear evidence of the dominance of intragroup relationships over intergroup relationships when the totality of bonds is taken into consideration.

Summary: Morphological Integration in Ontogeny

The study of ontogeny by comparative analysis of morphological integration at a series of ontogenetic stages in the albino rat reveals much of interest and significance, in spite of rather severe limitations imposed by the nature of the experiment. First, it is apparent that analysis of ontogeny by the approach of morphological integration produces interpretable results. In spite of limitations, knowledge of the ontogeny in the rat is enriched. Phases of strengthening and weakening of total integration are revealed, and it seems evident that these are related in part to relative rates of increase in dimensions of the structures involved. Differentiation and development of a series of F-groups are recorded. These and other observations pave the way to additional research, for example, a study of the relationships of the phases of development to the physiological aspects of ontogeny, experiments upon the effect of modifications of environment and heredity on ontogeny, or an extension in earlier, prenatal stages to relate histogenesis and early morphogenesis to the stages now known.

Definitive generalizations can hardly be made from this single study. In view of the many close resemblances of ontogenetic phenomena in placental mammals other than those of morphological integration, it seems probable that most, if not all, of the general aspects of morphological integration in the ontogeny of the rats would be encountered in other members of this large group. It would appear highly probable that the pattern of alternating phases of high and low integration exists throughout the mammals and that these phases are related to growth phenomena and thus, ultimately, to physiological factors manifest in changes of dimensions. Pervasiveness of changes in integration throughout the organism, striking in the rat, may be a general phenomenon, but it seems highly likely that it might be less evident in cases in which rapid differentiation or consolidation of a system occurs during the ontogenetic stages studied. In such an event it seems probable that an imbalance between the level of integration in the affected system and that of other systems would exist. This might well be the general situation very early in ontogeny where histogenesis predominates, but it also would be expected when rapid major modifications occur in postnatal ontogeny, for example, at the onset of sexual maturity. Perhaps the most important single contribution that knowledge of morphological integration in ontogeny can offer is to be found in the comprehensive knowledge of the interactions of the parts of the organism which, in the course of ontogeny, result in the integration of the whole. This in turn makes possible a more realistic examination of the relationships of genetic and epigenetic factors to morphogenesis.

STUDIES OF MAMMALIAN DENTITIONS

The dentition of mammals, especially the cheek dentition, has provided the greatest single source of information upon the evolution of mammals throughout geological time. Early orders of mammals, from the Mesozoic and very earliest Cenozoic, are known largely from teeth and jaws, and even in more recent parts of the Cenozoic knowledge of most groups would be poor in the absence of the data supplied by preserved dentitions. Inasmuch as dentitions comprise single systems, highly differentiated from all others and internally highly complex, they appear to be ideally suited for analyses from the standpoint of morphological integration. As will be pointed out, there are both practical and theoretical difficulties in making such studies, but these are not insurmountable. Two studies of teeth have been made in the course of our investigations of morphological integration. The first involved a time-stratigraphic sequence of species of the Eocene condylarth, *Hyopsodus*. This study concerned only the lower molar dentition. The results were of considerable interest but served mainly to raise questions which could be answered only by a more comprehensive study in which both upper and lower series could be treated together. The use of a living species was indicated, since it was necessary to have upper and lower jaws of the same individuals for the sample. *Aotus trivirgatus*, a South American monkey, proved suitable for this study (see Figs. 54 and 55). Only the molar dentition plus the lower fourth premolar were used in the second study to insure that the same measurements could be taken on each tooth in the series. This is not a necessary restriction, but it was felt that it would serve to simplify a situation that, at best, is extremely complex.

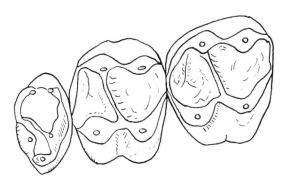

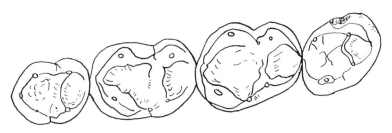

Fig. 54.—Sketches of the crown views of the upper molars (*above*) and lower molars and the fourth lower premolar (*below*) for *Aotus trivirgatus*. The upper third premolar (M³) is highly variable in this species. The illustration shows only one of several patterns.

A great variety of measurements can be taken on molar and premolar teeth. The number available upon a single tooth in series in which the teeth are somewhat similar along the tooth row is roughly multiplied by the number of teeth, and, as a consequence, even a minimum of desirable measures grows to proportions that are handled with difficulty. Selection of measures is essential and is best carried out in accord with the concept of representative measures (see pp. 9–10). We have used four general types of measures in our work, fully recognizing that other types are possible and might, for various purposes, be

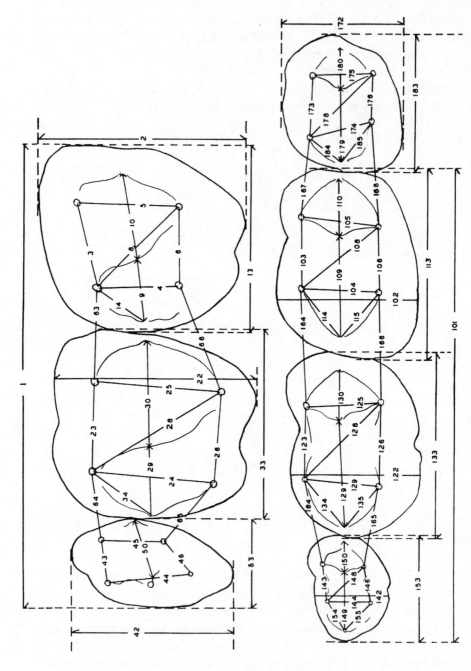

Fig. 55.—*Aotus trivirgatus*. Diagrams of measurements of upper (*above*) and lower (*below*) molars and lower fourth premolar. (See text for explanation of system of measurement.)

more effective than the ones that we have selected. The types used in our studies are as follows: (1) measures of general, over-all dimensions of the tooth row; (2) measures of gross dimensions of individual teeth; (3) measures of features of the crown surfaces of the teeth, often called "within-tooth"measures later; and (4) measures between crown features of adjacent teeth, often called "between-teeth" measures below.

The only measure of tooth-row dimension that has proved to be of specific value is the maximum length of the tooth row. Some expression of width of the tooth row would be desirable, particularly for use in partial correlations, but no adequate measure of this feature has been found. Offhand, the width of one of the molar teeth would appear to fill this need, but, at least in the cases studied, width measures of the successive teeth have not been found to be sufficiently highly correlated to justify the assumption that the measurements of one width dimension are representative.

Measures of gross dimensions of individual teeth should ideally include length, width, and height. Length and width offer no difficulties, but height measures pose difficult, usually insoluble, problems. The factor of wear commonly introduces significant inaccuracies even in the very best of materials. In addition, there is the problem of what constitutes height. Although it is usually possible to define a reliable position for the base of the crown, no single apical feature gives adequate expression of over-all tooth height in most molars and premolars. It appears that various heights of cusps, styles, and so forth on a particular tooth are not, in general, highly correlated with each other, although this cannot be accurately assessed in view of wear. Valleys, cinguli, and similar structures rarely provide accurate points for measurement. For these reasons we have not used height measures in our work, although we recognize their importance and realize that the use of such measures can be meaningful where restrictions of accuracy are less stringent.

An almost unlimited number of measures of the crown surface is available even in relatively simple mammalian molars and premolars. Height measures of these features suffer from the factor of wear just as do the gross features of height. Wear may also affect direct measures between cusps and crests. We have restricted our measurements to distances between crown features by making all measurements in directions normal to the vertical axes of the teeth. Measures between adjacent teeth have been similarly made. These, of course, suffer from the additional hazard of tooth displacement after death.

Measures between upper and lower teeth in a defined occlusal position would undoubtedly prove of great interest. It has been our experience, however, that repeatable measurements of this type cannot be obtained.

The studies that we report below were frankly exploratory, for we had little idea of what to expect in the way of results, and there is no very adequate basis for interpretation of the results that were obtained. Although dental systems are highly functional and in most cases composed of differentiated, functional subsystems whose basic adaptive characteristics are evident, interpretation of the detailed interrelationships of small characters is, to say the least, beset with difficulties. The situation is quite different from one in which a complex of moving elements must act in delicate harmony to perform even its most rudimentary tasks. Here nice adjustment is mandatory. It would appear off-hand that a rather wide range of variation of details in dental pattern would have relatively little effect upon the functional activity of, for example, the grinding dentition in a herbivorous species or the shearing functions of carnassials in a carnivore. Yet it is well known that the variability of most small characters, taken one at a time, is relatively low within a species, as emphasized, for example, by Simpson and Roe (1939). Even the minor details of crown pattern usually provide an excellent basis for the differentiation of species and the establishment of phylogenetic relationships. It is difficult to believe that these small differences in crown patterns can, in themselves, be adaptively significant. They do appear, however, to evolve under the influence of selection. It may well be that small differences observed in dental characters of related species represent differences in pervasive organizational patterns, as implicit in the application of the general concept of morphogenetic fields to dentitions as made by Butler (1937, 1939*a*, *b*, 1941) and Patterson (1949), among others. That selection has been operative in dental evolution seems established beyond any doubt, but the morphological level and, consequently, the genetic level at which it has operated are less certain. Evidence from studies of the genetics of dentitions is virtually non-existent, and until the situation is remedied the uncertainties inherent in purely inferential interpretations cannot be removed. If, however, it is possible to gain additional insight into the scope of the selective unit, some progress can be made toward an understanding of the subordinate dental characters that are so important in studies of fossil mammals. Investigations directed toward this end fall within the domain of the concept of morphological integration.

The two studies that have been made fall far short of the goal of complete and satisfactory answers. Analyses and interpretations have proved to be extremely difficult in view of the absence of any guiding principles. Once the patterns of morphological integration have been obtained and interpretation is required, we begin to tread on extremely dangerous ground. This becomes twice hazardous when it is recognized that the patterns that emerge are based on low-level correlations and that much is left "unexplained." Some progress, however, has been made, and we feel that it is sufficiently interesting to report. We ask that the conclusions be viewed in the spirit in which they are presented. An underlying order is revealed, and tentative conclusions concerning its meaning are offered as a stimulant to additional research and thought along the lines of our experiments.

DENTAL STUDIES OF *Aotus trivirgatus*

Samples and Measurements

Appendix H includes details about the sample of *Aotus trivirgatus* and a table of the measurements that were used in the study reported below. In all, eighty-three measurements, illustrated in Figure 55, were taken on each individual of a sample of eighteen specimens. These measurements covered the upper and lower molars and the fourth lower premolar. So far as possible, the same measurements were taken on each upper molar in the series, and analogous measurements were taken on the lowers. Where the series of measurements is incomplete, as for M^3, the missing measures either could not be obtained or were not repeatable with sufficient accuracy. In all cases measurements were made on only one side of the dentition, for distributions of commensurate measures on the same teeth on opposite side consistently fail to show significant differences. In each individual, measurements were made above and below on the same side, left or right as preservation dictated.

It is exceedingly difficult to keep the meanings of the eighty-three measurements in mind, and we are fully aware that this contributes to the difficulties that readers will have in following through the necessarily complex details of analysis. The numbering system used was designed to alleviate this difficulty as far as possible. It is essential to understand and know this system. Each tooth was measured as shown in Figure 55. For M^1, for example, measurements 2, 3, 4, 5, 6, 8, 9, 10, 13, and 14, were taken, and measures 7, 11, 12, and 15 were established but were dropped later, either because they were represented by another measure or were not repeatable with sufficient accuracy. Thus measures 2 through 14 apply to the first upper molar; measures 22 through 34 apply to M^2; and 42 to 54, to M^3. Measures 2, 22, and 42 thus designate homologous measures on M^1, M^2, and M^3, respectively. Lower molars 1 to 3 are similarly numbered, with some additions in the series, but in the 100's rather than the 10's. $P_{\overline{4}}$ is designated by measures 172 through 185. Measures 102, 122, 142, and 172 designate homologous measures on $M_{\overline{1}}$, $M_{\overline{2}}$, $M_{\overline{3}}$, and $P_{\overline{4}}$, respectively. These measures are *analogous* to similarly numbered measures in the upper molars. It should be noted that there is a labial-labial and lingual-lingual correspondence of the measures of the upper and lower series. The meanings of measures may thus be summarized as follows:

1. Length of total tooth series
   No. 1, total length of M^1 through M^3
   No. 101, total length of $P_{\overline{4}}$ through $M_{\overline{3}}$
2. Widths and length of individual teeth
   Nos. 2, 22, 42, for M^1, M^2, M^3, respectively
   Nos. 102, 122, 142, 172, for $M_{\overline{1}}$, $M_{\overline{2}}$, $M_{\overline{3}}$, $P_{\overline{4}}$, respectively

3. Measures between crown features within individual teeth

Nos. 3, 4, 5, 6, 8, 9, 10, 14, for M^1, and equivalents for M^2 (20 and 30) and M^3 (40 and 50)

Nos. 103, 104, 105, 106, 108, 109, 110, 114, 115 for $M_{\overline{1}}$, and equivalents for $M_{\overline{2}}$ (120 and 130), $M_{\overline{3}}$ (140 and 150), and $P_{\overline{4}}$ (170 and 180)

4. Measures between adjacent teeth

Upper molars

Nos. 63, 66, between $M^{\underline{1}}$ and $M^{\underline{2}}$

Nos. 64, 65, between $M^{\underline{2}}$ and $M^{\underline{3}}$

Lower molars

Nos. 163, 166, between $M_{\overline{1}}$ and $M_{\overline{2}}$

Nos. 164, 165, between $M_{\overline{2}}$ and $M_{\overline{3}}$

Nos. 163, 166, between $P_{\overline{4}}$ and $M_{\overline{1}}$

Procedures

Measurements were made to the nearest 0.01 mm. (permissible error ± 0.02), using calipers fitted with special needle points. All measurements were made by the senior author. Simple intercorrelations for the complete array of measurements were carried out. Partial correlations for the "within" measures of the individual teeth were performed with length and width held constant, $r_{xy.lw}$. We formed ρ-groups at the levels $\rho \geq .82, .86, .90, .93$ ($P = 0.01$) and later, when more complete grouping was required, at $\rho \geq .66$ ($P = 0.1$).[7]

Once groups were formed, the study was carried out in a series of progressive steps. It was found impossible to handle the totality of groups at the low levels of ρ simultaneously. At $\rho \geq .66$, for example, there were 713 non-contained ρ-groups, many with ten to fifteen measures included. The steps were as follows:

1. Assessment of intensities of correlation, using $I\rho$
2. Studies of gross measures and within-tooth measures
   a) Upper molars alone (simple and partial correlation)
   b) Lower molars and $P_{\overline{4}}$ alone (simple and partial correlation)
3. Upper and lower teeth considered together, cross-series relationships
4. Between-teeth measures
   a) Each measure as a group-former alone
   b) Paired measures as group-formers (as 63 and 66, 64 and 65)
   c) Comparisons and interrelationships of pairs

This procedure in general is followed in the account of the analyses given later in this section. The following factors were considered to have contributed to the existence and nature of integrative patterns: (1) factors of size and shape of individual teeth and total series of teeth, both above and below; (2) factors pertaining to homologous and analogous measures; (3) factors related to the general concept of morphogenetic fields, expressed in gradients of intensity and specific composition of integration along the tooth rows and between upper and

[7] Grouping was carried out by Eban Matlis, an assistant in NSF Project G-183. We wish to express our special appreciation to him for this work, which was immensely time-consuming and difficult.

lower teeth; and (4) factors of "interadaptive" relationships of upper and lower molars. It is needless to point out, of course, that these four sets of factors are not independent and that, to a degree, they are to be considered only as morphological expressions of underlying genetic and physiological factors.

To reach the objectives of this study, very detailed analyses were required. As far as possible, only the critical data and interpretations have been presented in the pages that follow. Tabular presentation has been used wherever possible. In spite of all efforts toward conciseness, the mass of data precludes the brevity and generality commensurate with our principal aims in this book. We do not feel that we can legitimately dispose of our responsibility to readers especially interested in mammalian dentitions without a coherent and reasonably detailed presentation of the evidence basic to the conclusions. For the more general reader, who may not wish to work his way through the complex treatment, a short account of the analysis and interpretations is included following the section on analysis.

TABLE 53

$I\rho$ FOR COMBINED UPPER AND LOWER DENTAL
SERIES IN *Aotus trivirgatus*
(Total 83 Measures)

Level of ρ	I_ρ
$\rho \geq .93$	.0000013
$\rho \geq .90$	.0000076
$\rho \geq .86$	.0000302
$\rho \geq .82$	.0000721
$\rho \geq .66$	.0000929

Analysis

The outline of procedural stages given on page 184 is followed in the discussion of analysis of the data. Interpretations are kept to a minimum, for these are included in context in the summary that follows.

INTENSITY OF INTEGRATION, $I\rho$, FOR TOTAL SERIES AND INDIVIDUAL TEETH

Table 53 lists the values of $I\rho$ for the upper and lower dental series (M^1 to M^3 and $P_{\overline{4}}$ to $M_{\overline{3}}$) considered together as a single system. All measures are considered. Integration is very low as compared to that found in various systems of the rat, squirrel, and frog (Table 38; Figs. 50 and 51). The coverage, of course, cannot be considered commensurate, but the discrepancies are so great that they leave no question but that they are real. Thus, although the molar system is highly functional, integration is distinctly limited. The situation appears quite the contrary, however, when individual teeth are studied separately, as shown by the values of $I\rho$ in Table 54.

The principal points of interest in Table 54 are as follows: $I\rho$ is not greatly different in the three upper molars but is somewhat higher in M^3 than in the other two molars. This occurs in spite of the fact that this is clearly the most variable tooth in the series if characters are considered one at a time. Integration in the lower teeth is consistently higher than it is in the uppers. Even the least integrated lower molar, $M_{\bar{3}}$, has a higher value of $I\rho$ than $M^{\underline{3}}$, the most highly integrated upper molar. The region of highest integration above and below is in a different position in the tooth row, at $M^{\underline{1}}$ above and at $M_{\bar{2}}$ below.

Additional insight into integration may be gained from Table 55, which shows the percentage of all possible intergroup bonds between each pair of

TABLE 54*

$I\rho$ FOR INDIVIDUAL TEETH OF *Aotus trivirgatus*

Tooth	$I\rho$	Tooth	$I\rho$
$M^{\underline{1}}$........	.01192	$P_{\bar{4}}$.........	.02429
$M^{\underline{2}}$........	.01075	$M_{\bar{1}}$........	.02667
$M^{\underline{3}}$........	.01633	$M_{\bar{2}}$........	.06479
		$M_{\bar{3}}$........	.01785

* Between-teeth measures omitted. $\rho \geq .66$; $P = .1$.

TABLE 55*

PERCENTAGES OF POSSIBLE BONDS REALIZED BE-
TWEEN UPPER AND LOWER TEETH
IN *Aotus trivirgatus*

	$M^{\underline{1}}$	$M^{\underline{2}}$	$M^{\underline{3}}$
$P_{\bar{4}}$.........	6.4	3.6	12.4
$M_{\bar{1}}$.........	10.5	3.6	13.1
$M_{\bar{2}}$.........	11.4	7.7	14.4
$M_{\bar{3}}$.........	8.6	4.6	9.8

* The index G_I; $\rho \geq .66$; $P = .1$.

upper and lower teeth. It will be noted that a special use of the index G_I is made here. It is the measures of each tooth, not of an F-group, that define the groups between which bonds pass.

The most highly integrated lower tooth, $M_{\bar{2}}$, is more completely bonded to *each* of the upper molars than is any other lower tooth, and $M^{\underline{3}}$, similarly, is more fully bonded to *each* lower than is any other upper tooth. Since the phenomena of bonding expressed in the table are independent of $I\rho$, these relationships are of more than passing interest. At present, however, such observations can do little more than provide points of departure for other types of studies, for morphological analysis can aid primarily in suggesting that fundamental factors of organization are involved.

STRUCTURE OF INDIVIDUAL SERIES, UPPER AND LOWER SEPARATELY

Upper molars.—All the large ρ-groups—those containing five or more measures—are formed about a "core" of either length or width measures. The primary "length" group consists of measures 1–13–33–45–53. Measure 1 is the total length of the molar series, and 13, 33, and 53 the lengths of the individual teeth. Measure 45 passes from the metacone to protocone of $M^{\underline{3}}$ and is technically a width measure, although, in view of partial rotation of the transverse axis of $M^{\underline{3}}$ with respect to the other molars, it falls somewhat along the longitudinal axis of the tooth row. Measure 45 also occurs in the "core" width group, which is constituted as follows: 2–3–4–5–22–23–24–25–45. Two longi-

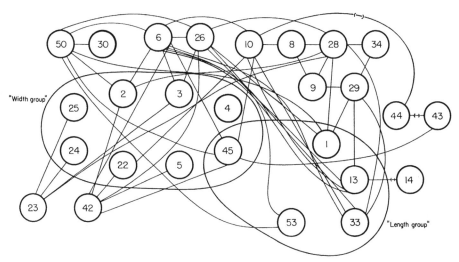

Fig. 56.—"Core" groups (length group and width group) in upper molar series of *Aotus trivirgatus*. Bonding is complete within circled "core" groups. Positive bonds are not entered, to reduce confusion of diagram. Negative bonds are indicated by (−). Note frequency of bonding of homologous pairs of measures (as 8, 28, etc.).

tudinal measures occur in this group, 3 and 23. Measure 42, width of $M_{\overline{3}}$, is not in ρ, although bonding to the group occurs. This absence and the presence of 45 in the "length" group both appear to relate to the rotation of the third molar noted above.

The two "core" groups are formed as follows at the successive levels of ρ studied:

ρ ≥ .93 . . . . Only elements of width group, pairs, present
ρ ≥ .90 . . . . Width group and length group, both represented by pairs of measures
ρ ≥ .86 . . . . Both length and width groups present in their entirety
ρ ≥ .66 . . . . Both groups complete and various additional measures added as ρ is lowered
            from the ρ ≥ .86 level

At ρ ≥ .66, bonding is very complex. To illustrate this, a diagram of the array of bonds is presented in Figure 56. In addition to the "core" groups, pairs of homologous measures tend to be increasingly evident as ρ is lowered. These are indicated in Figure 56, and occur at higher levels as follows: at $\rho \geq$ .86, 6–26, 8–28, 5–25, and 3–23 and at $\rho \geq$ .82, 4–24 and 30–50 in addition to the above. It will be noted that few measures of M^3 occur. These homologous pairs do not themselves act as group-formers, but they do merge into large groups, develop bonds between pairs, and become bonded to measures not in the "core" ρ-groups as the level of ρ is lowered. Various measures other than those of the homologous pairs and the "core" groups are present at $\rho \geq$ .66. They do not, however, develop strong patterns of bonds.

Lower molars and $P_{\overline{4}}$.—Only a single "core" group is present, with composition as follows: 102–103–106–108–126–128–143–145–146–148–173–178–183. Strongly bonded to this group but missing one or more bonds at $\rho \geq$ .66 are the following:

One bond absent..........	150 (178)[8]
Two bonds absent.........	125 (148, 173); 130 (103, 143, also 150)
Three bonds absent.......	122 (103, 143, 178, also 130)
Four bonds absent........	129 (128, 143, 146, 173, also 150)
	113 (108, 143, 145, 178)
Five bonds absent........	133, 175, and 176

Although both length and width measures are included in this "core" group, it is striking that measure 101, the length of the total lower series, is not in the group and not highly correlated with other measures of the lower series. At $\rho \geq$ .66 it is correlated with 113, 153, and 173, but at higher levels of ρ it is related to only a few measures in a somewhat haphazard way. The principal regularity in the lower series, based on the major ρ-group and highly bonded measures, relates to the existence of series of homologous measures, as follows: 106–126–146, 108–128–148–178, and 103–173. If the bonded measures are considered, additional relationships appear: 125–145, 102–122, and 113–133–183 (note that 113 and 133 are not bonded to 153).

The basic organization of the upper and lower series appears to depend primarily upon different relationships. Length and width are predominant in the upper series, with homologous measures distinctly secondary, whereas, in the lower series, homologous measures and their interrelationships provide the principal integration, and lengths and widths are distinctly secondary and evident only at very low levels of ρ. Over-all length, measure 101, is not important in the correlations in contrast to measure 1.

RELATIONSHIPS BETWEEN UPPER AND LOWER TOOTH SERIES AND INDIVIDUAL TEETH

The best single statement of the relationships between the individual teeth of the upper and lower series is provided by the modified index G_I, based on the percentage of possible bonds between measures of pairs, as shown in Table 55. This does not, of course, indicate *which* measures are involved. Two types of

---

[8] Numbers in parentheses denote the missing bonds.

approach can provide somewhat different aspects of the relationships. Examination of the ρ-groups that include both upper and lower measures shows some aspects of relationships; examination of the bonds that exist between measures of the upper and lower teeth reveals other aspects.

The ρ-groups.—Many large ρ-groups that include measures of both the upper and lower series are formed at ρ ≥ .86. The following are representative:

```
1–2–113–126–146–148–150–153–173–176–183
1–13–102–113–122–136–128–130–133–148
1–50–53–102–108–126–143–146–173–178
33–102–122–130–142–183
```

Two facts stand out: (1) only the lower series is represented by a large number of intercorrelated measures in these "joint" ρ-groups; and (2) the measures of the upper series all express gross dimensions. (Note that measure 50 is virtually an expression of the length of M³.) In smaller ρ-groups, of which 3–6–106–126–128–130 is characteristic, some upper and lower intercusp measures occur.

TABLE 56

PERCENTAGES OF BONDS BETWEEN PAIRS OF UP-
PER AND LOWER TEETH OF *Aotus trivirgatus*
INVOLVING SIZE MEASURES

	P₄	M₁	M₂	M₃
M¹.....	66	55	40	58
M².....	62	62	41	50
M³.....	68	65	58	33

They tend to include homologues and their corresponding analogues. Length and width measures of the lower series do *not* relate to the measures of the large ρ-groups, but integration in the *lower* series is strongly related to length and width measures of the *upper* series.

Bonds between teeth (ρ ≥ .66).—Bonds that involve at least one measure of length or width comprise at least 50 per cent of the total number of bonds present in nine of the twelve possible pairings of teeth (see Table 56 for percentages of possible bonds realized). The relationships are graphically portrayed in Figure 57. In this figure, a tendency is evident for the highest percentage of bonds that involve length or width to occur in teeth with the least topographic proximity. There seems to be no similarity between this distribution and that in which the percentage of bonds existing between pairs of upper and lower teeth is shown (see Table 55). The bonds that exist are listed in Table 57. Each pair shows the presence of analogues; bonds of measures along the same axis, longitudinal or transverse with respect to the tooth row, tend to predominate slightly. There is, however, no striking relationship.

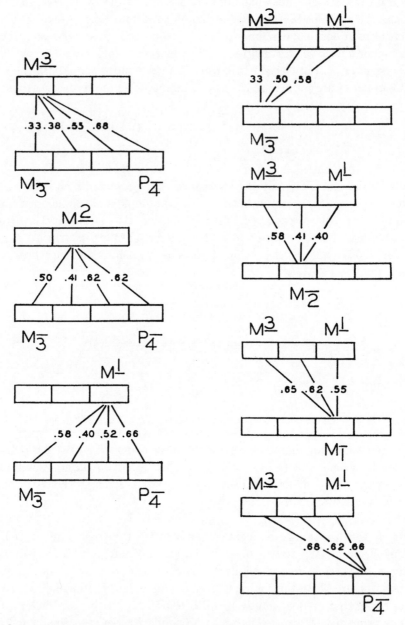

FIG. 57.—Percentage of possible bond developed between individual upper and lower molar pairs in *Aotus trivirgatus*, lower P$\overline{4}$ included.

Both the ρ-groups and the bonds between pairs show length and width to be important in the relationships of the upper and lower teeth. The total length of the upper series predominates when large ρ-groups are considered. Very slight bonding of upper and lower crown details occurs in large ρ-groups. These shows a higher level of occurrence when pairs of teeth are considered, but the pattern is irregular. Length and width in general tend to assume the most important position in teeth that are most widely separated along the tooth rows.

PARTIAL CORRELATION

The importance of the role of length and width in integration both between tooth rows and within rows suggests that they may mask important details

TABLE 57

BONDS BETWEEN PAIRS THAT INCLUDE ONE UPPER AND ONE LOWER TOOTH
IN *Aotus trivirgatus*

M¹ and P₄

2-173	3-184	6-183
2-176	5-175	8-183
2-178	6-173	13-175
2-179	6-176	13-183
2-183	6-178	

N¹ and M₁

2-102	5-106	10-110
2-106	6-106	10-114
2-113	6-104	13-102
3-102	6-105	13-104
3-106	6-106	13-108
3-110	6-108	13-108
3-113	8-110	13-114
		14-102

M¹ and M₂

2-126	4-122	6-128
2-128	5-123	6-130
3-122	5-124	6-133
3-123	5-126	8-123
3-126	5-128	8-124
3-128	5-130	13-122
3-130	6-125	13-129
3-133	6-126	13-130

M¹ and M₃

2-146	3-148	13-146
2-148	3-154	13-148
2-149	5-148	13-149
2-150	6-145	13-150
2-153	6-146	13-155
2-154	6-150	
3-144	10-155	

M² and P₄

22-184	26-184
23-172	28-172
25-184	30-172
26-174	33-184

M² and M₁

22-110	
22-114	25-113
24-104	29-115
25-110	30-113
	33-102

M² and M₂

22-123	26-129	30-134
22-124	28-123	33-122
23-134	29-129	33-129
24-123	30-122	33-130
25-123	30-123	
26-128	30-130	

M² and M₃

22-144	25-149	30-142
22-154	25-154	30-154
25-144	26-146	33-142
		33-145

M³ and P₄

42-173	42-185	50-185
42-174	44-185	53-173
42-175	50-173	53-175
42-178	50-175	53-178
42-179	50-178	53-179
42-183	50-183	53-180
		53-183

M³ and M₁

42-102	45-110	50-113
42-104	45-113	53-102
42-106		
42-108	45-114	53-103
42-115	50-102	53-105
43-103	50-105	53-108
43-104	50-106	
45-102	50-108	

M³ and M₂

42-122	45-126	50-130
42-125	45-130	50-133
42-126		
42-128	50-122	53-122
42-130	50-128	53-125
43-123	50-125	53-126
45-122	50-126	53-128
45-123	50-128	53-130

M³ and M₃

42-145	45-145	50-145
42-146	45-149	50-146
42-150	45-150	50-148
42-153	45-153	50-150
	46-144	
45-155	50-143	

and patterns not dependent upon their effects. As noted above, these dimensions cannot be held constant for the whole tooth row, since no reliable measure of width is available. Consequently, the teeth have been treated separately to determine $r_{xy.lw}$ for each tooth. By standard procedure, ρ-groups derived from this analysis ($\rho \geq .70$, $P = 0.1$) were formed. The ρ-groups for each tooth, r_{xy} and $r_{xy.lw}$, at equivalent values of ρ, are shown in Table 58. In Figure 58 ρ-groups based on partial correlation are illustrated diagrammatically.

A number of points need special emphasis. First, the important length and width in the integration of each tooth are shown in Table 58. The sizes and compositions of groups, beyond the loss of length and width measures, provide the evidence. Second, with few exceptions, the groups formed pertain to the central portions of the teeth. All triplets in M¹ and M² involve measures between cusps, and only one additional measure is present in any group, No. 34. For $M_{\bar{1}}$ and $M_{\bar{2}}$ again almost the whole relationship is found in the measures between cusps. $M_{\bar{2}}$, however, shows one pair, 110–114, that involves posterior measures. Both $M_{\bar{1}}$ and $M_{\bar{2}}$ have a "4–5–8" group, a "3–4" group, and a "6–8" group.

Both the upper and lower third molar show important characteristics. The high integration of $M^{\underline{3}}$ under simple correlation (see Table 54) is clearly due to the factor of size, which accounts for the apparently anomalous situation of high

TABLE 58

ρ-GROUPS FOR INDIVIDUAL MOLAR TEETH OF *Aotus trivirgatus*

	r_{xy}	$r_{xy \cdot lw}$
$M^{\underline{1}}$......	2–5, 8–9, 8–10 2–3–4, 2–3–6, 3–4–5 6–10–13	3–4–5
$M^{\underline{2}}$......	22–25, 22–33, 25–30, 28–34 24–33, 24–34, 23–24–25 23–25–28, 26–28–33	29–30, 29–34 23–24–25 23–25–28 24–25–34
$M^{\underline{3}}$......	42–43, 43(–)44, 　43–45, 42–45–50, 　45–50–53	43(–)44
$M_{\bar{1}}$......	109–115, 110–114, 104–105–108 102–103–104–106–108, 　102–103–106–108–109 102–103–106–109–113	103–104, 103–106, 106–108 110–114, 104–105–108
$M_{\bar{2}}$......	122–125–126–128–130–133 122–125–126–129–130–133 123–124–125–126–128 123–125–126–128–130	123–124, 126–128 124–125–128
$M_{\bar{3}}$......	148–149, 144–153–154 150–153–154, 146–148–150–153 143–145–146–148–150	143–148, 148–149 143–145–146–148 143–145–150 144–154–155

* Based on r_{xy} ($\rho \geq .66$, $P = .1$) and $r_{xy \cdot lw}$ ($\rho \geq .70$, $P = .1$).

integration in the most variable tooth in the series. Variability in individual characters is a function of size rather than of the crown structure per se. $M_{\bar{3}}$ reveals a strong differentiation of the talonid portion, 144–145–155, from the rest of the tooth. It is this portion of the tooth that makes contact with a single upper tooth above, $M^{\underline{3}}$, whereas the more anterior portions relate to the posterior part of $M^{\underline{2}}$ in occlusion. The meaning of this relationship in interadaptive characters is taken up in the next section.

Except for the differentiation in $M_{\bar{3}}$, there is no evidence of integration of the parts of the upper and lower teeth that meet in occlusion. Rather, each tooth has central integration, which involves measures from structures related to different teeth above and below. In other words, the highly integrated portions of the teeth, as revealed in the absence of between-teeth measures in partial correlation, appear to alternate in position as the teeth are brought into tight occlusal relationship. Except for the case of M$\underline{3}$, there is considerable evidence in the patterns of integration that homologues, either in pairs or larger groups, play an important part in integration, with modifications in intensity along the tooth row.

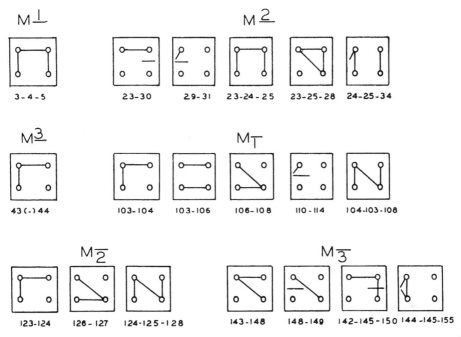

Fig. 58.—Diagrammatic representation of ρ-groups formed for each tooth of *Aotus trivirgatus*, for $r_{xy \cdot lw}$. Each tooth repeated for each separate ρ-group. The solid lines show the measures in a ρ-group. Thus there is one ρ-group in M$\underline{1}$, six in M$\underline{2}$, etc.

RELATIONSHIPS OF MEASURES BETWEEN ADJACENT TEETH AND WITHIN TEETH

Measurements 63, 64, 65, and 66 in the upper series and 163, 164, 165, 166, 167, and 168 in the lower cross between adjacent teeth (see Fig. 55). They will be referred to as "between-teeth" measures. They were taken in an effort to determine how the segments of the dental length so covered related to the segments of the opposite row that roughly corresponded in occlusion. The relationships between measures and teeth are approximately as follows: 63 and 66 equivalent to M$_{\overline{2}}$; 64 and 65 equivalent to M$_{\overline{3}}$ (exclusive of talonid); 163 and 166 equivalent to M$\underline{1}$; and 164 and 165 equivalent to M$\underline{2}$. Since 167 and 168 do not have equivalents, they will not figure importantly in the considerations. If an integration were based on these crude spatial relationships, we should expect groups including measures on M$_{\overline{2}}$ and 63 and 66, M$_{\overline{3}}$ and 64 and 65, and so forth. As the results cited below will show, this is *not* the case, and no such simple relationship holds.

The complete patterns of ρ-groups and bonds formed with measures of the 60 and 160 series are extremely complex, largely as a result of rather low correlation that precludes the formation of a few large groups. Even the most complete bonding, at $\rho \geq .66$, suffers from this difficulty. Patterns at higher levels are, of course, less complete. They show, in skeleton form, the same general patterns as those at $\rho \geq .66$ but are difficult to interpret without knowledge of the grouping at the lowest level. The essential patterns of morphological integration are shown in the large ρ-groups. For the sake of simplicity, we shall pay only passing attention to measures not in these groups. Little is lost by this practice because the small groups and measures bonded to the large groups, in this case, give no insight into integration, which is not seen in the large groups. Considerations are based on the level $\rho \geq .66$, for maximum completeness.

The most striking single feature of the correlations of the 60 and 160 series is the existence of a great many negative values of r. This is shown diagrammatically in Figures 59 and 60, in which the largest ρ-group formed by each between-teeth measure is illustrated.

These figures will be used as the basis for discussion of the analysis, but, it must be emphasized, the illustrations are deceptively simple. They directly show neither group intersections nor that many measures in the different groups have cross-bonds nor that there are other measures not shown which are strongly bonded to one or more of the ρ-groups. The total situation is too complex to be presented diagrammatically. Essential points, however, are brought out in the discussion.

Bonding between measures in the 60 series and the 160 series is shown in Table 59, with the values of ρ given. To the extent that significant correlations exist, the between-teeth measures show a consistent relationship, with positive correlations between measures of the same series, upper or lower, and negative correlations between measures of opposite series. "Expected" correlations, 63–66, 64–65, 163–166, and 164–165, are all significant. Of the paired measures above and below, 64–164 and 65–165 are significantly correlated but 63–163 and 66–166 are not. This sort of irregularity is characteristic throughout the groups formed by the between-teeth measures and consistently plagues analyses with difficulties and doubts. There is no real basis for a decision as to which of many possible reasons are responsible for this situation. Interpretations have been based on patterns that show consistencies, and, since these give only partial solutions, conclusions must be considered tentative.

We shall consider the between-teeth measures of the upper molars first. The irregularity of the complete groups formed by these measures is striking (see Fig. 59). Actually it is less real than appears in the diagram, for, when measures strongly bonded to the groups are considered, as discussed later, more intelligible patterns emerge. In general, measures of the 60 series are negatively correlated to measures of the upper teeth and positively correlated to measures of the lower teeth. There are certain noteworthy exceptions. Measure 64 is bonded negatively to 101, total length of the lower series, and to 154. It will be recalled

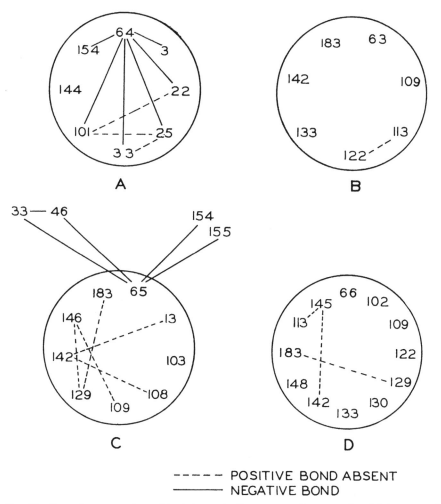

POSITIVE BOND ABSENT ----

NEGATIVE BOND ———

Fig. 59.—Largest ρ-groups formed by between-teeth measures of upper molars (sixty groups of measures). Groups in each case circled, and all measures within circles correlated at ρ ≥ .66, except those connected by dashed lines. Negative bonds are indicated by solid lines.

that 101 was found to be related to upper molar lengths rather than to those of the lowers. Measure 154 is a talonid measure of $M_{\overline{3}}$. This structure has been shown to separate under partial correlation into an integrated group distinct from central measures of $M_{\overline{3}}$. In Figure 59, *A*, it is shown to correlate with various measures in the upper dentition and to be related positively only to 144, at the anterior margin of the talonid, and to 184, which is homologous to 144 on $P_{\overline{4}}$. In relationship to between-measures of the upper series, 154 behaves as if it were closely associated to upper teeth.

Measure 65 is positively correlated with measure 64. It forms a tight positive ρ-group with very few bonds missing; with eight measures of the lower series.

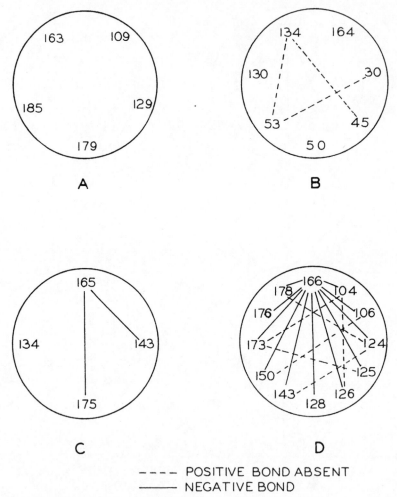

---- POSITIVE BOND ABSENT
———— NEGATIVE BOND

Fig. 60.—Largest ρ-groups formed by "between" measures of lower molars of *Aotus trivirgatus*. (Symbols as in Fig. 59.)

Measure 13, length of $M^{\underline{1}}$, is positively correlated with 65 and is in the ρ-group. This is an example of the irregularity of relationships between the measures of the 60 series and the lengths of upper teeth. In general, where a length measure is of a tooth involved in the 60 measure, the correlation between it and the 60 measure is negative; otherwise it is positive (see, for example, measurement 33 [length of $M^{\underline{2}}$] in the group of measure 64). Measure 65 forms a triplet with

154 and 155 and with 33 and 46. In the first instance, the measures are of the talonid of $M_{\overline{3}}$ (see discussion of measure 64), and in the second the measures are of the upper teeth.

Measure 63 forms a strong positive ρ-group that includes only measures of the lower teeth. The only negative bonds are to measures 29 and 165. Measure 66 similarly forms a ρ-group that consists of positively related measures. This measure has no bonds to upper measures except to 63.

The four measures between upper molars show the following general features. Each tends to form a rather large ρ-group. In these groups, measures of

TABLE 59*

LEVELS OF ρ AND SIGNS OF THE CORRELATIONS FOR MEASURES BETWEEN
TEETH IN *Aotus trivirgatus* (60 AND 160 SERIES)

Meas.	64	65	66	163	164	165	166
63	$+$ $\rho \geq .78$	$+$ $\rho \geq .66$	$+$ $\rho \geq .76$	0	0	$-$ $\rho \geq .66$	0
64	. . . .	$+$ $\rho \geq .90$	0	0	$-$ $\rho \geq .68$	$-$ $\rho \geq .85$	0
65	. . . .	. . . .	0	0	0	$-$ $\rho \geq .90$	0
66	. . . .	. . . .	. . . .	0	0	0	0
163	. . . .	. . . .	. . . .	. . . .	$+$ $\rho \geq .67$	0	$+$ $\rho \geq .68$
164	. . . .	. . . .	. . . .	. . . .	. . . .	$+$ $\rho \geq .78$	0
165	. . . .	. . . .	. . . .	. . . .	. . . .	. . . .	0
166	. . . .	. . . .	. . . .	. . . .	. . . .	. . . .	. . . .

* Values of $\rho < .66$ considered insignificant (r not greater than 0, $P = .1$) are entered without sign as 0.

lower teeth are positively correlated with measures of the 60 series and with each other, except for measures of the talonid of $M_{\overline{3}}$ and the total length of the lower series. Measures of the upper series, other than those in the 60 category, are negatively correlated with measures of the 60 series, but, on the whole, there are few significant correlations. They are confined to measures 64 and 65, except for a negative bond between 63 and 22.

Examination of Figure 60 shows that measures of the lower teeth, correlated to those of the 60 series, fall into three categories, summarized in Table 60: (1) length measures, (2) width measures, and (3) measures of the anterior and posterior parts of the lower molars.

Some evidence of the complex group intersections is present in Table 60. Coupled with the correlations between members of the 60 series (Table 59), these produce a complexly integrated series that is difficult to grasp in its entirety. In addition to the measures shown in the table, three which do not fall in the categories noted occur in the 65 group, namely, 103, 108, and 146; one measure, 145, is present in the 64 group. These are measures of the central parts of teeth. In large part, then, tooth size, length and width, and dimensions of the fore and aft parts of the lower molars are involved in the integration.

The 160 series of the lower molars (not including 167 and 168) shows some similarities and some differences as compared to the 60 series just analyzed (see Fig. 60). Measure 163 has only positive correlations with members of its ρ-group, as does 164. Measures correlated with 163 are all of measures of the

TABLE 60

CATEGORIES OF MEASURES OF THE LOWER MOLARS CORRELATED TO
MEASURES OF THE UPPER MOLARS IN *Aotus trivirgatus*

MEASURE	LENGTH MEASURES	WIDTH MEASURES	ANTERIOR AND POSTERIOR MEASURES OF MOLARS	
			Anterior	Posterior
64........	101 (−)	....	....	$\begin{cases}144\ (+)\\154\ (-)\end{cases}$
65........	183 (+)	142 (+)	....	$\begin{cases}109\ (+)\\129\ (+)\\154\ (-)\\155\ (-)\end{cases}$
66........	$\begin{cases}133\ (+)\\183\ (+)\\113\ (+)\end{cases}$	$\left.\begin{matrix}102\ (+)\\122\ (+)\\142\ (+)\end{matrix}\right\}$	130 (+)	$\begin{cases}109\ (+)\\129\ (+)\end{cases}$
63........	$\begin{cases}113\ (+)\\183\ (+)\end{cases}$	$\left.\begin{matrix}122\ (+)\\142\ (+)\end{matrix}\right\}$	130 (+)	109 (+)

lower molars encompassed by the dimensions covered by the between-teeth measures. The two lower measures of the 164 group are of similar character. These are measures that are negatively correlated with measures of the 60 series above. Of the four upper measures in the 164 group, one is a length measure (53), two pertain to anterior parts of upper molars (30, 50), and one is a width measure of M$^{\underline{3}}$.

The 166 group is made up of a suite of measures of the lower dentition, positively correlated with each other and negatively correlated to 166. Only one of these, 150, is not a measure of central portions of one of the lower teeth. Intersects, except those that result from correlations of the measures of the 160 series, are not extensive; bonding, however, which is not shown in the figures, is very extensive. This bonding is an expression of the relatively high correlation of the lower molars (see p. 188) and produces an extremely complex situation. It does not alter the basic patterns, but merely involves more measures in the array. For this reason and to avoid unnecessary complexity, this bonding is not figured.

Size relationships are not evident in groups of the 160 series. Only in the case of the 164 group are measures of the upper teeth found in these groups. The upper and lower dentitions thus appear to be somewhat differently organized with respect to measures between molars, as were the measures on single teeth (see pp. 188–91). Joint groups involving measures of the 60 and 160 series are small, as can be deduced from collation of Figures 58, 59, and 60, and they add nothing to the general picture. In large part the organization of the lower teeth, viewed from the standpoint of the lower between-teeth measures as group-formers, fails to show any marked relationship to the upper series. The general plan involves positive relationship of measures within the spans of the 160 measures and negative correlations of the central parts of the teeth with measures of the 160 series. When, however, the lower series is viewed from the aspect of the measures between teeth of the upper series, the 60 measures, and the relationships of the lower between measures to analogous measures above, the various relationships that have been detailed above come into a proper perspective.

We have now presented the important elements of the data that have come from analyses of the dentitions of *A. trivirgatus*. It is not always clear-cut and definitive. The complexities make interpretations difficult. It has been necessary to simplify in places to maintain some semblance of clarity. With full recognition of these difficulties, we shall attempt to synthesize the results in a brief summary and to follow this with a condensed interpretation.

Summary of the Data

Integration based on the total suite of eighty-three measurements used in the study of the cheek dentition of *A. trivirgatus* is low when compared to integration found in such functional systems as the limbs or axial structures of various vertebrates. Integration in individual teeth, however, is relatively high. The intensity of integration, based on single teeth, is higher in the members of the lower series, $P_{\overline{4}}$ to $M_{\overline{3}}$, than in the members of the upper, $M^{\underline{1}}$ to $M^{\underline{3}}$. In the upper series, $M^{\underline{3}}$ is the most highly integrated tooth; in the lower series, $M_{\overline{3}}$ occupies this position. There is no apparent relationship between the occlusal contacts of the upper and lower teeth and the intensity of integration of the teeth in the two series. In this case, as in others noted later in this summary, the organization of the upper and lower series appears to have a marked degree of independence. Some interdependency, however, is found when bonds between measures of upper and lower teeth in pairs are considered. Each of the upper molars is most highly bonded to $M_{\overline{3}}$, and all the lowers are most highly bonded to $M^{\underline{3}}$.

The details of the integrative composition of the teeth require study through analysis of a comprehensive suite of measures. The great number of measures needed inevitably introduces complexities that are difficult to visualize. For simplicity, the upper and lower series have been considered alone in initial analysis. All large ρ-groups of the upper molars include a high proportion of measures of length and width. Factors in the ρ-groups, which express maximum relationships of measures, are clearly related to over-all size, expressed in length of tooth row and in length and width measurements of individual teeth. In the upper series, however, the effects of lengths and widths are separate at high levels of integration. A less important factor in integration in this series is found in the tendency for homologous measures along the tooth row, exclusive of length and width, to be correlated.

Only a single "core" ρ-group is present in the lower molars and the fourth premolar. It is composed of a series of interbonded homologous measures. Over-all size relationships are of little importance. This is a second indication of the difference in the organization of the upper and lower cheek teeth. When the upper and lower series are considered together, however, a high integration of the length and width measures of the upper series and within-teeth measures of the lower series is revealed. The integrated homologous measures of the lower series are thus strongly related to general size features of the upper series. In this instance, as in others, the upper dentition appears to maintain a dominant position in integrations that involve both upper and lower teeth.

Partial correlation applied to each tooth with length and width of the tooth held constant ($r_{xy.lw}$) confirms the importance of size in integration. The resulting groups consist very largely of measures of the central parts of the teeth, both above and below. In general, M^1 and M^2 are similarly organized, with the latter having a somewhat more complete integration. The most highly integrated upper molar under simple correlation, M^3, retains only a single bond, 43–44. This strong negative bond, negative in both simple and partial correlation, reflects the reciprocal difference in positions of the cusps of M^3, a highly variable tooth, in different individuals. Except for this one relationship, the integration of M^3 is entirely size-dependent. In the lower series $M_{\overline{1}}$ and $M_{\overline{2}}$ resemble each other in the patterns of integration. Principal groups represent the central portions of the teeth. $M_{\overline{3}}$, however, possesses a well-defined talonid group which meets M^3 in occlusion.

Since the upper and lower teeth overlap one another in occlusion, M^1 over the posterior part of $M_{\overline{1}}$ and the anterior part of $M_{\overline{2}}$, and so forth, it is essential that the relationships of between-teeth and within-tooth measures be known. Such a study in *A. trivirgatus* reveals many interesting aspects of dental organization. It must be understood, however, that the completeness of patterns, desirable for definitive conclusions, has not been obtained for any single measure, and that it is only the combination of pairs, triplets, and quadruplets, as they relate to other measures, that gives fairly clear outlines of the organization. No simple one-to-one correspondence of opposing "between" and "within" areas of the teeth occurs. The integration patterns, on the contrary, relate to the total molar series and appear to be strongly size-dependent. If the situations indicated variously by the different between measures are collated, a general pattern of relationship emerges, as described in the following paragraph.

Many negative bonds between within-tooth and between-teeth measures occur. In some instances, large ρ-groups of within-tooth measures bonded negatively to between-teeth measures but positively to each other are present. The labial measure between $M^{\underline{3}}$ and $M^{\underline{2}}$ and the lingual measure between $M_{\overline{1}}$ and $M_{\overline{2}}$ form such groups. Negative bonds, with rare exceptions, occur between measures of intertooth distances and within measures which fall into the following categories: (1) measures in the same series, i.e., upper within and between measures, or lower within and between measures; and (2) measures of upper intertooth distances and measures of the talonid of $M_{\overline{3}}$. With reference to the first category, there is evidently no selective pattern of types of measures involved.

Positive ρ-groups are formed by the between-teeth measures not cited in the preceding paragraph, along with some negative bonds that fall into category 2 above. It may be that the differences of bonding noted for the various between measures have some significance, but it is not evident from the sample what this might be. Large ρ-groups that involved positive bonds are formed by the lingual measure between $M^{\underline{1}}$ and $M^{\underline{2}}$ and the labial measure between $M_{\overline{2}}$ and $M_{\overline{3}}$, the analogues of the between measures that have large arrays of negative bonds. Lesser positive ρ-groups are formed by the other between measures. Positive bonds in the ρ-groups are in large part related to between-teeth measures and within-tooth measures of teeth in opposing series. Those formed by between-teeth measures of the upper series, with few exceptions, involve the following types of measures: (1) measures of lengths and widths (of lower teeth); (2) measures of the anterior and posterior areas (of lower teeth). Very few measures of the central parts of the teeth show any positive correlations to between measures of the opposite tooth row. The between measures of the lower series are positively bonded to measures of category 2 above but are not bonded to size measures.

A considerable part of the total picture is omitted when the between measures are considered one at a time. There is a rather extensive bonding between these measures (see Table 59). Some concept of the great extent of integration that is evident when these measures are combined can be gained by visualizing the intersects of the ρ-groups shown in Figures 59 and 60. In addition, however, there is extensive bonding of measures in the ρ-groups but not in the intersects and also of measures not in the ρ-groups but bonded to many measures of the ρ-groups as well as to each other. These arrangements are too complex to be meaningful in diagrams. The full array of ρ-groups in which this information is contained—713 in all at $\rho \geq .66$—reveals the basic patterns only upon very extensive and tedious analysis. In lieu of full presentation of the evidence, we request the reader's indulgence to accept the statement that the intersections, bondings between ρ-groups, and additional bonded measures do not alter the basic concepts available from what has been presented, but clarify the situation in two ways. First, they show integration to be more extensive than that indicated by the ρ-groups formed by the between measures. A more complete patterning is thus developed than is indicated. Second, they extend markedly the positive and negative bonds from the between measures and positive bonds between the within measures in about 95 per cent conformity with the patterns shown by the ρ-groups.

Interpretation

In the introduction to this report on the study of the cheek teeth of *A. trivirgatus*, the hypothesis that several factors entered into the formation of patterns of morphological integration of mammalian molar teeth was set forth. These factors involved over-all size, homologous and analogous dimensions, morphogenetic field control, and "interadaptive" features specifically related to within and between measures of the opposing series of teeth. The experiment was designed expressly to "test" this hypothesis, and thus we shall attempt to evaluate the results from this point of view. The factors are, in a direct sense, strictly ostensive features of morphology, but into each may be read something of the developmental background that may have led to their existence.

The evidence clearly shows that over-all size, both the total length of the tooth series and length and width measurements of individual teeth, is fundamental to the great majority of gross patterns of integration. This may, of course, be stated in terms of growth as follows: growth of the teeth in their formative stages occurs in such a way that there results a harmonious pattern of a great many of the dimensions of the teeth. Although there is no dynamic functional control of this growth, in the sense that patterns are imposed by dynamic interactions during ontogeny, it is evident that the sizes of dentitions in general are not unrelated to the size of the organism in which they develop and more specifically to the dimensions of adjacent bones. Rather meager data, however (see, for example, *Sciurus*), have not indicated significant correlation between dimensions of the jaw elements and the cheek dentition. The factor of over-all size in the tooth-row series and in the individual teeth thus appears to operate in integration specifically in relationship to the teeth alone.

This result is not unexpected and has an evident relationship to the very general concept of morphogenetic fields. The somewhat unexpected result of the studies, however, arises in the differences of the factor of size as expressed in the upper and lower series. Length of the tooth row and the included teeth are broadly integrated with each other and with other measures in the upper dentition. Width measures in the upper series are similarly correlated with other tooth measures and are integrated with length measures at low levels of ρ. Neither length nor width measures are critical in the associations of the within-tooth measures of the lower series. The striking fact is that it is the lengths and widths in the upper series that enter into large ρ-groups of the lower series, when all within-tooth measures are considered together. The within measures of the upper series, however, do not appear in these ρ-groups to any notable degree. The factor of over-all size in morphological integration, while important, is not a simple, single influence. From a strictly morphological point of view, the impression is gained that the lower dentition is dependent upon size factors that are detectable as such only in the upper series. There is no evidence of a reciprocal relationship. What this means with respect to genetics, epigenetics, and the ontogenetic process leading to the final formation of the teeth, we are not prepared to say. Speculation on these points, in the absence of the critical data of genetics and growth of the teeth, can be nothing but a series of wild guesses.

Importance of homologous measures in the integrative patterns is evident in both the upper and the lower series. Analogous measures, on the contrary, even length and width measures, are of little or no importance. The factor related to analogues may, of course, be used to infer a common genetic background for the structures involved. Whether this would be considered specific, as related to individual patterns of each tooth, or repetitive expression in serially homologous structures seems at the present state of knowledge to be largely a matter of personal opinion. The way that the data were treated allowed no decision on this matter. It appeared possible, as work progressed, that correlation of homologues might be related primarily to size factors. If this were the case, however, size dependency was on gross dimensions of the upper series, for both upper and lower homologues. The homologous measures in the lower series formed patterns of integration not dependent upon gross dimensions of their own series. Partial correlation performed on each tooth, with length and width held constant, demonstrates that those measures that form integrated patterns of homologues in the series persist in integrated groups in individual teeth with the effects of length and width removed. We may infer, on admittedly somewhat inadequate grounds, that this may apply between teeth.

In view of the widely divergent views on homologues of cusps of the upper and lower molars of mammals (see, for examples, summaries in Gregory [1934], Butler [1939a, b, 1941], and Patterson [1949]), it is difficult to discuss "homologues" between upper and lower molars. None of the patterns revealed in this study, however, shows a consistent upper-lower relationship that suggests homologues to be important under any of the schemes that have been proposed.

The concept of morphogenetic fields as applied to adult structures, such as the teeth of *Aotus*, seems to us to be somewhat tenuous and in a sense descriptive rather than related to genesis. If the concept of a field is considered strictly in the sense of phenotypic description, there results a way of thinking that views large units, which is all to the good. It does not seem to us, however, that there is sufficient evidence to draw inferences to fundamental causation of the phenomena involved. Selection certainly has resulted in the shifts in the molar "field," as in its extension in many herbivorous mammals or its restriction in various carnivores. We may assume that there is a genetic basis, expressed through morphogenesis. What may have been the nature of genetic modification that selection favored remains largely unknown.

Can we think of morphological integration in the sense of the concept of the morphogenetic field? Offhand, this seems a natural consequence of this type of study. Unless, however, the field concept can give additional meaning to the results, little is gained and much may be lost by adopting the terminology of a concept based largely upon work in a separate subdiscipline of biology. We feel that at present it would be a mistake to attempt to cast the results of our work on *Aotus* into this framework. Once it has become possible to relate the observed morphological relationships experimentally to the processes that resulted in their existence, the situation will, perhaps, be different.

Finally, we shall consider briefly the "interadaptive" features of the molar teeth. Under this term we embrace the concept that the cusps and other crown features of the teeth that are brought together as the functional activities of the teeth are carried out will show a higher degree of integration with each other than with other structures, if there has been a selection related to joint performance of their functions. We must point out immediately that such an adjustment is not necessarily independent of over-all size relationships and might, in fact, be in large part related to it. The data, however, show that in *Aotus* there is no instance in which the over-all length of the tooth row is positively correlated with between-teeth measures used to study this relationship. On the other hand, these measures do relate to gross size-measurements of individual teeth, in general with positive correlation with those of the opposite series and negative correlation with those of the same series.

There does exist an integration of between-teeth measures of one series, upper or lower, with the measures of the opposite series that lie in the areas of contact. There is no integration of the between-teeth measures and the central measures of teeth of the opposing tooth row. The integration that does exist, however, is weak and includes measures along the tooth row, as well as some in the area immediately opposed by a particular between measure. In *Aotus*, at least, the factor of interadaptation, while operative in the total pattern of integration, is neither strong nor definitive. It may well be that this is the result, in part at least, of the type of dentition present in *Aotus*, which is relatively simple and much less specialized than that of the great majority of mammals. One fact points in this direction. The trigonid of $M_{\overline{3}}$ relates in occlusion predominantly to the total of M$^{\underline{3}}$. Here we might expect a strong interadaptive situation that involves only within-tooth measures. The measures of the talonid of M$_{\overline{3}}$ are integrated in the between-measure ρ-groups in the same way as are the within measures of the *upper molars*, not as are the within measures of the other lower teeth. By indirection the sort of relationship proposed appears to exist. Perhaps in mammals with more specifically functional occlusal relationships throughout, analogous cases of greater interadaptation exist. A great deal more work must be done on this aspect of integration in mammalian dentition before the general importance of this factor can be evaluated.

There is little basis at present for speculations concerning the genetic and developmental aspects of the factors observed or for casting morphological integration in the framework of the concept of morphogenetic fields. We may feel quite certain that the patterns revealed are the result of adaptation under natural selection, but in the present state of knowledge it seems futile to speculate upon the nature of the genetic structures upon which selection has acted.

LOWER MOLAR DENTITION IN *Hyopsodus*

Eight samples of the lower molar series of the small Eocene condylarth *Hyopsodus* have been studied in the course of investigations of morphological integration. The principal purpose of the study was an analysis of the role of integration in dental evolution at the species level. Results from this point of view are taken up in chapter viii. For the present we shall be concerned with the phases of the work that are appropriate to a comparison of the lower molar series of *Hyopsodus* and *Aotus* and with one sample that shows the behavior in integration in a spurious assemblage. The data are taken in large part from Tables 68–75, included in chapter viii. The taxonomy of the samples of *Hyopsodus* and other relevant information on the genus are available in Appendix E.

The measures of the lower molars of *Hyopsodus*, as shown in Figure 61, give approximately the same coverage as those used for *Aotus* (Fig. 55). Similar coverage was possible because the lower molars in the two genera have basically similar morphological patterns. The molars of *Hyopsodus*, in fact, are so similar to those of some of the lemuroid primates that the genus was once assigned to this order. The resemblances of the dentition in *Hyopsodus* and *Aotus* permit much closer comparisons between members of separate orders than are usually possible. Only the general aspects of morphological integration can be compared, however, for the details show a wide divergence even within the genus *Hyopsodus*, and we have no information upon the range of the various features in genus *Aotus*.

Comparisons of Hyopsodus and Aotus

The index of morphological integration in the lower molar series of *A. trivirgatus* at $\rho \geq .51$ ($P = .01$) is .0014. A range from .0002 to .0030 is found in

Hyopsodus (exclusive of *H. paulus* "mutants") at this level of ρ. *Hyopsodus miticulus* from the Gray Bull member and *H. mentalis* from the Lysite (and Lost Cabin) member of the Wasatch formation, early Eocene, have indexes of .0017 and .0014, respectively, and resemble *A. trivirgatus* closely in this respect. Without knowledge of more than one species of *Aotus*, it is impossible to know what part of the genus range the value of .0014 of *A. trivirgatus* represents. It does seem evident, however, that the morphological integration in the lower molars of *Aotus* and *Hyopsodus* is of the same order of magnitude.

Morphological integration of the lower molar series in *A. trivirgatus* has been shown to depend in large part on ρ-groups that involve interbonded homologous measures. Length and width measures play a relatively minor role

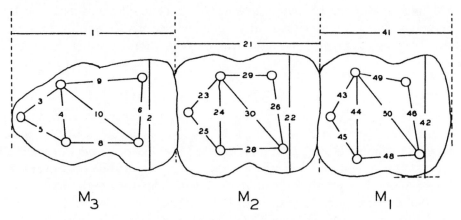

Fig. 61.—Measurements of lower molars of *Hyopsodus* used in studies of the various species. Same measurements used for each species. (See text for system of measurement.)

in group formation. This is not true in *Hyopsodus*, for in each species over-all length of tooth row and the lengths and widths of the individual molars do enter into the large ρ-groups. This feature is least developed in *H. paulus*, in which the index of integration for the lower molar series is .0002. Where integration is relatively high, as in *H. miticulus*, gross dimensions are prominent in the ρ-groups (see Table 69). *Aotus trivirgatus* finds its closest counterpart in over-all integration patterns in the molars of *H. paulus*, but here the indexes, based on commensurate levels of ρ, are very different, .0002 for *H. paulus* and .0014 for *A. trivirgatus*. In both there is little dependence of groups on length and width measures. One marked difference appears. In *H. paulus* there are very few cross-bonds between teeth, whereas in *A. trivirgatus* such bonds predominate (cf. Table 68 and Fig. 56). It should be noted that in these two, where the level of ρ is such that only values of *r* not significantly different from zero are excluded, $P = 0.1$ and 0.01, respectively. In other species of *Hyopsodus* in which cross-tooth bonds are prominent, length and width measures are important. There appears thus to be a fundamentally different plan of integration in the lower molars of the two genera as these are viewed without reference to the upper.

Individual lower molar teeth in *A. trivirgatus* show length and width measurements to be prominent in the ρ-groups (Table 58). The same is true for the various species of *Hyopsodus* (Table 68). Also, $M_{\overline{2}}$ was shown to be the most highly integrated tooth under simple correlation in *A. trivirgatus*. Values of $I\rho$ in Table 74 show that *different* lower molars play the role of the most highly integrated in the various species of *Hyopsodus*. This feature is unstable in the species and, if lower molars are considered alone, relates to the importance of the size factor in the individual teeth. Where size measures are important in the formation of ρ-groups, integration tends to be relatively high; where they are not, integration tends to be relatively low.

The results of partial correlation bear on this point. Table 61 shows the change in $I\rho$ in *A. trivirgatus* as length and width are held constant. All values

TABLE 61

$I\rho$ AT COMMENSURATE LEVELS OF ρ IN
LOWER MOLARS OF *Aotus trivirgatus*

	$I_{\rho xy}$	$I_{\rho xy.lw}$
$M_{\overline{1}}$............	.0267	.0059
$M_{\overline{2}}$............	.0648	.0024
$M_{\overline{3}}$............	.0179	.0170

of $I\rho$ drop, but for $M_{\overline{3}}$ the decrease is negligible. Size is not important in the bonds and ρ-groups that enter into $I\rho$ for this tooth, but it is very important in the other two molars. The highest integration in the series, r_{xy}, is in $M_{\overline{2}}$. This is reduced to the least under partial correlation $r_{xy.lw}$. It will be recalled that $M^{\underline{3}}$ of *A. trivirgatus* showed the highest value of $I\rho$ under simple correlation of the upper molars but that it retained only a single bond under partial correlation, $r_{xy.lw}$.

A somewhat similar relationship between size-dependent integration and integration after partial correlation to remove the effect of size occurs in the species of *Hyopsodus* (Table 75). The effectiveness of size as a factor of integration differs markedly in the species of *Hyopsodus*. Clearly there is danger in drawing conclusions from the values of $I\rho$ of the individual teeth under simple correlation. Removal of the effects of size appears to provide a better basis for evaluation of relative integration. The results, however, take on a specific meaning, as shown in the studies of *Hyopsodus* and *Aotus*, only when they are studied either in conjunction with the upper molars (*Aotus*) or with information about the composition of the ρ-groups formed after partial correlation (*Aotus* and *Hyopsodus*).

It was noted in *A. trivirgatus* that, after partial correlation, the ρ-groups of M_1 and $M_{\bar{2}}$ contained only central measures of the teeth (Table 58). Those of $M_{\bar{3}}$, on the contrary, included both central and talonid measures. The occlusal relationships of the upper and lower third molars impart meaning to this condition (see pp. 188–89). Central integration is present in the lower molars of the species of *Hyopsodus*, consistently in $M_{\bar{2}}$ and variously in $M_{\bar{1}}$ and $M_{\bar{3}}$. Talonid integration is found in $M_{\bar{3}}$ of five of the seven valid species of *Hyopsodus* but also occurs variously on $M_{\bar{1}}$ and $M_{\bar{2}}$. The importance of shifts in patterns in the course of phyletic evolution of the species of *Hyopsodus*, as taken up in chapter viii, suggests that adjustments are evidence of relatively small shifts and that comparisons of patterns between widely separated genera, like *Hyopsodus* and *Aotus*, can be misleading.

Comparisons between these two genera cannot be carried further in the absence of knowledge of the upper molars in *Hyopsodus*. It is evident from the information available that the general similarity in dental morphology is reflected in a broad similarity of patterns of integration but that there are fundamental differences. The wide range of difference in the details of integration in the species of *Hyopsodus*, however, is a warning signal against heavy reliance upon generalization to a genus from a single species. They argue strongly against using a "typological" approach for comparisons of genera and more widely separated taxonomic categories.

Morphological Integration and Hyopsodus paulus "Mutants"

The specimens referred to *H. paulus* "mutants" by Matthew represent a middle-size range of the specimens of *Hyopsodus* from the Bridger C. As pointed out in Appendix E, the ways in which this "species" was assembled and defined indicate that it is probably spurious and consists of specimens that represent more than one species of the Bridger C. Certain aspects of the morphological integration in the sample drawn from this assemblage have implications of rather general importance.

The index of morphological integration, derived from the sample, is .0046 for the total molar series, the highest of any found in the species of *Hyopsodus* studied. *Hyopsodus powellianus* is second highest, with $I\rho = .0030$ (Table 68). For the individual molars, $I\rho$ of $M_{\bar{1}}$ in *H. paulus* "mutants" is higher than that in any other species studied (.0107). Values for $M_{\bar{2}}$ and $M_{\bar{3}}$ are among the highest, although each value is exceeded by commensurate values in two other species. *All* ρ-groups for *H. paulus* "mutants" include length and width measures, a situation not found in any other sample studied. Length and width are thus very important in the integration revealed in the indexes of this spurious species.

Under partial correlation ($r_{xy.lw}$) the value of $I\rho$ drops to zero for each tooth (Table 74). With the elimination of the effects of gross dimensions, there remains no integration. The behavior of $I\rho$ in this case, where the sample consists of members of a genus but not of the same species, is that to be expected under a situation where the size factor is pervasive in integration throughout the genus, whereas species differ in integration largely in patterns that are not size-dependent. In simple correlation, only size-dependent groups will tend to form. The number of bonds, relative to the number of possible bonds of the measures, will tend to be high, relative to those in the species, since each size measure will be bonded to all size-dependent measures and these, in turn, to each other, as a result of their common size-dependency. In this situation, ρ-groups will be large relative to the number of measures and few in number. A relatively high value of $I\rho$ will result. Under partial correlation, a drastic lowering of the index of morphological integration will occur when measures of size are held constant. It will drop to zero in the event that no integrations common to the members of the various species and not size-dependent have been included. This appears to be the case for *H. paulus* "mutants."

The need for caution in the use of $I\rho$ in simple correlations, noted in the comparisons of *Aotus* and *Hyopsodus*, is emphasized by the problems that may be raised by spurious samples or by samples representing supra-specific categories. It is essential in all cases to examine the ρ-groups that contribute to the index for the existence of measures that may express ubiquitous factors. Testing these factors by partial correlation is a necessary safeguard to errors in conclusions such as those drawn from the teeth of the various species of *Hyopsodus* in the next chapter.

CHAPTER VIII

Morphological Integration and Evolution
I. Empirical Studies

The goal of all of the studies that have gone into the preparation of this book has been the establishment of a basis for a study of evolution in which some major portion of the organism is considered to be the evolving unit. Some of the possible roles that such studies might play in elaboration of the theory of evolution were outlined in the closing pages of chapter ii. Now that the details of the techniques of study or morphological integration, as well as theoretical considerations and examples of analyses, have been presented, we are in a position to explore some of these roles and to attempt to provide answers to some of the questions posed earlier. It would be pretentious to imply that, at the present time, more than a meager part of the whole potential of such studies can be treated. We can, however, make a beginning.

At present, theory is still far ahead of the sound empirical basis which eventually must exist as a foundation for extrapolation. To assure that these two areas are not confused, we propose to treat separately the empirical studies and the more theoretical aspects of evolution as related to morphological integration. The present chapter is devoted to actual studies that have been made, with consideration of theoretical aspects only as they stem directly from the problems treated. A more general synthesis will be the subject of the final chapter of the book.

The two studies presented in this chapter show ways in which morphological integration behaves during evolution. All the various analyses stress the development of means of treatment and the interpretation of changes in integration that are observed. The groups of organisms that the materials represent are not per se of primary concern. We are, for example, less interested in new information about *Pentremites* than we are in the principles that emerge from a study of it, principles that appear to have some general import. Similarly, whatever is revealed about dentitions in *Hyopsodus* is secondary to the determination of whether or not any information on dental evolution, not otherwise available, can be obtained by the type of analyses that are used. Analyses stop short of detailed biological interpretations that pertain specifically to the evolution of the organisms in question.

To date, four studies have been made on materials suitable for analysis of evolution. The results of two of these have been published previously, one on captorhinomorphs (Olson and Miller, 1951*b*) and the other on the extinct amphibians *Diplocaulus* and *Trimerorhachis* (Olson, 1953). The other two, on *Pentremites* and *Hyopsodus*, are reported for the first time in this chapter. The first two studies were preliminary and in large part set the stage for the more ambitious analyses that followed. We shall treat them only in summary fashion to show the basis for certain principles more fully developed in this and the next chapter.

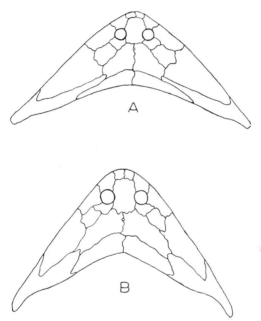

Fig. 62.—Dorsal aspects of skulls of the amphibian *Diplocaulus*. A, *D. magnicornis;* B, *D. recurvatus*.

RESULTS OF EARLY STUDIES

The study of captorhinomorph reptiles revealed a semblance of order in ρ-groups formed by means of correlation and showed that these groups were different in the various species. Some suggestion of the relationship of groups derived mathematically to biological groups of measures was found, and it was determined that size, expressed by a gross measure, was an important agent in integration.

A profound modification of pattern and intensity of morphological integration in the development of one species from another was revealed in the study of *Diplocaulus magnicornis* and its direct descendant, *D. recurvatus*. Measure-

ments of the dermal pattern of the dorsal surface of the skull and of the aper-
tures for the sensory organs of the head provided the data. The only important
gross morphological difference observed between skulls of the two species was
one of shape, related to the development of the posterolateral "horns" (Fig.
62). As in the captorhinomorphs, it was made evident by partial correlation
that much of the pattern was dependent upon a factor of size. When length
and width measurements were "held constant" in *D. magnicornis*, there remained only a small "core" ρ-group, which included measures intimately associated with the brain area and the sensory organs. With the change in shape
in the development of *D. recurvatus*, the co-ordinated action of length and
width, evident in *D. magnicornis*, was eliminated, and length and width operated more or less independently as agents of integration. Thus the changes of
many characters were nicely related to one major feature and brought into a
simple, coherent framework for interpretation.

Another point of some interest was noted in a comparison of the integration
patterns of the skulls of the amphibians *D. magnicornis* and *Trimerorhachis
insignis*. These two are commonly placed in separate subclasses and thus have
only a very remote common ancestry. Both showed complex, highly integrated
patterns of the dermal elements and sensory openings of the skulls. Under
partial correlation, with skull length and width held constant, there remained
in each a small residuum of integration related to the brain case and sensory
organs. There appears to be a very stable pattern of morphological integration
here, which persisted through the vicissitudes of a highly diverse evolution.
Although the evidence is tenuous on this point and no further work has been
done to test it, the implications are sufficiently important to warrant more
critical and more properly conceived research.

Essentially, the early work provided a series of tentative ideas, to be looked
into and tested by the use of more suitable materials and more refined techniques. The studies of *Hyopsodus* and *Pentremites* represent a beginning in this
direction. Other work along both similar and different lines is in progress or
under consideration.

PENTREMITES

Data from the six samples of the genus *Pentremites* used in the study reported
below are presented in Appendix G. Only essential points are summarized in
the present discussion. The samples are all from the Chester series of the Mississippian system of Illinois. They were obtained from the shale and limestone
formations of the series. These formations occur in more or less cyclical alter-
nation with sandstones. A generally similar environment of deposition appears
to have obtained for each of the samples, although it is highly probable that
detailed sedimentological and paleontological analyses might reveal important
differences. In the absence of this information, interpretations cannot take

cognizance of adaptations as they relate to the different environments. We hope that it may be possible to remedy this deficiency in the future, but for the present we can note only that its existence is recognized. The specimens of the samples appear to have been deposited at or near the sites in which they lived, for they are usually fairly complete and occur in concentrations that suggest the general colonial aspects of their existence. The sediments, furthermore, suggest relatively quiet waters during periods of deposition.

The individual samples consist of specimens that are either godoniform or pyriform in the shape of the calyx (Fig. 63). The oldest sample, from the Renault, is godoniform. The pyriform type does not, to the best of our knowl-

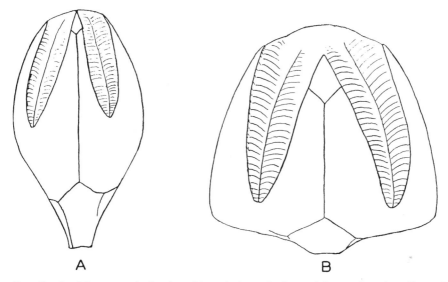

A   B

Fig. 63.—Semidiagrammatic sketches of lateral views of calyces of *Pentremites*. *A*, pyriform; *B*, godoniform.

edge, occur in this formation. Both godoniform and pyriform samples have been taken from the Paint Creek formation, next youngest limestone in the series. Individuals of both types occur at the same localities. The same is true for the Golconda, which is the first limestone encountered above the Paint Creek. In the youngest formation of the series, the Glen Dean, only godoniform individuals occur. The pyriform type apparently does not occur above the Golconda.

As is generally the case, it is necessary to make some assumptions to establish a framework for investigation. In this instance, in which we wish to study morphological integration in an evolutionary series, it is essential to derive an understanding of relationships from data other than those of morphological integration. These data come primarily from the gross morphology of the specimens and the stratigraphic relationships of the samples, augmented, of course, by principles generally applicable in studies of stratigraphic series. It is assumed, in this regard, that the Renault godoniform population is ancestral to both the godoniform and the pyriform populations of the Paint Creek. All the evidence available from stratigraphy and morphology supports this assumption. The two types occur in the next fossiliferous bed above the Renault. The morphology of the Renault sample is permissive of such a relationship, and there is no other known possible ancestor of the pyriform type of *Pentremites*. While the assumed relationship seems the most logical one, it cannot, of course, be conclusively demonstrated to be correct.

On the basis of the morphology of the godoniform and pyriform types, the fact that they occur together, and the temporal duration of each type, it is assumed that they were sexually isolated. Presumably, at the time of division of the ancestral population after the Renault and prior to the Paint Creek, some sort of isolation took place, probably a geographic isolation. We assume that sympatric speciation did not occur. The coexistence of the two in the Paint Creek and Golconda times must represent a later movement into the same habitat. We further assume that the samples are in each case representative of the populations of their times. This is undoubtedly an oversimplification, for surely there was local variation in the populations beyond that which has been taken into account. We feel, however, that this was probably insufficient to invalidate the broad outlines of the evolutionary flow shown by the samples. A similar assumption is made with respect to temporal variation. The samples have been drawn from a series of temporal stages. It must be assumed that evolution progressed along a somewhat regular course rather than with wide fluctuations and that the samples represent this course adequately. Each of these assumptions appears to be the most logical that can be made under the circumstances of occurrence. Results, of course, must be judged with the recognition that they have been made.

The evolution of any series of organisms can be viewed in many ways, and this is true even within the limits of morphological integration that we have established. Since we propose to evaluate the role of morphological integration in evolution to make clear the types of phenomena that can be observed and interpreted, it has been found profitable to consider the various aspects of morphological integration separately, prior to a general analysis of the whole picture. The first step, after data have been gathered, is the calculation of the coefficients of correlation. Thereafter, the following order of procedure, which we follow in the presentation, is practical in this problem and others of similar nature:

1. Analysis of the general level of total integration based on the index of morphological integration, $I\rho$

2. Determination of the gross patterns of integration through study of groupings of measures at successively lower levels of ρ

3. Study of the individual morphological systems and their interrelationships through the use of basic-pair analyses
4. Study of "factors" of integration by use of partial correlation
5. Collation, interpretation, and summary of results

Each of these steps involves comparisons of populations appropriate to the particular way in which the data are formulated. Each of the first four yields information of a somewhat different nature. The final step of collation and summary is directed primarily by the objectives of the study. In the case at hand, it involves an effort to draw together the various aspects that show how morphological integration has fared in the course of a particular set of evolutionary events. Since $I\rho$ represents the closest approach to the total picture of evolution of morphological integration, the evidences from the other sources can best be considered as they relate to it and as they give insight into complexities that contribute to the final value of $I\rho$.

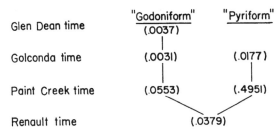

Fig. 64.—Values of $I\rho$ for *Pentremites*, arranged stratigraphically and phylogenetically

1. $I\rho$, the Total Integration

The index of morphological integration has been developed to provide a single numerical expression of the level of integration of a population. The value is dependent, of course, upon the measures that have been taken and the level of ρ (with a given level of confidence) selected. The same measures were taken on each specimen of *Pentremites* and were designed to represent the calyx as fully as possible. To this extent they represent the totality, and, since all samples have been similarly treated, results are directly comparable. Also, $I\rho$ is evaluated at the level of ρ for each sample that includes only those values of r that are significantly different from zero.

A total of twenty-nine measurements has been taken on each individual, as shown in Figure 31 and described in Table 20. The measurements represent dimensions of the respiratory system, the digestive system, the deltoid system, the basal system, and over-all size and shape in terms of selected gross dimensions. Figure 64 shows the values of $I\rho$ for each of the samples of *Pentremites*. The populations are arranged stratigraphically and in phylogenetic relationship. It will be noted that the index rises sharply from the Renault *Pentremites* to the two Paint Creek populations, presumed to be derived from the first.

Between the Paint Creek and the Golconda there is a marked drop in both lines. The Glen Dean godoniform *Pentremites* approximates the level of $I\rho$ found in the Golconda population.

The series of events recorded in the changes of $I\rho$ provides a general and simple expression of the course of morphological integration in the segment of evolution of the *Pentremites* that has been studied. It is, however, based upon a very complex analysis and takes on increasing meaning as the various facets of this analysis are examined more in detail. Interpretation is possible without further exploration, but it can be markedly enhanced by additional information. Here, as in the following sections of the study, interpretation of the meaning of the observed patterns will be deferred until all phases have been considered.

2. *Evolution of ρ-Groups Based on Gross Determinations*

The number and size of ρ-groups contribute to the value of $I\rho$ in such a way that we may make a rough estimate of these values from $I\rho$ alone. The index, however, tells nothing of the composition of the groups. A first approximation to this feature—the qualitative aspects of the groups—may be made by a study of the formation of bonds at successively lower levels of ρ, as described in chapter iv. In this study we are dealing with all measures and maintaining the concept of the totality expressed in $I\rho$, in contrast to the later sections, in which restrictions reduce the number of measures considered.

The results of the study are given in Table 62 for the populations of *Pentremites*. The ρ-groups are not formally established, but it is inevitable that any orderly assemblage of measures will be apparent. This, then, represents an informal approach to recognition of F-groups from the data of ρ and one that may be carried out rapidly. It is suitable for a general estimation but, of course, is inadequate as a basis for additional analytical work that requires precise formulation of ρ-groups. General statements concerning F-groups can be made, as has been illustrated in Table 62. It is important to recognize, however, that the bonds *between* measures of the F-groups have a strong masking effect as the level of ρ is lowered. This represents the initiation of merging of groups and is portrayed in like manner in formally constructed ρ-groups.

It is evident in the summary in Table 62 that, at very high levels of ρ ($\rho \geq .98$) in the Renault population, bonding occurs between measures of the basal structures (24, 25, 26, 27, 28) and that some bonds occur between these measures and those of gross size (8, 23, 31, 32, 34). Specifically, the general statement depends upon the existence of the following groups: 8–23–31 (size), 8–31–32 (size), 24–25–26–28 (basal), 24–25–28–34 (basal and size), 25–26–28–31 (basal and size), 25–28–31–32–34 (basal and size), and 27–28–31–34 (basal and size). Even at this level the two groups of measures show some overlap, that is, there are ρ-groups that include measures of both the basal complex and the size.

Extension of this summary approach to lower levels of ρ gives a crude but useful basis for comparisons between populations. We thus arrive at a qualitative description of the evolution of the integrated groups of measures and the

TABLE 62

Survey of Evolution of *Pentremites* Based on Informal Analysis

		Godoniform Line	Pyriform Line
Glen Dean	$\rho \geq .98$	Weak basal group	....
	$\rho \geq .97$	Same, slight increase in consolidation	....
	$\rho \geq .93$	Strong integration of basal group, some merging with size group, which is weak; deltoid measures independent	....
Golconda	$\rho \geq .98$	Size group only	Size group, weak deltoid group
	$\rho \geq .97$	Size group consolidated, merging with deltoid	Basal and one crown measures brought into size group
	$\rho \geq .95$	Crown comes in as unit; anal measures discretely bonded; size group further consolidated	Most crown and measures of digestive structures bonded; basal, deltoid, and size measures highly consolidated
	$\rho \geq .93$	Same as $\rho \geq .95$, except basal group appears as independent unit	Very high degree of consolidation
Paint Creek	$\rho \geq .98$	Size group present; basal measures weakly related	Size group strong; basal group weak
	$\rho \geq .97$	Consolidation, but no new groups	Crown and basal measures come into size group; deltoid measures form somewhat separate group; digestive measures form partially separate group
	$\rho \geq .95$	Size group consolidated; basal group increases, remaining independent	Most measures merge into common basal size group
	$\rho \geq .93$	Crown measures come in, in part; size group and basal group increase but are largely independent; anal measures come in as unit	Consolidation nearly complete
Renault	$\rho \geq .98$	Basal and size group, partially merged	....
	$\rho \geq .95$	Basal and size group consolidated; deltoid group brought in	....
	$\rho \geq .93$	Further consolidation; addition of digestive group	....

relationships between groups. The combination of such information with the data from $I\rho$ may provide all that is desired about the nature of evolution in terms of morphological integration. We can at this stage, for example, make the following statements:

1. With the split of the parent population into two populations, one pyriform and the other godoniform, total integration rises sharply to each of the derived populations.

2. During the evolution of each of the derived lines, there is a sharp drop of integration from the initial stage to the one that follows.

3. Different patterns of ρ-groups occur in the two derived populations of the Paint Creek—the pyriform type and the godoniform type. This is well shown in the independence of the basal and size shape groups in the godoniform population as contrasted with the close association of the two in the pyriform population.

4. The patterns of the groups established in the two lines after the split tend to persist in the two lines with, however, a weakening of the associations of various measures as $I\rho$ decreases.

For some studies such an analysis may be sufficient. Certainly it provides a reasonable basis for examination of the various facets of evolution of *Pentremites* for investigation, for instance, of relationships of changes in the morphological systems to gross conditions of sedimentation or, more precisely, conditions under which the populations are presumed to have lived. Often, however, it will be necessary to pry more deeply into the integrative relationships, and this is particularly true in the present instance, in which we are interested in all aspects of the behavior of morphological integration in evolution. Thus we shall proceed to more technical aspects of the study by the use of basic-pair analyses.

3. *Basic-Pair Analysis*

Basic pairs for each of the populations have been determined by the method described in chapters iii and iv. The pairs and the levels at which they form are shown in Table 63. As is to be expected, the basic pairs differ in the various populations. Either the existence of different groups or slight shifts in emphasis within the same groups can cause the differences. Sampling fluctuations, of course, may also be responsible for some indeterminate part of the observed differences. It is the group-forming capacity of the basic pair rather than its precise composition that is of primary concern in comparisons of populations except where the minute details of each group must be considered. On the basis of this potential, the basic pairs in the table fall into the following primary categories:

1. *Size.* Here the basic pairs serve to form groups that consists of measures of gross size. Measures 8–31, 8–32, 8–35, 31–32, and 31–34 all act in this way. Occasionally, single measures other than those of gross size appear in the groups. These are measures not brought into groups by other basic pairs. They may have importance in very detailed analyses but for general comparisons may be ignored. It should be noted that measure 34 (base of the ambulacrum to base of the calyx) has properties of both a gross size measure and a basal measure. It tends to be closely associated with both groups and acts as a link between them. The measures that are referred to *size* by the groups formed by these basic pairs are 8, 23, 31, 32, 35, (34).[1]

[1] The size group may be used to illustrate what is meant by the "group-forming capacity" of a basic pair and to show how similar groups are formed by the various pairs listed above. Groups formed

2. *Deltoid* (size) *group*. Basic pair 29–30. This pair tends to exist independently, not to form large groups, above levels of ρ at which it merges with other basic pairs. Its failure to be present in some populations arises from the fact that it is merged with other basic pairs at the highest level (of ρ) of occurrence.

3. *Basal group*. Basic pairs, 25–26, 26–27, 24–28. This group generally tends to form independently of others at high levels. Only in the Golconda godoniform population is it merged

TABLE 63

BASIC PAIRS IN *Pentremites* POPULATIONS

	$\rho \geq .98$	$\rho \geq .97$	$\rho \geq .95$	$\rho \geq .92$	$\rho \geq .90$	$\rho \geq .86$	$\rho \geq .80$
Glen Dean godoniform.....	26–27	29–30	....	....	....	....	1–2
Golconda pyriform	8–32, 24–28, 31–34	....	....	....	....	....	....
Golconda godoniform.....	29–30	31–35	....	....	1–2	....	....
Paint Creek pyriform.......	5–7, 8–35,–24–28, 29–30, 31–32	....	....	....	....	....	....
Paint Creek godoniform.....	9–31	26–27, 4–7	....	....	....	....	....
Renault..........	8–31, 29–30, 25–26, 27–34	1–2	....	....	....	....	....

with another basic-pair group at the time of appearance. Measures included are 24, 25, 26, 27, 28, (34).

4. *Spiracle group*. Basic pair 1–2. These two measures relate to the spiracles and appear to be independent, except at very low levels of ρ, of other measures that have some relationships to these structures. The failure of 1–2 to appear as a basic pair in some populations expresses the fact that it is merged with another basic pair at the time of appearance.

---

at levels of ρ that are comparable in the various populations with respect to intensity of integration are shown below. The size measures are underlined to distinguish them from others that are incorporated, as basic-pair groups merge or as non-size measures are taken into the groups. Basic pairs of other groups are marked by double underlining.

RENAULT
Basic Pair (Size) 8–31

$\rho \geq .98....$ 8–31–32, 8–23–31
$\rho \geq .97....$ 8–23–25–26–28–31–32
8–25–26–27–28–31–32–34

PAINT CREEK GODONIFORM
Basic Pair (Size) 8–31

$\rho \geq .98....$ 8–31–32–35
$\rho \geq .97....$ 8–32–31
8–9–31–32–34
8–9–31–32–35

PAINT CREEK PYRIFORM
Basic Pair (Size) 31–32

$\rho \geq .98....$ 8–23–31–32–34
8–9–32–31–32–35

GOLCONDA GODONIFORM
Basic Pair (Size) 31–35

$\rho \geq .95....$ 8–31–35
23–31–35
$\rho \geq .93....$ 8–23–35
8–28–29–30–31–32–34–35

GOLCONDA PYRIFORM
Basic Pairs (Size) 8–32, 31–34

$\rho \geq .97......$ 23–31–34
8–29–30–32–35
$\rho \geq .95......$ 8–29–30–32–35
23–24–25–26–28–31–34
8–24–28–32–35

5. *Anal group*. Basic pairs 5–7, 4–7. The group formed by these pairs includes measures that relate to the anal opening, plus one that expresses the distance between a oeltoid and a spiracle. Only a small group, 4–5–6–7, tends to form. Measure 14, width of the side plate, is brought in, in some cases. The group relates in general to the digestive system. It tends to merge at high levels of ρ. Measures 1, 2, and 3 are commonly associated with it at the level of merging.

There are two basic pairs, 27–34 and 28–34, in one or more of the populations that are not included among those listed above. The dual role of measure 34 in the size and basal group has been noted. There is a strong tendency for the basal and size groups to merge in some populations (Fig. 66). A high degree of interdependency is found between some measures of the two groups. In the Renault population, 27–34 and 8–31 are both present. Measures 27 and 34 form a basic pair that tends to link the basal and size group at high levels of ρ, as follows:

$\rho \geq .98$.....29–30
8–23–31
8–31–32
24–25–26
27–34

$\rho \geq .97$.....1–2
29–30
8–11–23–28–31
8–25–26–27–28–31–32–34

In the Glen Dean population, however, the situation is rather different, for the size group is weak and appears for the most part only as some of its measures are attached to the basal group by 28–34. There is no discrete basic pair of a size group. The condition is shown by the following list of basic-pair groups:

$\rho \geq .95$.....26–27
28–34
29–30

$\rho \geq .92$.....5–14
23–28–31–34
24–25–28–34
26–27–28–34
28–31–32–34
29–30

$\rho \geq .86$.....5–14
9–21–24–25–26–27–28–31–32–34
9–23–35–26–27–28–31–32–34
12–28–31–34–35
29–30

We shall consider that each of the basic pairs in the five groups listed gives evidence of the existence of that group with which it has been associated. For the general objectives we are following, this is valid practice. Were our purpose the study of *Pentremites* from the perspective of morphological integration, the shift in composition would be examined to determine what importance it might have with respect to differences between populations. If we pursue the first noted procedure, the essential information is readily consolidated into graphic or tabular form. This has been done, with the results shown in Figures 65 and 66 and Table 64 for the six populations of *Pentremites*.

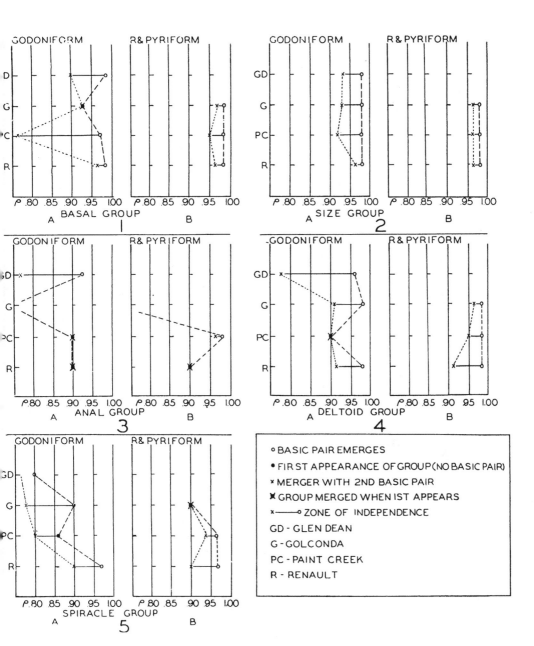

Fig. 65.—Graphs of evolution of the basic pairs of the various F-groups of *Pentremites*. For each F-group the Renault basic pair is entered in both the godoniform and the pyriform graph. Above Renault, godoniform and pyriform patterns graphed separately.

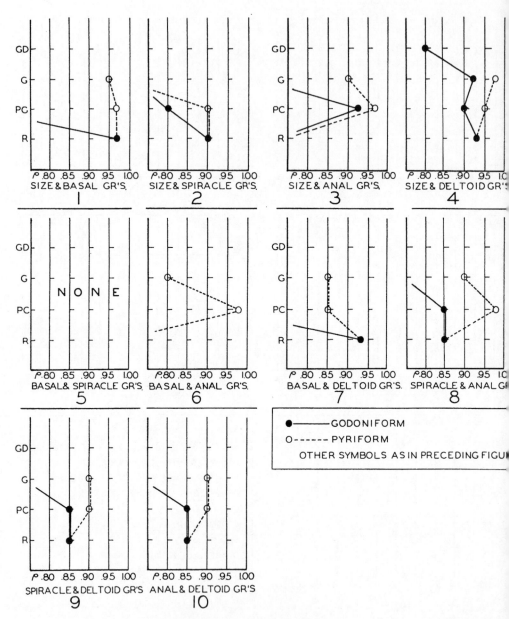

Fig. 66.—Graphs of mergings of basic pairs in the evolution of *Pentremites*. Pairs of F-groups plotted arately. Godoniform and pyriform *Pentremites* entered on same graphs.

Once the construction of the graphs is understood, the information that they contain and the comparisons that may be made from them are essentially self-evident. Combination of the information in the figures and table gives all critical information about the basic pairs and the groups that they form except that of the details of composition of groups in various populations. For our purposes these details are unimportant, but for studies of evolutionary series per se the changes in group composition must be considered.

The basal group can be used to illustrate the interpretation of the graphs. In Figure 65 the ρ level of emergence of the basic pair for each of the populations is shown by the circle. The flow of the level through the temporally suc-

TABLE 64*

PERCENTAGE OF F-GROUPS FORMED BY BASIC PAIRS OVER THE
ZONES OF INDEPENDENCE IN *Pentremites*

F-GROUP	HORIZON	GODONIFORM		PYRIFORM	
		Zone of Independence	Per Cent F-Group	Zone of Independence	Per Cent F-Group
Basal.........	Renault	$\rho \ge .96$	18	. . . .	. . . .
	Paint Creek	$\rho \ge .80$	100	$\rho \ge .95$	100
	Golconda	None	. . . .	$\rho \ge .97$	13
	Glen Dean	$\rho \ge .92$	52	. . . .	. . . .
Size-shape....	Renault	$\rho \ge .97$	60	. . . .	. . . .
	Paint Creek	$\rho \ge .92$	100	$\rho \ge .97$	100
	Golconda	$\rho \ge .93$	50	$\rho \ge .97$	60
	Glen Dean	$\rho \ge .93$	30	. . . .	. . . .
Anal.........	Renault	None	0	. . . .	. . . .
	Paint Creek	None	0	$\rho \ge .97$	75
	Golconda	None	0	None	0
	Glen Dean	$\rho \ge .80$	75	. . . .	. . . .

* The lower level of the zone is indicated; upper = 1.00. Deltoid size and spiracle basic pairs do not form large enough groups to be considered. They are not entered.

cessive series is indicated by the heavy dashed line. The godoniform and pyriform populations are plotted separately with the Renault population included in each diagram for ease of comparison. Where no basic pair of the basal group is present, the first appearance of measures of the group is marked by a dot.

In Figure 65, (*1*), *A, B* the basal group of the Renault population is present at $\rho \ge .98$. The zone of independence[2] is small, with the lower boundary at $\rho \ge .97$. Thus only the group formed at $\rho \ge .98$ is discrete. For the godoniform population of the Paint Creek the ρ level of emergence is $\rho \ge .97$. There is a wide zone of independence. A further drop in the level of emergence occurs to

[2] The "zone of independence" is defined as that band of ρ over which a basic pair forms a ρ-group which consists of measures that pertain only to the basic pair in question. This zone thus exists only above the level of merging of two or more basic pairs. Technically, the zone ranges from the specified ρ level of merging to $\rho = 1.00$. For practical purposes, the line in the graphs is drawn only from the level of emergence of the basic pair to the level of merging with a second basic pair.

the Golconda population, and there is no zone of independence. Finally, in the Glen Dean the level of emergence has risen to $\rho \geq .98$ again, and there is a moderate zone of independence, to $\rho \geq .90$.

Slight shifts in level of emergence, as from $\rho \geq .98$ to $\rho \geq .97$, are of little importance and may be due to sampling and measurement errors. The noted fluctuations of the zones of independence, however, are much too great to be attributed to such causes.

The two pyriform populations show a marked contrast to the godoniform populations of equivalent age. There is no change in the level of emergence between the two pyriform populations, and the zone of independence is small in both cases. The most striking difference between the populations at the same horizon is to be seen in the Paint Creek just after the split of the Renault population. A very strongly independent basal group has emerged in the godoniform population, whereas this has not taken place in the pyriform.

The same process of rather simple analysis of a complex situation can be used for each of the other groups. From each analysis comes the basis for statements about each of the primary morphological systems as the populations are followed through the split of the ancestral stock and the phyletic development thereafter.

The graphs shown in Figure 66 present supplemental information related to the merging of the different groups. The highest level of merging, of course, corresponds to the level shown in Figure 65 for each group. In Figure 66 both the godoniform and the pyriform populations are plotted on the same graphs. In Figure 66, (*1*), for example, it is shown that there is no merging of the basal and size groups in the godoniform line after the Renault but that there is high-level merging in the pyriform line. This same general situation holds for most other pairs of groups, but in Figure 66, (*7*), an exception is seen in the basal and deltoid groups. Complete independence of the basal and spiracle groups is shown in Figure 66.

The third set of data, Table 64, adds information on the percentage of the total F-group incorporated into the basic-pair ρ-group above the level of merging of two basic-pair groups. This percentage is based on the relationship of the total number of bonds realized above this level to the total number of bonds present in the fully integrated F-group. Since the deltoid group and spiracle group are composed of only a very few measures, percentages are misleading and have not been entered.

The use of the data in Table 64 is strikingly illustrated by a comparison of the Paint Creek godoniform and pyriform populations with reference to the basal group. The level of merging in the former is below the level $\rho \geq .80$, whereas in the latter it is at the level $\rho \geq .95$. In both, however, 100 per cent of the bonds are present in the zone of independence. Thus in both lines the group is fully realized as an independent unit, but in the godoniform population independence is maintained at all levels studied, whereas in the pyriform population it is lost at a high level. In contrast to this full formation of groups are such cases as the size group of the Golconda godoniform population, which merges at $\rho \geq .93$ but in the zone of independence realizes only 50 per cent of the total possible bonds.

The co-ordinated information of Figures 65 and 66 and Table 64 provides a basis for study of the evolution in the perspective of the principal systems of the body as they exist independently and as they are variously related to each other. For a detailed study of morphological changes this may be extended to include information upon the composition of the individual groups as they emerge, form independently, and merge with other groups.

For present purposes these data add to the information derived from $I\rho$ and study of unreduced ρ-groups by pointing specifically to the integrative characteristics of the individual systems that contribute to the over-all integration. It is evident that, in the course of evolution, there are marked shifts both in the independently integrated systems and in contributions of systems to total integration by virtue of intersystem association. The shifts are, in general, greatest at the time of the split of the ancestral population, for, in the phyletic lines that follow, there is a lower level of change. It is noteworthy that change proceeds in each of the two lines in a fairly orderly fashion but that the courses in the two lines are very different. A basis for comparison of rates of change of the systems and system relationships is present in the results of the analysis. The rates for each system are seen to be quite different from one another within one line, and the rates of the same system are different in the two lines. If, from another point of view, studies are directed toward the nature of changes that accompany modifications of environment, the data provide a point of departure for analysis of the systems that may be thought to be closely related to environmental changes—the digestive system, for example. Finally, it should be noted that the rather extensive modifications revealed exist in spite of the fact that the only readily evident morphological change is that between the ancestral population and the pyriform line of the two derived populations. This change involved marked modification of the form of the base of the calyx.

None of the studies outlined to this point has taken into consideration the effect of factors that may account for a considerable part of the integration observed. It has, however, become apparent that such may well exist. In the considerations of the coalescence of basic-pair groups and in the general analysis of ρ-groups, the size group in particular stands out as a possible integrating unit. To test this, we must bring the information available through the use of partial correlation into the study.

4. *Partial Correlation and the Factors of Integration*

When there is a strong reason to believe that some measures represent an important factor of integration, partial correlation can be used to good advantage to explore the hypothesis, as explained in chapter iv. It has been noted above that there is good reason to believe that size operates in this way in *Pentremites*. The most evident source of this belief is to be found in the data of the basic pairs. There is, in addition, a second source which, since it does not require a basic-pair analysis, is more easily used. This is to be found in a simple evaluation of the number of ρ-groups in which a particular measure or pair of measures occurs at a selected level of ρ. If some measure or pair of measures occurs with high frequency, say in 75 per cent of the ρ-groups formed, it is reasonable to infer that it may have an important action as an integrating force. If such a measure, or such a pair, can be considered to represent some factor, then the effectiveness of this factor can be evaluated. In essence, this is an estimation of the contribution of this factor to the totality of integration indicated in $I\rho$, provided that the same levels of ρ have been used in the estimation of $I\rho$ and determination of the frequency of occurrence of the measures in question.

The probable effectiveness of the factor of size has been noted in the basic-pair analysis. Measures 31 and 32, gross size measures, may be considered to be representative of this group. They have a high frequency of occurrence, both singly and as a pair in the ρ-groups of the various populations, except at very high levels of ρ in some cases. Examples are as follows: In the Renault population, at $\rho \geq .90$, pair 31–32 occurs in eleven of fifteen ρ-groups. In the Paint Creek pyriform population, at $\rho \geq .95$, pair 31–32 occurs in eight of eleven ρ-groups. In the Glen Dean godoniform population, at $\rho \geq .80$, pair 31–32 occurs in forty-one of fifty-two ρ-groups.

These examples are typical of all populations in the series. Correlations in which these two measures are held constant will show a marked drop in integration, measured, for example, by $I\rho$, in the event that the size factor, which they are considered to represent, is in fact an important agent in integration. In addition, the groups of measures that exist independently of the effects of size will be more clearly revealed than by any other form of analysis consistent with the model.

We shall consider, first, the effects of partial correlation, $r_{xy.31,\ 32}$, upon the value of $I\rho$. This is shown, with values for $I\rho_{xy}$, in Table 65, in which the level of ρ is such for each population that the values of r significantly different from zero $(P = .01)$ are included.

The strong drop of $I\rho$ in each of the populations supports the hypothesis that the size factor is important in total integration. The effect in the different populations is not, however, equal, and this important feature requires evaluation. The shift in each may be shown simply to give the relative effectiveness in the populations by the ratio $I\rho_{xy.31,\ 32}:I\rho_{xy}$. Since the value of $I\rho$ is lowered in each case by partial correlation, the relative effectiveness can be considered to decrease as the value of the ratio approaches 1.00. The values obtained from this ratio are shown in Table 66. The general changes in effectiveness of the factor of size in integration are self-evident in Table 66. It is shown, for example, that the increase in integration from $I\rho_{xy}$ from the Renault to the two populations of the Paint Creek is in some part related to increase in the effectiveness of the size factor. Thereafter, in the godoniform line the effectiveness of size is reduced, whereas it remains about the same in the pyriform line. Additional interpretation, of course, requires data on group composition and group relationships and will be deferred until the final section, after group composition under partial correlation has been taken up.

TABLE 65*

$I\rho_{xy}$ AND $I\rho_{xy\cdot lw}$, *Pentremites* POPULATIONS

		Renault	Paint Creek	Golconda	Glen Dean
Godoniform.....	$\{I\rho_{xy}$	.0379	.0553	.0031	.0037
	$\{I\rho_{xy\cdot 31,\ 32}$	.00037	.0003	.000065	.000058
Pyriform........	$\{I\rho_{xy}$	....	.4951	.0177	....
	$\{I\rho_{xy\cdot 31,\ 32}$	....	.0017	.000055	....

* ρ is such that all values of r significantly greater than zero ($P = .01$) are included.

TABLE 66*

VALUES OF $I\rho_{xy\cdot 31,\ 32}/I\rho_{xy}$, *Pentremites*

	Renault	Paint Creek	Golconda	Glen Dean
Godoniform.......	.0096	.0054	.0210	.0163
Pyriform..........	....	.0034	.0031	....

* Based on levels of ρ such that all values of r significantly greater than zero ($P = .01$) are included.

As in the case of simple correlation, a rather complex series of intersecting ρ-groups emerges when measures 31 and 32 are held constant. Typical examples of large ρ-groups of the various populations are shown in Table 67. In the groups in Table 67 there occur a few measures of the size group. Two, for example, occur in the Renault complex. Thus, in this population, 31–32 does not perfectly represent the totality of this group. Measure 34 occurs in both the Golconda and the Glen Dean godoniform lines. It is associated with the basal and deltoid groups. It will be recalled that this measure (34) was assigned to both the size group and the basal group (p. 218). A combined ambulacral-deltoid suite of measures forms the major group in the Renault population. Basal integration does not play an important part in the absence of the effect of size, and it will be recalled that a high dependency between these two groups was indicated in other analyses. In both the godoniform and the pyriform populations of the Paint Creek, however, the basal group forms a dominant system. In the latter there are, in addition, groups with measures of the spiracular and deltoid systems. The organization is clearly different in the two, over and above the differences seen in simple correlation. It may also be recalled that the total integration, $I\rho$, was very high in the pyriform population and relatively much lower in the godoniform. This, along with the greater effectiveness of size as an integrating factor in the former (Table 66), brings the two populations somewhat more into conformity in $I\rho_{xy.31,\ 32}$, as shown in Table 67. There still exist in the pyriform population groups not present in the godoniform. The persistence of the basal group in the godoniform line is evident in the table. In the Golconda pyriform line, in which size is an extremely important integrating factor, the integration in its absence is extremely weak. This is in sharp contrast to the fairly high value of $I\rho$ (.0177) as shown in Figure 64.

TABLE 67*

Representative ρ-Groups for $r_{xy.31,\ 32}$, *Pentremites*

	Godoniform	Pyriform
Renault.........	8–22–29–30–35	....
Paint Creek......	24–25–26–27–28	14–25–26–27–28
		3–4–5–7–11–14
		3–4–8–11–39–30
Golconda........	27–28–29–30–34	9–28–29
Glen Dean.......	26–27–29–30–34	....

* ρ such that all values of r significantly greater than zero ($P = .01$) are included.

Basic-pair analysis may be used after partial correlation, just as it is used in simple correlation. Complexities of intersects are thereby removed, with the results for the complete arrays of ρ-groups shown in Figure 67. In this figure, negative bonds are entered. These may have real meaning in partial correlation in contrast to the situation under simple correlation (see chap. iv). Of particular interest in this regard is measure 34. It is negatively correlated to all measures to which it is bonded except those of the basal group. It may be concluded that the measure is affected by the size factor in the same way as measures of the basal group but that this does not hold for other measures to which it remains bonded. The joint nature of the measure in its relationship to both the basal group and the size group is graphically illustrated.

A second point to be noted is that a number of basic pairs that emerged under simple correlation persist in this capacity under partial correlation (1–2, 25–26, 29–30, 4–7, 28–34). Evidence of a partially independent existence of all the systems in the absence of the factor of size is thus at hand. The group-forming capacities are not greatly altered.

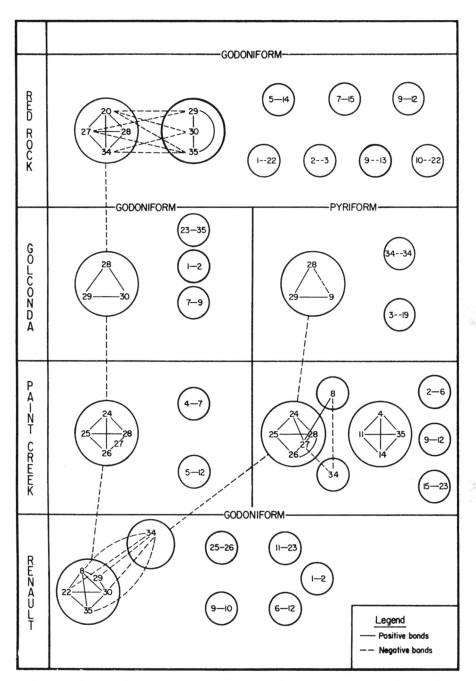

Fig. 67.—Diagrams of evolution of godoniform and pyriform lines of *Pentremites* based on basic-pair analysis of ρ-groups formed after partial correlation, with length and width of calyx held constant ($r_{xy.lw}$). Principal basic-pair ρ-groups circled. *Solid lines*, positive correlation; *dashed lines*, negative correlation.

It is the large ρ-groups that are most readily followed through evolution and offer the best opportunity to witness change. The patterns will be seen to be essentially refinements of the patterns revealed without the use of basic pairs (Table 62). From these groups a clear idea of the evolution of the dominant systems that are not size-dependent is available.

Complete analysis, to give the maximum information about a particular suite of evolving organisms, would require additional partial correlations based on other measures and pairs as they represent important F-groups. This may be done in a cumulative way by holding constant 31–32 plus additional measures, as $r_{xy.31, 32, 29, 30}$, or by partial correlation in which another pair is held constant with the size group present, as $r_{xy.29, 30}$. The general model for a study of evolution under morphological integration involves this procedure for full quantification. Likewise, studies of particular organisms must often be carried to this level. For the purposes of investigation and demonstration of a process, however, this is not necessary, and for this reason we shall not add unnecessary confusion by carrying the present investigation to this degree.

5. *Collation of Evidence of Morphological Integration in Evolution of the Pentremites*

In the study of the evolution of the *Pentremites* morphological integration has been observed under two evolutionary modes—speciation and phyletic change. The speciation (or splitting) involves the division of a parent population into two descendant stocks; the phyletic change follows two lines, one through three stages and the other through only two. Only a limited aspect of each mode is seen. The split involves a sufficient period of time between populations that the event is fully consummated by the first time that its results are observed. The phyletic change is at a very low categorical level, for, by morphological definition, the populations involved in each of the lines differ only at a level that is subspecific.

The changes of morphological integration that are shown by any of the four approaches used are much more intense in the case of speciation than they are during the phase of phyletic evolution. In both cases, however, they appear to be orderly and subject to reasonable interpretation. Because of the quantitative and qualitative differences found in the two modes, we shall consider them separately in this summary.

Upon the development of the godoniform and pyriform populations of the Paint Creek from the Renault population, there is a marked rise in the value of the index of integration, $I\rho$. The rise to the godoniform population, which resembles the ancestral population in gross morphology more closely than does the pyriform population, is the lesser of the two. A part of the increase in each of the lines can be attributed to an increase in the effectiveness of the factor of size as an integrative force, as shown by the ratio of $I\rho_{xy.31\text{-}32} : I\rho_{xy}$ (see

Table 66). The higher integration in the pyriform population is accompanied by an increase in the dominance of size. Qualitative analysis of bonding in the ancestral and descendant populations gives indication that the relationships of the basal and size groups of measures play an important role in the changes of integration. The size F-group is weakly developed in the Renault population but is strong in both the godoniform and the pyriform Paint Creek populations. Most measures of the calyx are rapidly incorporated into the size group as the level of ρ is dropped in the pyriform population, but the basal group of measures remains independent in the godoniform population. This difference is clearly a major item in the observed differences between the ratios of $I\rho$ as determined, respectively, by simple and partial correlation.

A more precise examination of the nature of the change can be made by use of basic pairs, by analyses that bring the individual groups under direct scrutiny. The size of the zones of independence of the basal and size groups is particularly pertinent. The basal group shows a great increase in this zone from the Renault to the godoniform Paint Creek population, and the size group shows a moderate increase. There is either no increase or slight increase in the zones of independence in the pyriform population. Several groups of measures are sorted by the use of basic pairs. These represent a basal, size, deltoid, spiracular, and anal group. Of these, the basal and deltoid groups are structural F-groups; the spiracular represents respiratory function; and the anal represents the digestive function.

The data concerning mergings of the basic pairs show the importance of the size and basal groups in contributions to $I\rho$, but they also indicate that other groups of measures enter in. In the pyriform population of the Paint Creek, the basal and anal groups, size and deltoid, size and anal, and spiracular and axial groups merge above the level of $\rho \geq .95$. In each instance, merging of these groups is at a lower level in the equivalent godoniform population.

The increase in $I\rho$ with the development of the two populations from the ancestral population is somewhat suggestive of the phenomena of decrease in variability, based on the coefficient of variability, V, that tends to occur under similar circumstances (see, for example, Simpson, 1953). In one sense, these phenomena are somewhat analogous. As in the case of variability, it is possible to envisage a rather diverse population, perhaps with two or more subpopulations, in which there is a diversity of integrative patterns. With such diversity, whether scattered more or less uniformly in the population or different between subpopulations, the index of integration for the whole population will tend to be lower than for a sample in which only a part of the range is present. The different patterns will presumably have different adaptive values. Isolation and the development of separate populations, perhaps under somewhat different environmental circumstances, could result in differential selection with a resultant pattern of integration more coherent in the derived populations than in the ancestral population. The type of rise of integration seen in the *Pentremites* in the increase of $I\rho$ may well have resulted from this general process. The analysis of the contributions to the change of $I\rho$ by the various integrative complexes, then, opens the way to a detailed understanding of the factors that are involved in the change and to interpretation of their meanings with respect to the problems of evolution that are under study.

There is a sharp drop of $I\rho$ in both the godoniform and the pyriform lines after the Paint Creek stage. Very little gross morphological change is evident in either line, so that the change in $I\rho$ appears to be a very delicate indicator of modifications not readily seen by other means. The decrease presumably represents an increase in the diversity of integration in the successive populations, again somewhat analogous, inversely, to changes in variability. The pyriform line, which, to the best of our knowledge, did not persist beyond the Golconda, shows a higher index than the equivalent godoniform population. It is striking in this case that the factor of size is extremely important in the pyriform integration and that there is almost no size-independent integration of systems. The Golconda godoniform population, on the contrary, shows a decrease in size-dependent integration, with the basal, deltoid, spiracular, and digestive systems all present with some degree of discreteness in the absence of the size factor. A roughly similar pattern of integration persists in this line into the Glen Dean. It seems possible that this difference may have an important bearing upon the success and failure of the two groups.

The single case analyzed cannot, of course, give sufficient evidence for any general conclusions, either about the full evolution of *Pentremites* or, much less, about the course of morphological integration in speciation and subsequent evolution in general. From what has been discussed, however, we can speculate about a possible series of events which, while largely hypothetical, may be stimulating as a direction in which further research may be carried out. Figure 68 shows a possible course of events based upon the case of *Pentremites*. In this scheme the Renault population represents a stage just prior to the development of two species. Morphological integration is relatively low. The two derived populations represent the Paint Creek godoniform and pyriform types with high integration. There follows a diversification within the populations derived from these two groups, with a drop in the values of $I\rho$. In one line, extinction follows, possibly because of the non-adaptive nature of much of the integration of the F-groups, which is size-controlled especially in the Golconda. The other line in a hypothetical stage reaches the condition in which appropriate circumstances may result in separation into two distinct populations.

HYOPSODUS

Seven samples of the small Eocene condylarth *Hyopsodus* provide the data used for the study of morphological integration in the evolution of mammalian molar teeth, the subject of this section. This genus was selected because of its relative abundance and the relative simplicity of the patterns of the molar teeth. Excellent as the materials are, relative to those available for most genera of extinct mammals, they nevertheless pose problems of sample size, taxonomy, and phylogeny. Discussion of these problems has been relegated to Appendix E, so that the text materials will not be needlessly cluttered with detail. The

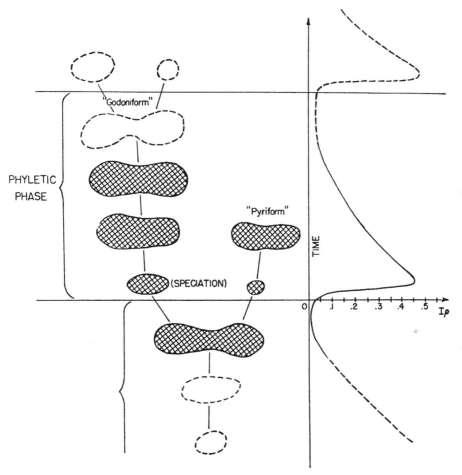

Fig. 68.—Suggested course of evolution in *Pentremites*, showing relationship of modes to $I\rho$

specific names that have been used, as well as their phylogenetic arrangement, are shown in Figure 69 and are explained in detail in Appendix E, where pertinent data for the samples are also available.

There are, of course, other species of *Hyopsodus*, both from the areas which furnished the collections studied and from other areas. None of these provided samples sufficiently large for the type of analysis needed for our studies. It was necessary, furthermore, to confine our attention to either the upper or the

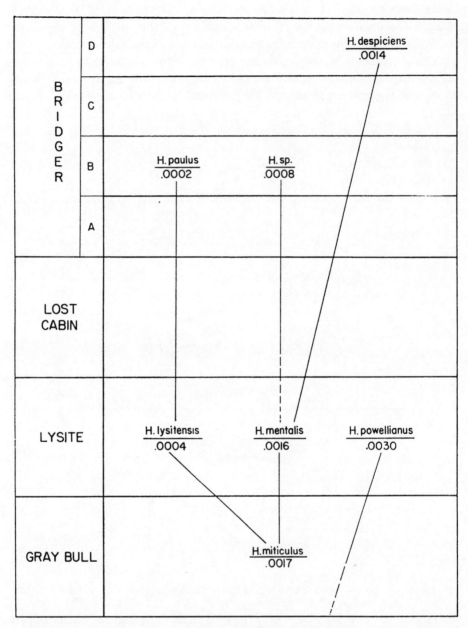

Fig. 69.—Phylogeny of the species of *Hyopsodus* as used in the text (see also Appendix E). Values below species names are the values of I_ρ based on levels of ρ such that all values of $r > 0$ are included.

lower molar series, because only rarely are the upper and lower jaws of individuals associated in preservation. We have chosen to use the lower teeth. The importance of the interdependence of upper and lower molars, as shown in the study of *Aotus* in chapter vi, places limitations upon the results that can be obtained by the use of one or the other. In spite of these it has been possible to gain considerable insight into a number of aspects of the evolution of the molar teeth in this genus and to compare the evolution of characters viewed in the standard way to the patterns of evolution of morphological integration.

The measurements used are shown in Figure 61. They are similar to those employed in the study of *Aotus*, and the rationale for their use is that discussed for *Aotus* in chapter vii. The problems attacked are, in a very general sense, comparable to those already discussed in the section on *Pentremites*. They differ in one important respect, for in *Hyopsodus* we are dealing with structures at a single ontogenetic stage. The difference is even greater than would be the case, were the structures the product of an ontogenetic history in which they performed some active role in the function of the organisms during growth. Although the systems of dental structures are obviously functional, they cannot be considered to have been determined in any way by functional activities that existed during their development. Much of the direct basis for inference concerning the meanings of correlation that applies in growing and mobile structure is absent in teeth. In this respect they present a special case. Some aspects of this case were considered in the discussions of *Aotus*. Others, particularly those relevant to evolution, are considered in the presentation and analysis of the lower molar dentitions of *Hyopsodus*.

The Data

The raw data, the measurements of the teeth, are available in Appendix E. The results of the various types of analyses are presented in Figures 69–77 and in Tables 68–75 and are applied variously in examples throughout the text. For clarification, brief comments on the sources of data and the part that they play in the study are included in the captions for the tables. The results are used in various ways, together and separately, in the interpretations, but in general the order of introduction follows the order of arrangement of the tables.

Some correlations between measures may, in one sense, be considered the result of "direct" genetic effects, not dependent, that is, on some broad integrating factor. Others, probably the great majority, may be "indirectly" affected and depend for their existence upon one or more major factors of integration. The most evident examples of dependency in teeth are found in measures that are size-dependent, as revealed in our various studies that involve partial correlation and determinable as well by various other means (see, for example, Wright, 1932, 1934). Such dependent correlations may be expected to be somewhat ubiquitous in a genus, whereas less dependent patterns of correlation may be more strictly confined to species or phyletic lines. The mere fact that patterns are not related to factors of size, of course, does not imply that they do not depend upon some other general controlling factor. In teeth, in partial contrast to hard structures that participate in ostensive functions during growth, such general factors must be considered basically genetic.

Close resemblance of the individual teeth in, for example, the lower molar series, tendencies for simultaneous changes along the tooth row, and commensurate modifications of upper and lower teeth of the same series argue strongly for common genetic influences that may be integrated under the broad concept of morphogenetic fields. Evidence favorable to the application of this concept in dental evolution has been advanced in particular by Butler (1937, 1939a, b, and 1941) and Patterson (1949). A search for evidence in a single species on the basis of morphological integration, as described for *Aotus trivirgatus* in chapter vii, yielded predominantly negative results. If the concept of the morphogenetic field does in fact have application in dentitions beyond a vague generality too tenuous to be of much use, it should, from the arguments advanced above, be revealed in studies of correlation. In spite of the negative evidence in *A. trivirgatus*, the hypothesis has been tested in the empirical studies of *Hyopsodus* and, as will be shown, appears to have some importance when applied to phylogenetic series.

The studies of *Hyopsodus* have been cast within the general framework of the preceding discussion, and interpretations depend to some degree upon the concepts advanced. As in *Pentremites*, the reliability of our conclusions is dependent upon the validity of the assumptions about taxonomy and phylogeny which have been made (see Appendix E) and upon the representativeness of the samples that have been used as the source of the raw data. The results of the study of *Aotus* have been used as a guide in interpretation in the absence of upper dentitions of *Hyopsodus*. It is assumed that the changes of lower dentitions participate in and are indicatve of the more complete patterns of modifications that involve both upper and lower dentitions.

Interpretation

The principal objective of this study, as in the case of *Pentremites*, is the determination of how morphological integration behaves in evolution. Here, however, we have a somewhat more detailed knowledge of a number of morphological changes determined independently of morphological integration. It is thus possible to go a step further to compare some details as revealed by the usual morphological practices and the use of morphological integration. The validity of both interpretations, of course, depends in large part upon the correctness of the phylogeny that is followed. The basis for the phylogenetic arrangement that is used is explained in detail in Appendix E. There it is concluded that one sequence, *Hyopsodus miticulus*, *H. lysitensis*, and *H. paulus*, expresses a valid line of descent but that the other sequence, *H. miticulus, H. mentalis,* and *H. despiciens,* includes species which, while rather closely related to each other, may not lie in a direct line but, rather, represent species that are somewhat divergent from a central phyletic series. *Hyopsodus* sp. "vicarius" is considered more closely related to the second series than to the first but to be somewhat more divergent than are the others from the line. *Hyopsodus powellianus* is considered to represent a third line, which developed in complete independence, once it had diverged from an (unknown) common ancestor. For convenience, the two major lines are hereafter referred to as the "paulus" and "despiciens" lines, with the name taken from the end member of each series.

MORPHOLOGICAL INTEGRATION IN *Hyopsodus* EVOLUTION

1. *The intensity of integration.*—It is only necessary to scan Tables 68–75 to see that no two species of *Hyopsodus* that have been studied have the same characteristics of morphological integration of the lower molar dentition. Separation of the species by standard procedures and by the use of morphological integration gives consistent results to the extent that no two populations shown to be different by the first are grouped together by the second. We infer, from the analysis of the data on samples from a population of pigeons treated in chapter iv that, were any of the samples in fact drawn from the same species population, differences of the magnitude observed would not exist. The extent of differences of morphological integration at different taxonomic levels needs much more extensive investigation than has been possible up to the present time. Data are insufficient for generalizations beyond the individual cases that have been studied.

From the standpoint of evolution, our present concern, the existence of ordered relationships of features of morphological integration is a point of central interest. As in the case of *Pentremites,* this may be approached from various different points of view and at various levels of detail. In general, Tables 68–75 have been arranged in the order used in interpretation, and we proceed from the comprehensive and general to the particular with an effort to interrelate the various levels and types of information in explanation of how each is developed in terms of the others.

The most general level of integration is shown by the index of morphological integration, based on simple correlation for the molar series of each of the species. Values of $I\rho$, illustrated in Figure 69, show an orderly array as entered in the two phyletic lines, "paulus" and "despiciens." The difference between the two lines is marked. The "paulus" line shows a marked drop in $I\rho$ from *H. miticulus* to *H. lysitensis* and a continuation of the decrease to *H. paulus.* There is essentially no change in the "despiciens" line. *Hyopsodus* sp. "vicarius," however, has a value of $I\rho$ distinctly lower than that of either *H. mentalis* or *H. despiciens.* If it was derived from either *H. miticulus* or *H. mentalis,* as suggested in the discussion in Appendix E, it represents a divergence of the "despiciens" line that, in the value of $I\rho$, is convergent with the "paulus" line. *Hyopsodus powellianus*, which clearly represents a third line, shows a very high (relatively) value of $I\rho$.

There is very little basis for interpretation of the meaning of the changes observed from the value of $I\rho$ alone. It is evident, however, that the course of events is different from that seen in *Pentremites*, in which there was a sharp

TABLE 68*

ρ-Groups of Dentition for Species of *Hyopsodus*

H. miticulus

5–44, 8–10, 23–43, 1–8–41, 1–26–41, 4–6–42, 4–6–10, 8–28–46, 6–42, 50, 9–29–43, 25–41–46, 29–46–49, 28–46–48, 42–48–50, 45–46–48, 1–41–42–50, 4–22–29–42, 5–24–25–41, 21–25–26–41, 21–22–41–42–50, 21–24–25–26–41, 4–21–25–41–42, 4–21–24–25–41, 1–2–21–22–42–50.

H. sp. "vicarius"

1–26, 2–42, 4–25, 8–10, 22–42, 24–26, 24–41, 25–41, 28–46, 45–50, 1–28–30, 3–21–23, 3–21–43, 3–23–44, 3–24–43, 5–6–24, 5–24–44, 10–21–43, 28–29–30, 1–5–6–45, 1–5–9–45, 1–6–21–28, 4–5–6–23–45, 4–5–9–23–45, 4–6–21–23–28, 5–9–23–44–45.

H. lysitensis

1–24, 6–44, 9–29, 21–46, 22–42, 26–48, 41–48, 42–48, 44–49, 3–4–24, 4–5–24, 23–24–25, 4–5–6–30, 4–5–28–30, 4–6–10–30.

H. despiciens

2–30, 4–5, 5–26, 8–41, 22–42, 43–44, 1–2–9–21, 2–9–21–41–49, 2–9–21–42–49, 2–21–29–41–48–49, 2–21–29–42–48–49, 2–28–29–41–48–49.

H. mentalis

1–2, 4–5, 24–29, 2–5–42, 3–4–10, 5–42–48, 6–41–49, 24–41–50, 24–41–44–49, 30–41–42–50, 6–30–41–42–48, 8–21–24–41–44, 21–28–30–41–42–48.

H. powellianus

4–45, 23–43, 26–45, 2–44–45, 2–6–42–50, 1–2–6–24–50, 1–2–6–28–50, 1–2–22–24–25, 1–6–9–24–49–50, 1–6–28–49–50, 1–9–21–25–41, 1–9–22–24–25–41, 2–6–10–28–30–48–50, 6–8–10–28–29–30–48, 6–10–28–29–30–48–50, 6–10–28–30–48–49–50, 6–28–29–30–44–48–50.

H. paulus

1–10, 1–30, 2–22, 4–6, 4–9, 4–24, 5–24, 5–25, 6–26, 8–30, 22–24, 23–28, 23–44, 24–30; 24–44, 25–45, 26–30, 28–41, 29–48, 30–46, 41–46, 43–46, 44–46, 44–50, 45–49, 46–50.

* The ρ-groups in this table have been determined from simple correlation of all measures of the lower molar series. The confidence level is $P = .01$, and the level of ρ for each sample is such that all values of r significantly different from zero at this level are included. The complete expression of molar integration, so far as present in the measures used, is found in the groups. The values of $I\rho$ (Fig. 69) are based upon these arrays.

rise in the value of $I\rho$ as we passed from the parent population to the two derived populations. The situations portrayed, however, are different in most respects, in growth and in the part of the total animal represented, so that such a difference is not unexpected.

The ρ-groups of Table 68 provide some evidence concerning the processes in operation in the changes of integration shown in Figure 69. Inspection alone indicates a prevalence of length and width measures of the teeth (1, 2, 21, 22, 41, and 42) in the ρ-groups. The relationships of the values of $I\rho$ to the percentage of all ρ-groups in a species in which measures of length and/or width occur give a rough estimate of the place that the length and width of individual teeth occupy. That there is a fair correspondence in the ordering of the two sets

TABLE 69*

$I\rho$ AND PERCENTAGE OF ρ-GROUPS, INCLUDING SIZE MEASURES IN *Hyopsodus*

Species	$I\rho$	Percentage of ρ-Groups with Size Measure(s)
H. paulus	.0002	23
H. lysitensis	.0004	33
H. sp. "vicarius"	.0008	50
H. despiciens	.0014	75
H. mentalis	.0016	78
H. miticulus	.0017	72
H. powellianus	.0030	79

* The relationship for $I\rho$ of the molar series ($P = .01$) and the percentage of the total number of ρ-groups in which one or more measures of length and/or width of molars are included. A rather close correspondence in ordering of the values for $I\rho$ and the percentages is apparent. This gives a crude estimate of the extent that size enters into the value of $I\rho$ and is used in lieu of partial correlation, which cannot be performed, holding width constant, in the absence of an expression of width for the total molar series.

TABLE 70*

ρ-GROUPS, *Hyopsodus miticulus* AND *H. paulus*, BASED ON PARTIAL CORRELATION, WITH TOOTH-ROW LENGTH HELD CONSTANT ($r_{xy.l}$)

H. miticulus	*H. paulus*
3–5	4–6
5–15	4–24
6–10	5–25
25–26	22–42
28–48	25–45
45–46	29–49
46–48	43–46
	46–50

* $P = .01$, with level of ρ such that all values of ρ greater than zero are included. A sharp drop in the size of ρ-groups (for comparison, see Table 68) has occurred from r_{xy}, and width measures have assumed a prominent position. Longitudinal measures are few. This type of reduction is characteristic of the various species studied.

of values is obvious from Table 69, and there can be little doubt that a major part of the difference between the species in $I\rho$ is dependent upon the effectiveness of the size of the teeth as an integrating factor. This is further emphasized by the reduction effected by partial correlation in which the length of the tooth row is held constant ($r_{xy.l_{ms}}$), as shown for *H. miticulus* and *H. paulus* in Table 70. At this point, however, a search for factors based on the complete molar series must end, for width cannot well be held constant for the total molar series, as discussed for *Aotus* in chapter vii.

To this point, we are able to make the statements that there is an orderly progression of total integration in evolution of the molar series in the species of *Hyopsodus* and that this progression is different in the two phyla studied. Further, a major part of the change may be explained in terms of modifications of the effectiveness of the factor of size in integration. Where size enters importantly, integration tends to be relatively high; and where it does not, integration is relatively low. The effectiveness of the factor of size is not related to absolute increase in size, for this occurs in both lines (see Table 70; Appendix E).

TABLE 71*

ρ-Groups of Single Molars Based on r_{xy} ($P = .01$) for Species of *Hyopsodus*

Species	M_3	M_2	M_1
H. miticulus	3–5, 8–10, 1–2, 4–6–10	21–22, 21–29, 26–28, 21–24–25, 21–25–26, 24–25–26	45–46–48, 46–48–49
H. lysitensis	3–4, 4–5–6, 4–6–10	23–24, 28–30	41–48, 42–48, 44–49
H. mentalis	1–2, 3–4, 3–10, 4–5	21–24, 24–29, 21–22 28–30	41–42, 42–48, 42–50, 41–44–49
H. paulus	1–10, 4–6, 49	23–28, 24–30, 26–30	44–46–50, 41–46, 43–46, 45–59, 47–50
H. sp. "vicarius" . . .	1–9, 8–10, 1–5–6, 4–5–6, 4–5–9	21–23, 23–38, 24–26, 28–29–30	44–45, 45–50
H. despiciens	4–5, 9–10, 1–2–9	21–29, 28–29	41–48–49, 43–44, 42–48–49
H. powellianus	1–6–9, 6–8–10, 2–6–10	22–24–25, 28–29–30	42–50, 44–45, 48–49–50

* Levels of ρ for the species are such that all values of r greater than zero are included and all not significantly different from zero excluded. This table is one of several in which the molars are treated singly, to provide a basis for partial correlation in which both length and width are held constant. Most of the work on single molars is based on $P = .05$. The present table is provided for comparison of the ρ-groups of the single molars with the ρ-groups at $P = .01$ for the total molar series.

For more precise information on the effect of size and a study of the integration that is independent of it, we must turn to analyses of the individual molar teeth, in which both length and width can be held constant. The ρ-groups portrayed in Tables 71 and 72 give a basis for determination of the frequency of size measures (1, 2 for $M_{\overline{3}}$; 21, 22 for $M_{\overline{2}}$; 41–42 for $M_{\overline{1}}$) in the integration of individual teeth. Although there is a tendency for relatively high frequency in most teeth, there is considerable difference between $M_{\overline{3}}$, $M_{\overline{2}}$, and $M_{\overline{1}}$ in single species and between homologous teeth of different species. This is best shown in Table 72, where $P = .05$. Both an estimation of percentages of ρ-groups with length and/or width measures to the total number of ρ-groups and a comparison of Tables 72 (r_{xy}) and 73 ($r_{xy.lw}$) provide bases for somewhat crude analyses of the effect of size. More precise statements, however, are possible on the basis of the entries of Table 73 and the values in Table 74, which are derived from the expression $I\rho_{xy.lw}/I\rho_{xy}$, with data from Table 75.

The relationship expressed by $I\rho_{xy.lw}/I\rho_{xy}$, although it appears as a single number, is the result of a complex series of manipulations and must be considered with this in mind. The values give an expression of the impact of the factor of size in the integration of the teeth in question. Values may range from

TABLE 72*

ρ-Groups of Single Molars Based on r_{xy} ($P = .05$) for Species of *Hyopsodus*

Species	M$_3$	M$_2$	M$_1$
H. miliculus	3–5, 1–2–6, 1–2–8, 2–8–10, 4–5–9, 2–4–6–10	21–26–28–30, 21–23–25–26, 21–22–24–25–26, 21–22–24–25–29, 21–25–26–28	44–45, 21–45–46–48, 43–45–46, 41–45–46–50, 41–42–46–48–50, 41–42–46–49–50
H. lysitensis	1–8, 8–9, 9–10, 1–2–3, 1–3–4–5, 3–4–5–6, 4–5–6–10	22–28, 23–24, 24–25, 24–28–30	41–48, 42–48, 44–45, 44–49, 44–50, 46–48, 46–50
H. mentalis	2–3–8, 3–4–10, 1–2–3–5, 2–3–4–5	23–30, 22–24–29, 21–22–24–30, 21–22–28–30, 22–28–29–30	43–44, 41–42–44, 41–49–50, 44–45–50, 41–42–48–40, 41–44–49–50
H. paulus	1–8, 1–10, 2–8, 3–5, 4–5, 4–6, 4–8, 4–9	21–23, 21–26, 22–24, 22–25, 22–29, 24–28, 24–29	43–50, 44–49, 45–50, 41–43–46, 44–46–50
H. sp. "vicarius"	1–9, 1–6–10, 1–8–10, 1–3–4–5–6, 3–4–5–6–9	21–23–24, 21–24–26, 21–26–30, 23–24–25, 23–25–30, 28–29–30, 21–23–28–30	41–50, 48–50, 49–50, 44–45–50
H. descipiens	4–9, 1–2–9, 4–5–10, 4–6–10, 4–8–10	24–29, 21–22–29, 21–28–29	43–44, 42–46–48, 42–46–50, 41–42–48–49
H. powellianus	4–5, 2–6–8–10, 1–2–6–9–10	25–26, 22–24–30, 21–22–24–25, 24–28–29–30	41–42–45, 42–44–45, 48–49–50, 42–44–48–50

* Levels of ρ for the species are such that all values of r greater than zero are included. The confidence level used here proved to be the most definitive for partial correlation, $r_{xy.lw}$, for the individual molars. The groups may be compared with those formed after partial correlation in Table 59.

zero[3] to indefinitely large. For all values less than 1, the factor of size is acting to *increase* integration. Where values are greater than 1, the factor of size is acting to decrease integration, that is, to reduce the correlations.

The values in Table 75 may be studied from two different aspects, one in which a particular tooth is considered as the unit and the other in which the

[3] The indeterminate ratio

$$\frac{0}{I\rho_{xy}} \overset{\text{def.}}{=} 0 .$$

Thus the ratio ranges from zero (as defined) to indefinitely large $I\rho_{xy.lw}/0$.

trend along the tooth row is the center of interest. Figure 70 presents temporal series based, respectively, on $M_{\bar{3}}$, $M_{\bar{2}}$, and M_T, with the values of $I\rho_{xy.lw}/I\rho_{xy}$ plotted against time and with connecting lines indicating the phylogenetic relationships. Each graph shows the distinctness of the "paulus" and "despiciens" lines. The greater effect of size in the "despiciens" is apparent in each instance. It should be noted also that, in general, the individual teeth reflect

TABLE 73*

ρ-GROUPS FOR INDIVIDUAL TEETH OF *Hyopsodus* $(P = .05)$ $r_{xy.lw}$

Species	M_3	M_2	M_1
H. miliculus	4–9, 8–10, 4–6–10	26–30, 24–25–26	46–48, 44–45, 45–46
H. lysitensis	3–4, 4–6–10	23–24, 28–30	44–48, 49–50
H. mentalis	3–4, 4–5	24–25, 24–29	43–44
H. paulus	4–6, 4–9, 9–10	24–29, 26–29, 23–28, 28–30	43–45, 44–49, 44–49
H. sp. "vicarius" ...	8–10, 3–4–5	28–29–30	44–45, 44–50, 48–49–50
H. despiciens	4–5	24–29, 28–29	43–44, 48–49
H. powellianus	6–8	28–29–30	44–45, 44–50, 48–49–50

* Levels of ρ for each species are such that all values of r significantly greater than zero are included. The material of this table forms the essential basis for the study of intensity of integration under partial correlation and for comparisons of group compositions.

TABLE 74*

$I\rho$ FOR INDIVIDUAL LOWER MOLARS IN SPECIES OF
Hyopsodus, r_{xy} AND $r_{xy.lw}$ $(P = .05)$

Species		M_3	M_2	M_1
H. miliculus	r_{xy}	.0289	.0741	.0567
	$r_{xy.lw}$	.0204	.0181	.0068
H. lysitensis	r_{xy}	.0319	.0069	.0054
	$r_{xy.lw}$	.0189	.0044	.0044
H. mentalis	r_{xy}	.0326	.0261	.0248
	$r_{xy.lw}$	.0044	.0044	.0023
H. paulus	r_{xy}	.0062	.0125	.0035
	$r_{xy.lw}$	.0068	.0079	.0555
H. sp. "vicarius"	r_{xy}	.0557	.0286	.0069
	$r_{xy.lw}$	.0181	.0204	.0189
H. despiciens	r_{xy}	.0187	.0093	.0233
	$r_{xy.lw}$	.0023	.0044	.0023
H. powellianus	r_{xy}	.0504	.0378	.0278
	$r_{xy.lw}$	.0027	.0024	.0189

* Based on data of Tables 72 and 73.

the ordering of values of $I\rho$ for the whole tooth row, shown in Table 69. Only in $M_{\overline{1}}$ of *H. paulus* and *H.* sp. "vicarius" does size act to reduce integration.

The case of *H.* sp. "vicarius" is worthy of special note. If its assumed derivation from *H. miticulus* or *H. mentalis* is correct, it must be concluded that, in the integrative characteristics shown in the diagrams, it represents a development that closely parallels that in the "paulus" line, especially as $M_{\overline{3}}$ and $M_{\overline{1}}$ change from *H. lysitensis* to *H. paulus*. If the placement in phylogeny is incorrect, which is, of course, quite possible, the diagrams are misleading. Were placement made upon the basis of the integrative features noted to this point, rather than on other morphological grounds, there would be a strong tendency to place *H.* sp. "vicarius" close to the "paulus" line.

TABLE 75*

VALUES OF $I\rho_{xy.lw}/I\rho_{xy}$ FOR INDIVIDUAL
TEETH OF *Hyopsodus* SPECIES

Species	M_3	M_2	M_1
H. miticulus	.71	.24	.12
H. lysitensis	.59	.64	.81
H. mentalis	.12	.17	.09
H. paulus	.91	.63	15.85
H. sp. "vicarius"	.32	.71	2.73
H. despiciens	.12	.47	.10
H. powellianus	.05	.06	.68

* These ratios show the relative effect of size in the integration of the individual molars of the species of *Hyopsodus*. The values may be viewed in two ways: as they are followed through the comparable molars in each of the species, as M_3 for all species and as the changes along the tooth rows in each of the species compared. These values form the basis for the graphs of Figs. 70 and 71.

The data viewed from the vantage point of trends along the tooth row are shown in Figure 71. Here it is the trend as traced by lines between points that is important. Most evident in the "paulus" line is a very strong, consistent decrease of the effectiveness of size in increasing integration forward along the tooth row, especially as shown by $M_{\overline{1}}$. A small irregularity appears in $M_{\overline{3}}$, for in both *H. miticulus* and *H. paulus* effectiveness of size is greater in $M_{\overline{3}}$ than in $M_{\overline{2}}$, whereas in *H. lysitensis* there is a slight decrease. The difference between the two phyla is evident from the graph. In the "despiciens" line the effectiveness of the size factor in increasing integration tends to decrease on $M_{\overline{2}}$ relative to both $M_{\overline{3}}$ and $M_{\overline{1}}$, and this is accompanied, between *H. miticulus* and *H. mentalis*, by a sharp increase in the effectiveness of the factor in $M_{\overline{3}}$. Resemblance of *H.* sp. "vicarius" to the "paulus" line is again evident.

The full array of data, as they pertain to intensities of integration and the factor of size as an agent of intensity, has now been discussed. We feel that there is clear evidence that intensities show regular changes in the evolution of the species in the two phyla; that the two phyla diverged in this respect; and

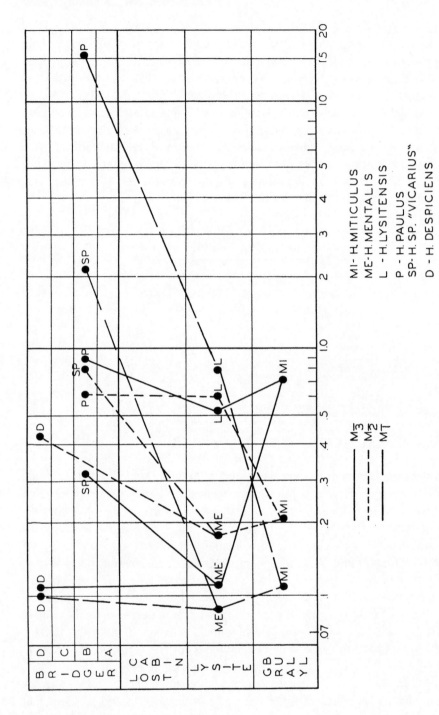

FIG. 70.—Graphs of lower molars of species of *Hyopsodus*, with each molar (M_1, M_2, M_3) plotted separately; values on abscissa (log scale) are for $I_{p_{xy}}/I_{p_{xy\cdot kv}}$. Phylogenetic pattern is that used throughout text.

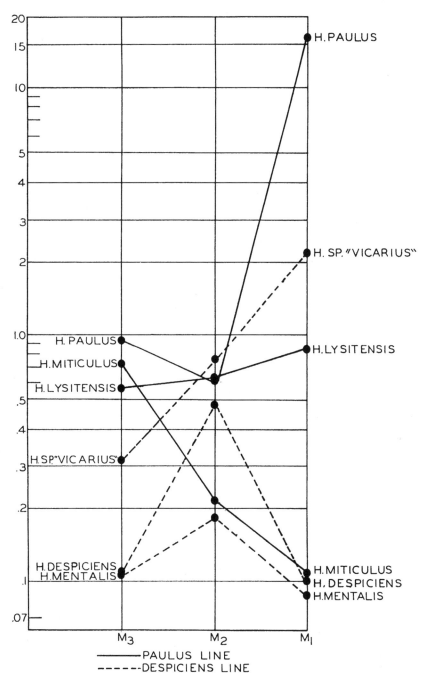

FIG. 71.—Graphs of trends of values of $I_{\rho xy}/I_{\rho xy \cdot lw}$ along the tooth row in species of *Hyopsodus*. Values on ordinate (log scale) are values of the ratio.

that the factor of size plays a very important role. Translation of this knowledge into the terms of dynamics of evolution, genetics, selection, adaptation, and so forth is quite another matter and not the goal set in this section. The stage is merely set for studies in this direction.

2. *Composition of the ρ-groups in evolution.*—One aspect of the composition of ρ-groups has already been covered in the study of intensities, the part that

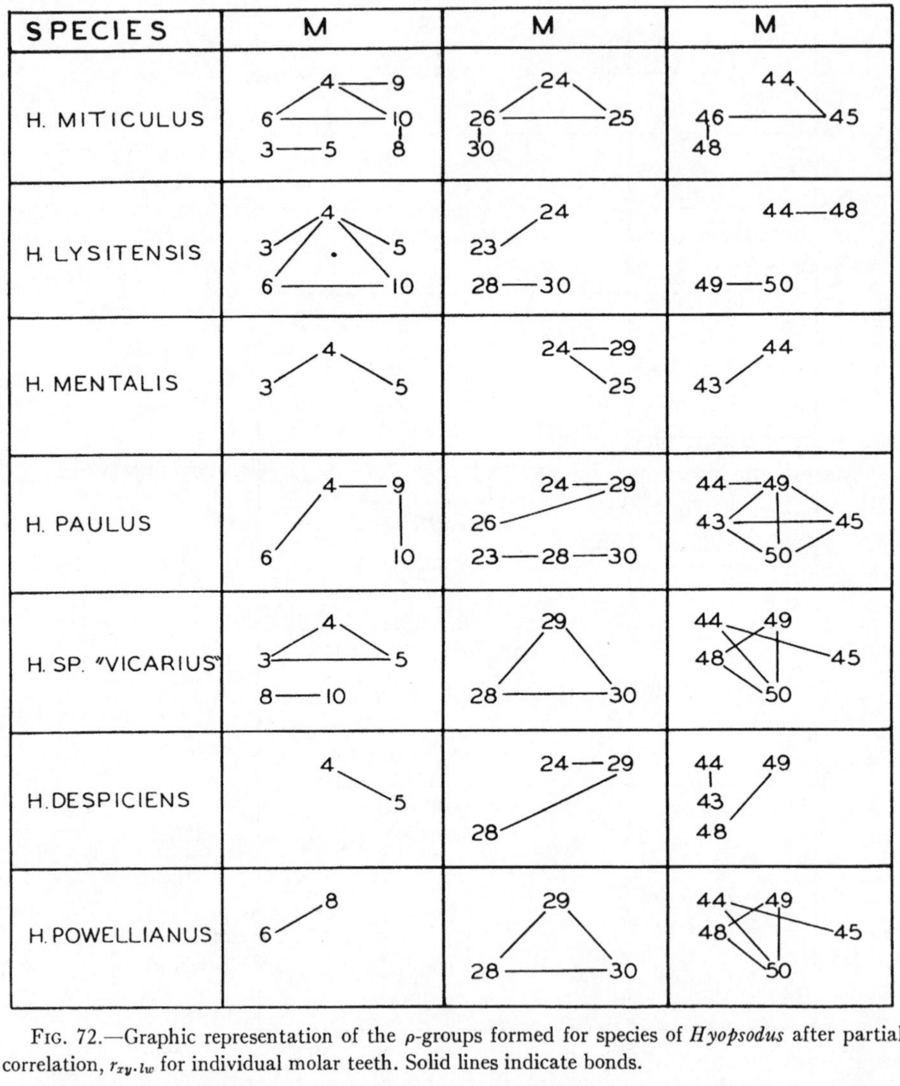

SPECIES	M	M	M
H. MITICULUS	4—9, 6, 10, 3—5, 8	24, 26, 25, 30	44, 46, 45, 48
H. LYSITENSIS	4, 3, 5, 6—10	24, 23, 28—30	44—48, 49—50
H. MENTALIS	4, 3, 5	24—29, 25	44, 43
H. PAULUS	4—9, 6, 10	24—29, 26, 23—28—30	44—49, 43, 45, 50
H. SP. "VICARIUS"	4, 3—5, 8—10	29, 28—30	44, 49, 48, 45, 50
H. DESPICIENS	4, 5	24—29, 28	44, 49, 43, 48
H. POWELLIANUS	8, 6	29, 28—30	44, 49, 48, 45, 50

Fig. 72.—Graphic representation of the ρ-groups formed for species of *Hyopsodus* after partial correlation, $r_{xy \cdot lw}$ for individual molar teeth. Solid lines indicate bonds.

size plays in their makeup. We may now turn our attention to the composition as it exists in the absence of the factor of size. The data are presented in Table 73. These can be more graphically shown by an arrangement in which the full array of bonds is portrayed, as in Figure 72. Although patterns of resemblances and differences do appear in this figure, a procedure somewhat more formal than mere scanning is useful in making interpretations of the patterns and their meanings.

Thus we shall direct our attention to what may be called *important associations* of measures. Such associations, as expressed by correlations, are defined to satisfy the following criteria:

1. The frequency of a bond[4] in the total array of groups of Figure 72 or Table 73 is greater than 1.

2. Pairs of bonds (such as 4–5 or 5–6) occur with a frequency greater than 1 in the total array.

3. Pairs of bonds (as in 2) are parts of some larger group, in the present instance, with one exception ($M_{\overline{1}}$ of *H. paulus*), of triplets. The existence of a triplet in one tooth satisfies this criterion.

This definition is, of course, arbitrary and designed to meet only the demands of the specific situation being considered. Essentially, it establishes the level of complexity at which we wish to work and formalizes the risk, in terms of assumptions, that we are willing to take. The primary unit under this definition is the triplet. We consider that the existence of a bond between two measures which occur in a triplet on at least one tooth is evidence of the partial existence of this triplet unit even in the absence of its complete formation.

The criteria are applied in the present case as shown in Table 76. All the pairs that occur together at least twice in Table 76 have a common measure and thus the potentiality for forming a triplet. From Table 76, it can be determined that the following occur in triplets:

3–4	4–10
3–5	8–9
4–5	9–10
4–6	

The triplets formed are as follows:

3–4–5
4–5–6
4–6–10
8–9–10

The only confusion in this particular situation attends the formation of the quadruplet 3–5–9–10 in $M_{\overline{1}}$ in *H. paulus*. For the moment we shall ignore this, for its position in the evolution of the patterns will become apparent in the development of the analysis. The triplets listed above are considered to comprise the bonds that produce the highest expression of integration. They

[4] Hereafter, to reduce confusion, homologous bonds in different teeth (as 4–24–44, or 6–26–46 for $M_{\overline{3}}$, $M_{\overline{2}}$, and $M_{\overline{1}}$, respectively) are considered the same and are indicated only by use of the right-hand digit. Thus, for example, 4–24–44 becomes 4–4–4, and any such bond is referred to by the number 4.

express the most persistent integrative patterns. These patterns are shown diagrammatically in Figure 73.

Three of the patterns appear to represent readily definable integrated units, as follows:

3-4-5: This is clearly a talonid integration. It can have no trigonid counterpart in the absence of the paraconid.

4-6-10: This group involves two measures normal to the long axis of the tooth row and the diagonal between the metaconid and hypoconid. It is an integration of measures across the molar teeth and the diagonal and will be called the "transverse" group.

8-9-10: This includes two measures roughly parallel to the long axis of the rooth row, plus the diagonal. It will be termed the "longitudinal" group.

The other triplet, 4–5–6, includes two transverse measures and a measure of the talonid. Thus there are both elements of the talonid and transverse integra-

TABLE 76

APPLICATION OF CRITERIA OF "IMPORTANT ASSOCIATION" TO *Hyopsodus* SPECIES

a) FREQUENCY OF OCCURRENCE OF BONDS POSSIBLE IN INTER-
CORRELATIONS OF MEASURES TAKEN

BOND	FREQUENCY			BOND	FREQUENCY		
	Pairs Only	Larger Groups	Sum		Pairs Only	Larger Groups	Sum
3–4	5	1	6	5–6	1	1	2
3–5	1	2	3	5–8	0	0	0
3–6	0	0	0	5–9	0	1	1
3–8	1	0	1	5–10	0	1	1
3–9	0	1	1	6–8	2	0	2
3–10	0	1	1	6–9	1	0	1
4–5	6	1	7	6–10	1	2	3
4–6	1	3	4	8–9	2	4	6
4–8	1	0	1	8–10	4	4	8
4–9	4	0	4	9–10	2	5	7
4–10	0	2	2				

b) FREQUENCY OF JOINT OCCURRENCE OF PAIRS OF BONDS
FROM *a* WHERE FREQUENCY IS GREATER THAN 1

Bond	3–5	4–5	4–6	4–9	4–10	5–6	6–8	6–10	8–9	8–10	9–10
3–4	1	3	1					1		1	
3–5			1	1	1			1		1	1
4–5			1	1	1	2	1		1	1	1
4–6				1	2	2	1	2		1	1
4–9									1	1	2
4–10								2	1	1	1
5–6							1	1			
6–8											
6–10										1	
8–9										4	4
8–10											3

tive groups. This triplet occurs in the series in only *H. miticulus*. It seems to act as an expression of partial combination of these two groups and thus may be called a "talonid-transverse" group, or, for simplicity of reference, the "mixed" group.

The patterns of integration based on these triplets are shown diagrammatically for each tooth of each species in Figures 74 and 75, for the "paulus" line and the "despiciens" line, respectively. Some points stand out in these diagrams. First, it is evident that three elements of integration occur in $M_{\overline{3}}$ of *H. miticulus* and that this is not the case for any other tooth in either series. When $M_{\overline{3}}$ is followed through the two series, it is clear that the 4–6–10 group, transverse integration, persists in $M_{\overline{3}}$ of the "paulus" line, whereas 3–4–5, talonid integration, persists in the "despiciens" line. In this there is some evi-

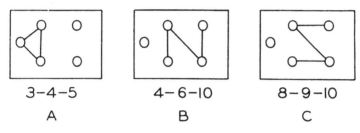

Fig. 73.—Diagrammatic presentation of three principal groups of associated measures in *Hyopsodus* after partial correlation, r_{xy}. *A*, talonid group; *B*, transverse group; *C*, longitudinal group. Solid lines indicate the measures that are correlated.

dence of a sorting-out of one or another group from the ancestral conditions in each of the lines. Beyond this point, however, it is more difficult to trace the evolution of the pattern without some guiding principle. It seems reasonable on the basis of this evidence to consider the lower molar series as a system under a single morphogenetic influence in which there exists a gradient of intensity of total integration *and* expression of this gradient in the composition of patterns.

Each species reveals longitudinal changes of intensity (in $I\rho$, Table 74) and longitudinal shifts along the tooth row of the composition of integration. Extension of this concept to the evolutionary sequences, so that gradient modifications are considered as part of the process of evolutionary development, reveals a generally coherent picture. The patterns of change from this point of view are shown in Figures 76 and 77 for the "paulus" and "despiciens" lines, respectively, with $I\rho_{xy.lw}$ entered to show the course of intensities (see also Table 75). The process of change of integrative patterns appears to be one of sorting out the three primary groups, a process that proceeds quite differently in the two lines. The three groups are present in $M_{\overline{3}}$ of *H. miticulus*. The transverse and mixed groups are present in $M_{\overline{2}}$, so that elements of both the talonid and the transverse groups are in evidence. Only the mixed group is present on

$M_{\overline{1}}$. All transverse integration is centered in $M_{\overline{3}}$ in the "paulus" line above the *H. miticulus* stage. Longitudinal integration spreads forward from $M_{\overline{3}}$ of *H. miticulus* to $M_{\overline{2}}$ and $M_{\overline{1}}$ of *H. lysitensis* and occurs on all lower molars in *H. paulus*. Emphasis in this line is upon transverse integration in $M_{\overline{3}}$ and upon a spreading of longitudinal integration through the whole series. Talonid integration, which is developed discretely only on $M_{\overline{3}}$ of *H. miticulus*, shows a forward spread to $M_{\overline{2}}$ in *H. lysitensis* and shows forward movement to be present only on $M_{\overline{1}}$ of *H. paulus*.

The "despiciens" line is somewhat less clear-cut, as might be expected in

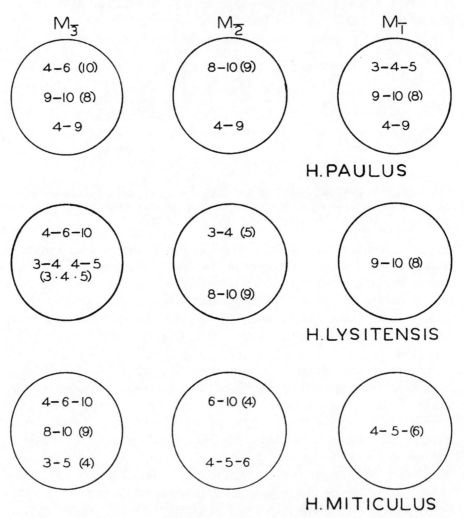

FIG. 74.—Diagram of pattern of each tooth of the species of the "paulus" line of *Hyopsodus*. Based upon partial correlation $r_{xy \cdot lw}$ and determination of important associative groups of measures. Bonds shown by solid lines. Full composition of groups represented indicated by number in parentheses.

view of the less direct line of descent of the species that has been postulated. It is evident, however, that *emphasis* is upon talonid integration. This is present in all molars of *H. mentalis* and in $M_{\bar{3}}$ and $M_{\bar{1}}$ of *H. despiciens*. Longitudinal integration does not occur above *H. miticulus*. The transverse integration of *H. despiciens* poses a problem. If *H. mentalis* is ancestral, it must be assumed

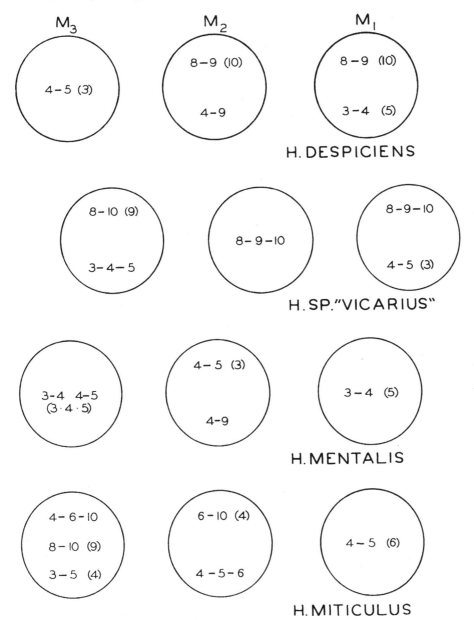

Fig. 75.—Diagram of patterns of species of *Hyopsodus* of the "despiciens" line. (Symbols as for Fig. 74.)

that *H. despiciens* has redeveloped this integration. This is, of course, not out of the question. On the other hand, it is quite possible that there existed an unknown intermediate stage (or stages) between *H. miticulus* and *H. despiciens* within the general group and that *H. mentalis* was not on the direct line between the two species.

The pair 4–9 occurs in both series, in $M_{\bar{2}}$ in *H. mentalis* and *H. despiciens* and in all molars in *H. paulus*. Its frequency suggests that it may have some impor-

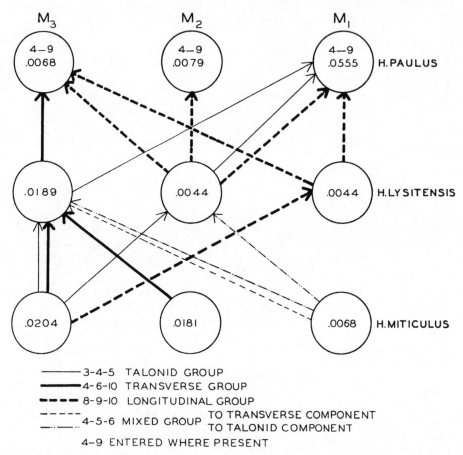

Fig. 76.—Diagrammatic summary of evolutionary shifts of the important associative groups of measures in teeth of the species of *Hyopsodus*, "paulus" line. $I\rho$ indicated for each tooth.

tance, even though it does not occur in triplets. It appears to represent a new development in both lines.

An event that may be of considerable importance is seen in *H. paulus*, in which a group is formed from measures 4–5–9–10. This results from the integration of components of the talonid and longitudinal groups which come together in this tooth. If it is assumed that orderly change is the rule and that groups do not appear sporadically without antecedents, this consolidation could not take place in the "despiciens" line. We may further speculate that the formation of such a new assemblage from pre-existing components could set the stage for continued development quite different from that which preceded its formation. No such course of development could arise from the "despiciens" line.

Finally, we may look briefly at *H.* sp. "vicarius," which has been considered a probable descendant from *H. mentalis*, and at *H. powellianus*, which is on a

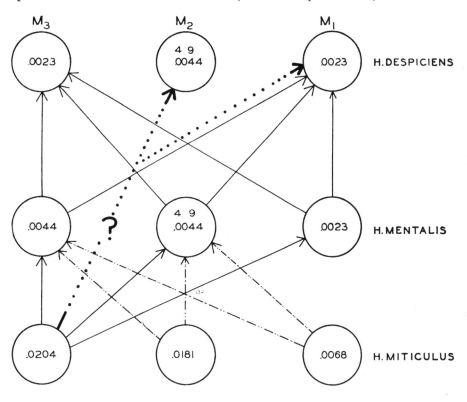

KEY AS FOR PRECEDING FIGURE

Fig. 77.—Diagrammatic summary of evolutionary shifts of important associative groups of measures in teeth of the species of *Hyopsodus*, "despiciens" line. I_p for each tooth indicated.

different line of descent. Data are presented in Table 73. *Hyopsodus* sp. "vicarius" is similar to *H. mentalis* in the presence of the talonid groups (3–4–5) on $M_{\overline{3}}$, and it also has this component in $M_{\overline{1}}$. All molars, however, are marked by the presence of the transverse group (8–9–10). There is no source of this group in *H. mentalis*. As in the case of *H. despiciens*, it is possible to suppose that this group developed independently. The alternative interpretation, that *H. mentalis* is not directly ancestral, is equally acceptable and, in view of other divergent characters in *H.* sp. "vicarius," perhaps somewhat more reasonable.

Hyopsodus powellianus is identical with *H.* sp. "vicarius" in $M_{\overline{2}}$ and $M_{\overline{1}}$, but there is marked difference between the two in $M_{\overline{3}}$. Neither talonid nor longitudinal integration occurs on this molar in *H. powellianus*. It seems certain that *H.* sp. "vicarius" and *H. powellianus* did not possess a common ancestor at any time later than the very earliest Eocene. This being the case, the identity of the patterns of the two anterior lower molars, which, in view of its absence in the "paulus" and "despiciens" lines, presumably was not present in the common ancestor, must be due to parallel development.

The compositional aspects of the lower molar dentitions in the species of *Hyopsodus* are less evidently arrayed in logical sequence than are the aspects of intensity of integration. When the factor of size has been held constant, however, we do arrive at an array of patterns that portrays a consistent difference between the two phyla and evidence of orderly change in each. The more consistent series of changes is found in the "paulus" line, for which evidence of ancestor-descendant relationships is fairly clear. From this line it would appear that changes in composition have a directional aspect in the evolution of successive species. In this result lies an answer, for a particular case, of how morphological integration behaves in the evolution of a series of species where the samples are drawn from a single ontogenetic stage. This instance, plus the evidence from *Pentremites*, provides a basis for the inference that orderly changes of integrative features are characteristic of organic evolution at the low level studied. This inference alone, of course, provides no basis for an answer to the more vital question of whether or not studies of morphological integration add anything to an understanding of evolution in particular cases or in general that is not to be obtained from other approaches. A different type of description is possible, but, if this is all, such studies are hardly worth the effort. It was a faith that there is more to be gained that initiated our work and kept it going. Nothing that has been found to the present argues to the contrary, and we propose to attempt in the final chapter a synthesis which involves consideration of positive evidence in this direction. In the special case of *Hyopsodus* we are dealing with a situation in which it is possible to assess the results of two types of study to see whether or not they are duplicative, complementary, or supplementary. This properly belongs in the present chapter, even though it is partly speculative, and it is now considered.

Matthew (1909, 1915) and, earlier, Osborn (1902) considered that the evolution of *Hyopsodus*, as recorded in the strata of the Eocene of western North America, was an example of *in loco* evolution. Matthew in particular regarded the known samples as several phyla that were progressing through time along distinctly parallel courses. He found it impossible to make clear-cut assignments of all species (including his subspecies) to one or another phylum, although some of the best-known species were arrayed in tentative relationships, as given in Appendix E. Regardless of the specific composition of the phyla,

Matthew made a strong case for parallel development so far as dental characters were concerned.

The parallelism in increase in size is abundantly clear. Both phyla treated in our discussions show increase in the length of the lower molar series, from a mean of 11.18 in *H. miticulus* to 12.37 in *H. paulus* and 14.31 in *H. despiciens*. *Hyopsodus powellianus*, which presumably arose from a small ancestor, which lay at the base of all the phyla, has a mean length of the lower molar series of 17.86. Parallelism also appears to have taken place in modifications of the talonid of $M_{\overline{3}}$ and the moralization of $P^{\underline{4}}$. Somewhat less well-defined, but apparently valid, is the parallel reduction of the paraconid in the lower molar series.

The modifications of morphological integration that occur within the series studied are *divergent* rather than *parallel*. This is particularly clear with regard to values of $I\rho$ for the total molar series, in which there is a sharp decline in the "paulus" series from *H. miticulus* and virtually no change in the "despiciens" series. A similar change is witnessed in the effectiveness of size as an integrating factor in the individual lower molar teeth. This factor maintains or increases its importance in the "despiciens" line but becomes less important in the other, particularly in $M_{\overline{1}}$, in which it acts to decrease rather than increase apparent integration. In the index for total integration, however, *H.* sp. "vicarius" approaches members of the "paulus" line more closely than it does the members of the line with which it was associated. If the phylogenetic position has been correctly assessed, there is parallelism in integration between this form and members of the "paulus" line.

The situation with respect to the composition of the ρ-groups is somewhat less clear-cut, partly, it may well be, because it was necessary to deal with teeth individually. In spite of this, there seems little doubt that the two lines of *Hyopsodus* analyzed in this study generally followed divergent rather than parallel courses as the patterns evolved in the series from the ancestral condition of *H. miticulus*. In this respect, *H.* sp. "vicarius" shows features of the "despiciens" line but, in spite of these, is somewhat divergent from the other species of this line. Resemblances between *H.* sp. "vicarius" and *H. powellianus* suggest parallelism between these two species.

Morphological integration and the general features of dental morphology that have been used to differentiate species and to express trend through time thus reveal opposite aspects of the evolution of the two phyla of *Hyopsodus*. Both, it would appear, must have changed under selective processes and, presumably, are a fairly direct reflection of the genotypes of the various species. The case for this interpretation in the gross morphological characters is reasonably straightforward. In addition to size increase in the teeth, which appears to be related to general increase of over-all size of the animals, the gross dental modifications appear to have produced an extended and more effective dentition for a herbivorous mode of life. Each of the changes may be interpreted as being of "general" value, regardless of precise differences in the feeding habits of the various species. If this is the case, the parallelism in the different phyla is readily explained. Increase in size, changes in patterns of the molar teeth, and extension of the molar form to P^4 suggest that the grinding dentition as a whole both above and below was the primary unit under selection.

Integrative features fall less readily into a pattern that is subject to simple explanation. The evolutionary events, however, are too regular and too consistent to be ascribed to chance modifications. If both the changes in general morphology and integration are adaptive and took place under selection, as seems reasonable in view of the regularity of change in both instances, it follows that different aspects of selection must have been in operation. Presumably each aspect is related broadly to function. The two types of changes must be complementary, but, while a balance is maintained between them, species that are fundamentally different in function must have developed, even in the absence of functional differences in the parallel characters. The parallelisms, while real, do not result in equivalency of function under this interpretation, but the divergences are those revealed by the study of integration. The two species developed from *H. miticulus* and their descendants can thus be considered adaptively different and not the product of essentially similar adaptive modifications that occurred, once isolation had produced the two species.

The lower molars alone, of course, cannot be considered an adaptive unit, for it is only in interaction with the upper teeth that their characters assume functional significance. The nature and intimacy of the relationships of the morphological integration of the upper and lower molar teeth were considered at some length in the study of *Aotus* (chap. vii). Without such information about the interactions for *Hyopsodus*, it is difficult to arrive at conclusions concerning the adaptive meanings of the changes that have been observed. This is particularly true as it relates to modifications of the index of integration, for the conditions in the lower series can give only partial information about the total selective unit. It is possible, for example, to place alternative interpretations upon the fact that total integration remained essentially stable in one line and dropped in the other. The populations that show reduction may merely be more diverse with respect to the characters that enter into the index. On the other hand, it may be that a higher integration of the opposing areas of the upper and lower teeth was taking place, with resultant lowering of the total integration both above and below. It is possible, further, that there was a trend toward the situation found in *Aotus* in which the gross size control of integration was centered in large part in the upper dentition. There is no answer from the values of $I\rho$ alone as to which, if any, of these explanations may be approximately correct.

Data on the factor of size, both for the total tooth row and for individual teeth, however, do not support the hypothesis that populations with relatively low values of $I\rho$ are merely composed of individuals with a wider range of integrative properties. It is shown in the discussion of *H. paulus* "mutants," an invalid species, in Appendix E that wide diversity, in which a ubiquitous factor of size is active in integration, is accompanied by a high value of $I\rho$ for the total molar series under simple correlation. When the factor of size is held constant under this circumstance, the value of $I\rho$ decreases to insignificance. These conditions are not encountered in the species of the two phyla. Either of the other two possible interpretations seems more probable, but there is little choice between them from the data on $I\rho$ and the factor of size.

The composition of the ρ-groups, after size has been held constant, furnishes some additional evidence. Transverse integration dominates on $M_{\overline{3}}$ in the "paulus" line with talonid integration restricted to $M_{\overline{1}}$ in *H. paulus*. Longitudinal integration spreads from $M_{\overline{3}}$ in *H. miticulus* to all molars in *H. paulus*. Talonid integration becomes dominant in the "despiciens" line. If, as the study of *Aotus* suggests, the areas of integration in the lower teeth imply a matching integration in the corresponding occlusal areas of the respective upper teeth, then we must conclude that different centers of high integration in occlusal relationships are characteristics of the two lines. Emphasis in the "paulus" line is upon the anterior part of the lower molars and the posterior part of the upper molar one forward in the series, as $M_{\overline{1}}$ to $P^{\underline{4}}$, $M_{\overline{2}}$ to $M^{\underline{1}}$, and $M_{\overline{3}}$ to $M^{\underline{2}}$. Emphasis in the "despiciens" line is on the posterior part of the lower molars and the anterior part of the corresponding molar in the upper series. Here, in contrast to the "paulus" line, a strong integration of $M_{\overline{3}}$ and $M^{\underline{3}}$ appears probable.

Presumably these different trends relate to the development of different actions in mastication. How these are to be translated into concrete interpretations related to feeding habits remains an open question. It seems probable that dietary differences play an important role in the divergence of the phyla, accompanied by a general improvement for a herbivorous mode of life. Until studies of diet and integration in modern forms are made, this will remain somewhat conjectural, and assessment of the qualitative differences in dietary habits is out of the question.

It is obvious that we are groping for explanations in attempts to interpret the data on integration in the absence of adequate information. The interpretation has been attempted more in illustration of the possibilities offered by study of morphological integration than in the hope of reaching definitive conclusions about *Hyopsodus*.

CHAPTER IX

Morphological Integration and Evolution
II. Summary and Theoretical Considerations

Opportunity for generalization has been in large part denied in chapters iii through viii by the necessity of sequential development of propositions, operational techniques, and illustrative examples of morphological integration. We are now in a position to bring together some of the most important aspects of the studies and to enlarge upon the scope of the work by generalizations and theoretical extensions. This final chapter is devoted to these ends, with emphasis upon the evolutionary aspects of morphological integration.

Evaluation of the place of morphological integration in the field of organic evolution may be approached from two somewhat opposed perspectives. One views the role of integration as predominantly a means of clarification and elaboration of the factors and phenomena of evolution within the framework of the current synthesis. The other involves consideration of morphological integration as a direct expression of organic evolution. Particularization becomes only a means to the end, namely, an operational conception of the organism as a whole and understanding of the totality as it moves and changes through time. These points of view are not necessarily incompatible, but preoccupation with one or the other may open diverse avenues of synthesis and interpretation.

The closing sections of chapter ii included general statements of the ways in which morphological integration could be related to evolution under the first point of view. It seems unnecessary to repeat these considerations in summary at this point, but, now that data, analyses, and interpretations have been presented, the various statements will presumably take on more precise meaning, and the scope and limitations of studies of morphological integration should fall more clearly into focus. The statements in chapter ii dealt in large part with generally well-understood aspects of evolution. The questions that were raised pertained to the evolution of morphological integration itself. We could do no more than raise such questions at that point, for the answers were dependent upon data and analyses not then presented. Even now, many must be left unanswered, and others can be treated only in a tentative way. The present chapter is devoted particularly to exploration of problems raised by the second point of view with the evolution of morphological integration as the center of interest.

TIME-ORDERED SEQUENCES AND MORPHOLOGICAL INTEGRATION

We have now reached the stage in the development of the concept of morphological integration at which it becomes possible to consider time-ordered sequences in some detail. This is the region of investigation where the paleontologist should reign supreme. The neozoologist has had at his command far more sensitive tools for investigation of the *cause* and *mechanism* than the paleozoologist. He does not, however, have the time sequences to which these tools can be applied. We have, in our opinion, *developed tools which are not only far more sensitive than any hitherto available to the paleontologist but approach those of the neozoologist in efficiency and sensitivity when applied to multivariate situations.* This holds for both extinct and contemporary animals but, in the absence of equally sensitive methods for dealing with the former, assumes greater importance in this area.

The literature on evolution is rich with discussions of modes of evolution. Investigations of the causality behind such complex phenomena, however, must proceed in the direction of finer and more delicate analysis if true progress is to be made. It is not enough to arrange and rearrange the fossil record at the level of species and genera or to reassess their higher taxonomic interrelationships. Detailed quantitative studies of temporally ordered anatomical changes and the attendant wealth of implications hold the key to knowledge that can be gained in no other way. What we have attempted in the pages of this book is a description and development of a quantitative approach to representation and analysis of anatomical changes. The specific aim from the beginning has been the analysis of the complicated total animal or organ in its anatomical changes through time. We feel that this goal has been reasonably well achieved. Now we propose to turn to a survey of time-dependent phenomena in the history of life, armed with a reliable and very sensitive recording device, the ρF-analysis.

Interpretation of such phenomena, beyond mere recording of events, brings the implications of causality into sharp focus. The search for causality in any complicated problem, however, is strictly a relative proposition. It is possible to explain a natural phenomenon in terms of properties more simple than the ones that make up the external expression of the phenomenon itself. But these simpler properties may require explanation in turn, and so on ad infinitum. The writers take the view that any gesture in the direction of ultimate organic causality can have no place in a book of this type. We are, however, seriously concerned with the notion of causality and with the idea of reduction of our problems to terms simpler than those encountered initially. We specify our goal in this respect as follows: *If, for a given problem in the context of this book, we are able to predict, by virtue of reducing the problem to a few quantitative terms and then noting regularities or lack of them in time-ordered sequences, we shall then claim that a step has been taken in the direction of a causal explanation.*

Time-dependent Sequences

The following model gives the basis for a preliminary discussion of time-dependent sequences. The correlation between two anatomical measurements is observed at various time levels. At any given time level only two outcomes, or states, are possible: either the pair of measurements is correlated, or it is not. The correlation is computed on a sample of individual animals from a population. The following notation will be used:

$$B = \text{bonded and implies} \mid \rho_{x_1 x_2} \mid \text{significantly} > 0 \; ;$$

$$U = \text{unbonded and implies} \mid \rho_{x_1 x_2} \mid \text{not significantly} > 0 \; ,$$

$$\text{where} \mid \rho_{x_1 x_2} \mid \text{is the correlation coefficient between}$$

$$\text{measurements } x_1 \text{ and } x_2 \; .$$

If the correlation between measurements x_1 and x_2 is observed at successive time intervals, the result is a time-ordered sequence, such as $B, U, U, B, B, U, B, U, U, B$. The runs test for random order may be profitably applied to this sequence. If the arrangement of B's and U's is not random, the alternative interpretations of the arrangements are: (1) if the number of runs is improbably small under the hypothesis of random arrangement, a clustering effect is suggested; or (2) if the number of runs is improbably large, a mixing effect is suggested (Feller, 1950, p. 56). Interpreted within our context, any of the three possible descriptions of a time sequence of this sort would have interesting and important evolutionary implications. It would be worth knowing (1) whether correlation between a pair of anatomical measurements is random through time; (2) whether the correlation alternates between long unbroken runs and runs of lack of association; and (3) whether the pattern fluctuates rapidly through time from correlated to non-correlated.

The application of tests of this sort requires extensive time-ordered sequences. At present our longest sequence includes only four separate time levels and is inadequate. Future investigations in the directions indicated depend upon the availability of temporally ordered materials from a relatively large number of time intervals. The work described in previous chapters has supplied biologically meaningful and sensitive relations that can be translated into numbers. The stage has been reached at which the building blocks have been fashioned and have been found to meet certain mandatory biological specifications. We can now survey a number of plans for construction of hypotheses that make use of these building blocks.

Although the runs test described above promises interesting results, more powerful models which include *prediction* are also available. Discussion of these will be confined to the types of prediction problems that appear, on casual inspection, to be appropriate to the models discussed.

Fundamental Considerations

I. INDEPENDENCE

We shall consider for illustrative purposes the bonding of a single pair of anatomical dimensions observed at time intervals. The question of independence will be taken up in this context, but the implications are general. As before, the observation of a pair of measures x_1 and x_2 at time t_k has only two outcomes —*B* (bonded) or *U* (unbonded). Thus a time-ordered sequence is formed which may, for example, take the following pattern: *B*, *B*, *U*, *U*, *B*, *U*, *U*, *U*, *B*, *B*, *B*, *U*. The question of independence is stated as follows: Does the outcome of time t_k depend on the previous result at time t_{k-1}? The answer to such a question may, in theory at least, be sought in biological and evolutionary context. Unfortunately, there is not enough direct knowledge in these fields to supply a definite answer, although it is possible to make an educated guess. An appropriate statistical test to aid in decisions has been suggested by H. T. David, of the Committee on Statistics, University of Chicago.[1] The generalization of the decision on the relation of the state at any time t_k to the state at time t_{k-1} is dependent upon whether or not the time process is *stationary*. This, then, is the second of the two fundamental properties which must be considered.

II. STATIONARY VERSUS NON-STATIONARY PROCESSES

According to Feller (1950, p. 365), a "stationary process" is one in which the forces and influences that determine the process remain absolutely unchanged over the time range being considered. It is, we suppose, possible to say that the evolutionary process in general is a stationary process by invoking the doctrine of biological uniformitarianism. Such a statement, however, is too vague and of little direct application in the more precise context of this book. What we must consider here is the specific process of change of bonding between a pair of anatomical measurements through time or change in *pattern of bonding*

[1] In the studies of the *Pentremites* described in chaps. vi and viii of this book, observations of the correlation between various pairs of anatomical measures were made at four time levels. The following sequence represents the association between measures 1 (spiracle width) and 9 (food groove length): *B*, *B*, *U*, *B*. We ask what the relationship is between *B* in the fourth entry and the previous entries. Let us suppose that replicate sequences of the four entries were made available by repeated sampling of the populations at the same four time levels. We would then subdivide the replicate sequences into eight categories formed by permutations of the *first three* entries. Now consider the category of sequences which have the form *B*, *B*, *U*, —, where the fourth entry may be either *U* or *B*. The proportion of those sequences which have *B* in the last entry to those which have *U* in the last entry is compared with the similar proportions computed for the other seven sequence types, e.g., *B*, *U*, *B*, —; *B*, *B*, *U*, —; etc. If the proportions are about the same in all eight categories, we may conclude that the entry in the fourth place in the sequence is independent of the previous entries.

when a number of measurements is considered simultaneously at various time levels.

We shall consider the forces and influences which affect the bonding process in terms of the following equation:

$$\rho_{t_k} = (\alpha) \text{ function} + (\beta) \text{ development} + (\gamma) \text{ residual} , \qquad (1)$$

where

$$\rho_{t_k} = \rho_{x_1 x_2} \text{ or } \rho\text{-group at}^2 \text{ time } t_k ,$$

so as to be consistent with earlier propositions.

This equation may be thought of as an expression of the total internal environment within which the various anatomical dimensions interact and react to give change in form and which we measure by the ρF technique. It is stated that the correlation at any time t_k between a pair of anatomical dimensions is equivalent to the sum of the factors of function and development plus a residual term. The term ρ_{t_k} is thus a parameter of the biological population structure at time t_k. Discussion of the relative importance of the three terms on the right side of the equation is based on evaluation of the three coefficients, α, β, and γ, and is reserved for a later section of this chapter.

We shall now extend consideration to closely related processes and again ask whether the notion of a stationary process applies. The *first*, which has been noted previously, is the correlation between a pair of anatomical measures observed at various time levels. As indicated, there are two possible outcomes, bonded (correlated) and unbonded (uncorrelated). The *second* involves a counting procedure and takes on values 0, 1, 2, . . . , n, depending upon how many changes from bonded to unbonded (or the reverse) have taken place from time t_{k-1} to time t_k, when the intercorrelations of all the measurements on an animal or organ are under consideration. Thus in the transition from time t_{k-1} to time t_k there may have occurred no changes or from one to n changes, where $n = (N^2 - N)/2$ and N is the number of measurements considered. A *third* process may also be useful for our purposes. This involves change of the index of integration $I\rho$ with time. In all three of these possible stochastic processes, consideration of the applicability of the notion of a stationary process depends upon the implications of equation (1).

In our opinion, the stationary process is applicable to each of the three processes described above. All depend directly or indirectly on equation (1). The effects of the individual terms change with time so that various states result, but the total effect of all the forces and influences as expressed in equation (1) holds for any time t_k!

The comments on the fundamental propositions may be summarized as fol-

---

^{2} In the following, ρ will be taken to mean either $\rho_{x_1 x_2}$ or a ρ-group or both, depending on the context.

lows: We are unable in our present state of knowledge to decide whether to treat the processes that have been discussed as independent or dependent. A simple statistical test to provide a basis for judgment has been suggested, but it requires data that are not yet available. We are convinced, however, that all three processes suggested above are stationary in nature and have presented our reasons for this conclusion. The two fundamental considerations (Feller, 1950, pp. 403–32), independence and stationary versus non-stationary processes, will determine which of the many powerful and useful aspects of stochastic processes will be the most promising for future study. The theory of recurrent events and Markov chains offer interesting possibilities. Poisson processes, or refinements, such as the pure birth processes (Feller, 1950, pp. 403–32; Yule, 1924, pp. 21–87) may prove valuable and interesting.

THE FUNDAMENTAL EQUATION

The fundamental hypothesis upon which the ρF analysis is based was presented in the first ρF paper (Olson and Miller, 1951) and has been discussed in the early pages of chapter iii. This hypothesis was restated in equational form in the last page of chapter iv and in modified form in chapter v. The equation states that

$$\rho = \alpha \text{ (function)} + \beta \text{ (development)} + \gamma \text{ (residual)}.$$

The coefficients α, β, and γ represent the relative contributions of the three factors; α, β, and γ are defined to sum to unity for convenience. In an earlier section of this chapter, the equation was used in a discussion of the applicability of the notion of stationary processes. We now turn to definitions and a discussion of the relative importance of the three factors[3] that contribute to $|\rho_{x_1 x_2}|$, the correlation between a pair of morphological measurements.

The stage may be set for discussion of the factors which follows, by specification of the *system* of interacting morphological dimensions that is formed when an interbreeding population is considered. The components of the interacting system are the individual measurements, and the interactions are represented by the correlation coefficients between measurement pairs, taken over a sample of the interbreeding population. The interaction of the totality of measurements is assessed by means of various formal methods of arrangement of the individual correlation coefficients. The system described operates under the influence of a very large number of factors, which we summarize in the following equation:

$$\rho = \alpha \text{ (function)} + \beta \text{ (development)} + \gamma \text{ (residual)}.$$

This conceptual model may be described as an open system in that factors that operate externally to the breeding population can affect the correlation of

---

[3] The three terms are called "factors" for convenience. Actually, they represent three distinct classes of factors, as defined in the text.

a given pair of measurements. A discussion of the relative importance and of the meaning that we attach to the three factors in the equation will be given later. First, however, it is necessary to elaborate certain aspects of the concept of causality.

The dictum that "correlation does not imply causation" is a relative statement and need not be taken literally. It may be rephrased more precisely as follows: When two properties, functionally very tenuously related but both subject to a very general collection of factors, have been found to be highly correlated, it is not to be concluded that there has been demonstrated a close causal relationship. An example is the high correlation between birth rate and the stork population in Sweden. On the other hand, there are many well-known examples in which a causal interpretation is far less tenuous, for example, the correlation between the number of chirps per minute of a cricket and temperature (Holmes, 1927). The difference between the two cited cases, both of which exhibit high correlation, may be considered to lie in the system in which the components operate. If the factors are fairly well known and if the components of the system operate under a suite of circumstances in which there is sound a priori knowledge, then the causal implications of correlation can be speculated upon without unreasonable difficulty.

We believe that the system dealt with in this book is well defined. The factors in the system that affect the components can be listed and discussed with varying degrees of sound a priori knowledge. We thus maintain that the correlations with which we deal *do* give insight in varying degrees into the classes of causal factors that operate on the system, both internally and externally.

With this in mind, we now turn to a more detailed examination of the equational statement of the fundamental hypothesis and a discussion of the factors and their weighing coefficients.

1. *Function*

We define *function* in the context of this study to include all *physicochemical interrelationships of the components of an organ, or part of an organ, directed toward performing any and all actions and reactions that operate after conception of the animal.*

2. *Development*

For our purposes we shall define *development* to include *all inherited growth patterns.*

Development and function, thus defined, represent two different sets of factors, the one inherited directly from the parents, the other operative after conception and superimposed upon the first. Reference to chapters iii and iv will provide illustration of the use of these concepts in a formal development of the use of the correlation coefficient in morphological integration and in application to specific cases. After conception the new individual is initially composed of a complex of biochemical substances in which are contained a set of building directions to be followed during the period of maturation. Superimposed upon this set of directions are the effects of function. The results may be partially and very crudely recorded in the growth curve of a single dimension, which we have called the "ontogenetic path" (chap. iv). The interaction of a single dimension with any other is recorded in the "joint ontogenetic path." From this basis the correlation coefficient is generated. That part of the degree of association expressed by ρ that is attained without direct response to *variations* of external factors we consider to be directly attributable to the factors of development and function as defined above.

The data on rat ontogeny discussed at the end of chapter iv suggest that the primary source of fluctuations of the correlation between a pair of measurements within a population is to be found in the earlier stages of ontogeny. The question which now arises is: How shall we assign these effects? If we take the position that functional requirements in the early stages result in wide variation in correlation during these stages, then we exclude the possibility that inherited chemical building plans call for readjustments at various thresholds in the maturation period and thus result in variation in correlation. For convenience and simplicity, we may take the position that the pattern inherited from the parents is set at conception. Individual variation, derived from β, arises from differences in genetic constitution of the individuals and from any latitude that may exist in the chemical building plan. Function, α, acts as an agent that restricts additional variation in the population.

It is intuitively reasonable to consider that the effects of "function" superimposed upon the growth patterns will vary for different anatomical dimensions. Thus, in the case of dimensions whose interrelationships are rigorously controlled by delicate and demanding functional duties, such as parts of the eye, most variability would come from inherited differences as defined by the term "development." "Function" would act in the direction of restriction of any further variability during the period of maturation. Indeed, we claim that when exceptionally high correlation is noted, preliminary consideration should be given first to the contribution of the term called "function."

It is, however, quite possible to speculate within our model that a "random" interaction of certain dimensions which results in very low correlation is due to the latitude allowed in the chemical building plan, uncontrolled during the period of maturation by any rigorous functional demands. An example of this is revealed in the dermal elements of the skulls of *Diplocaulus recurvatus* described in one of the reports on morphological integration published prior to the writing of this book (Olson, 1953).

3. *Residual*

The *residual* is a wastebasket term, erected to include all factors other than function and development. Sampling error and unassignable random effects are also included. For convenience we shall say that the factors of "function" and "development" operate within a system and that "residual" includes all factors external to the system—temperature, chemistry of the environment, and so forth. The importance of the residual term varies according to how wide a range of external factors has acted upon the sample of the population and how sensitive the pair of measurements (in terms of their covariance) under consideration are to the external factors. We feel that in most cases γ, the weighting coefficient for "residual," is small and may be neglected.

Obviously, there are important exceptions, and these tend to occur most commonly where the general economy of the organism is not seriously disrupted, perhaps even improved, by modifications in direct response to external stimuli during growth. Modifications of growth by external factors are commonplace. The effects of wave action on strand-line shelled invertebrates or the effects of temperature on size and proportions provide well-known examples. Such effects, however, may or may not be pertinent to the system under consideration. Only if the ways in which two measures vary with each other are affected, does a modification of the system occur. Mere size change of the organism as a whole or proportional changes within or between parts of the body do not *necessarily* result in alteration of the system. The requirements of "function" and the inherited patterns of "development" would appear to exert the major effects in the majority of cases, and the effects of the "residual" factor are generally strictly subordinate.

By definition, the coefficients a, β, and γ sum to unity. Thus, if for a given pair of measurements or for a ρ-group we are able to approximate the values of any two of the coefficients, we shall have a value for the third. It is hoped that further analysis of the study of ontogeny in the albino rat and similar experiments will give the needed additional insight into the relationship and that the results can then be applied to fossil materials.

In summary, it appears at the present state of knowledge that the coefficients a and β, in the majority of cases, contribute most heavily to the total. The relative importance of a and β depends upon the particular anatomical dimensions under consideration. Finally, it is hoped that further studies like the investigation of the rat ontogeny will aid in the separation of the effects of "function" and "development" upon the system expressed by ρF.

CONCLUSION AND RETROSPECT

The first chapters of this book summarized the concepts and studies that would make up the body of the text. As they may now be considered in retrospect, the task of summary at this time is in large part non-existent. We began with an idea, neither new nor novel, established a framework within which it could be investigated, performed the tests that seemed appropriate, and arrived at a basis for quantitative studies of the evolution of temporal series of organisms. We believe that this basis and the results obtained by its application are important in the new perspectives that they impart to evolutionary studies. Perhaps we have been immodest in our claims, but, even if immodest, we hope that this work has been provocative.

A great chasm, we feel, has existed between the reality of the evolutionary process, which involves the total organism and the population of which it is an element, and the study of the processes of evolution by particularization of its components. This has been not so much a gap of concept as one of methodology and operational theory. For as long as the theory of evolution has existed, it has been recognized that synthesis must consider the organism as an essential unit. The philosopher, of course, has followed an "organismic" course, but the scientist has suffered under the burden of restrictive difficulties. More and more, however, the organism and the interbreeding population in their internal and external relationships have become the focal point of study, in genetics and ecology in particular. For scientific analysis, however, it becomes necessary to take the population or the animal apart, to look at the components, and to isolate the particular. If it be granted that answers in the field of science cannot be gained without such dissection, we are faced with the direct question of how to put the parts together once more in such a way that we do not, in the process, lose sight of the components which entered into the synthesis.

There are many possible answers to this question. As the complex problems of synthesis are surveyed, the greatest success to date seems to have come from the use of stochastic models, exemplified particularly in the field of population genetics. Ecologists, in different ways, have been successful in entering the variables appropriate to their systems into models that have resulted in important syntheses. Our variables are the morphological characteristics of animals. Our problem, specifically, was how to express these variables and their relationships in such a way that the total organism or some part of it could be treated intelligibly without losing sight of the contributions of the parts. Only if an answer could be supplied, could we hope to tackle the problems of time series of organisms, as represented in the fossil record, from the point of view of totality that seemed to us so important. The use of some stochastic model had intrinsic appeal, and results in other areas of study were encouraging. The model presented in this book, based on the correlation coefficient, is our answer to the problem of synthesis.

The approach that has been used is by no means the only one that is possible, for even within our own work, preoccupied with one method of attack, other avenues become evident at many stages. To those who may ask Why did you not do this or that? we can only answer as follows: First, it may be that the suggested approach did not occur to us. Second, if we have considered one or another line and failed to follow it, the decision was based upon the consideration that it was not consistent with the basic model that we were using. We were unwilling to abandon the particular model, even in the face of alternative suggestions, because it gave promise of allowing the desired synthesis and, above all, permitted extensions that could be cast within the framework of powerful mathematical procedures that at once become available when the properly constituted data can be supplied. As discussed in this final chapter, the model that we have proposed is, we believe, internally consistent and does provide for important extensions along the lines suggested in the theoretical consideration of evolution. We do not claim that our way to synthesis is by any means the only way, but we do maintain that it is an extremely powerful way, and one that has been demonstrated to yield important results and to have a potential far greater than that realized in the studies reported in this book.

Work in the direction of realization of some of the more ambitious aims is in progress as this book is being written. We cannot hope, even with the generous aid supplied by the National Science Foundation, to carry our work into all the areas that are open to investigation. One of the principal aims of this book is to stimulate others to think along these lines, to investigate, to criticize, and to improve upon our work. The success or failure of our efforts will be measured in some part by the degree to which this is accomplished. It was our experience, early in this work, that communication on the subject of our studies was difficult and often unsuccessful. For this, of course, we must assume the responsibility. We hope that, in spite of the evident difficulty of following some of the detailed analyses, this problem may have been effectively reduced by publication of this book. The most difficult task, it has seemed to us, has been to cast thinking about morphology into the framework of the concept of covariance, for this requires a rejection of traditional thought, in which absolute dimensions, proportions, and the concepts of shape in terms of multiple variables are uppermost. Once the concept of covariance has become an integral part of the thought process, the interpretations that have been presented will fall into place beside more usual interpretations and lose what may appear at first sight to be an antagonistic position.

Appendixes

These appendixes are designed to supplement the data for experimental work presented in the text by giving pertinent materials which would interrupt the continuity of the text, were they included in the discussions of the results. The data include notes on the nature and composition of the samples, on how data were obtained, and on taxonomy, as well as raw data in the form of measurements and special items concerning collections where appropriate.

Appendix A: *Rana pipiens*

The sample of *Rana pipiens* consists of twenty adult and subadult individuals drawn from a breeding population in central Wisconsin. Eight males and twelve females are included. The specimens were skeletonized by Dr. Nicholas Hotton III, and disarticulated skeletons were used for measurement. Five specimens were dissected as a guide to measurement, and laboratory and field studies of functions were carried out. A total of fifty measurements was used, selected from a suite of nearly one hundred made by the senior author. Measurement was carried to the nearest 0.01 or 0.1 mm., as was appropriate. Measurements are given in Table 77. For descriptions of measurements and abbreviations see Table 12 and Figures 27 and 28.

Appendix B: Sciurus niger ruviventer

Thirty skulls, including cranium and mandible, form the sample of *Sciurus niger ruviventer*. The specimens are in the collections of the Chicago Natural History Museum. The thirty specimens include both males and females and were taken from a suite of fifty specimens studied earlier (Miller, 1950). Selection was made to include only those specimens upon which the total suite of measurements was available. The sample was collected from two adjacent counties in central Michigan. Measurements were made by the junior author, and the forty-five used in the present studies were selected from an original total of 235. Measurements were made to the nearest 0.01 mm. for all linear dimensions used. They are given in Table 78. For abbreviations and descriptions of the measurements see Table 18 and Figures 30 and 31.

Appendix C: Domestic Pigeons

Five samples of twenty individuals each of the domestic pigeon were collected from a breeding population on and near the campus of the University of Chicago, on the South Side of Chicago. Collections were made during January, February, and March, 1955. Each sample is made up of adult males and females. Specimens were grouped into samples by the order in which they were obtained; the first twenty in the first sample, the second twenty in the second, and so forth. The specimens were skeletonized by Phillip Harrison and were measured by Robert DeMar. Twenty-six measurements were taken on each individual, all to the nearest 0.1 mm., as follows (measurements for the samples are described in Fig. 78 and Tables 79–83):

MEASUREMENTS ON PIGEONS

1. Sk_l Length of skull tip of beak (bone) to middle of crest on occiput, giving maximum
2. Iow   Interorbital width, minimum normal to mid-line of skull
3. Sn_l Anterior end of narial opening to tip of bony beak
4. Oc_n Height of occiput, from top of foramen magnum to mid-point of ceast at top of occiput
5. J_l Length of lower jaw, maximum
6. Co_{al} Length of coracci along anterior border
7. Co_w Minimum width of coracoid in outer aspect
8. Gl_w Maximum width of glenoid fossa
9. Sc_l Length of scapula, maximum
10. Sck_l Length of keel on sternum, maximum
11. Sck_d Depth of keel, anterior at level of anterior margin of pillar
12. St_w Width of sternum, maximum
13. Sa_w Maximum width of dorsum of sacrum
14. Hu_l Maximum length of humerus
15. Hu_{dw} Distal width of humerus, dorsal orientation
16. Hu_{tl} Length of major trochanter of humerus, tip of proximal end to end of muscle scar, distally
17. Ra_l Maximum length of radius
18. Ul_l Maximum length of ulna
19. Ul_w Width of ulna, dorsal aspect, between feather scars nearest middle of bone
20.   (This measure dropped as inaccurate during course of study.)
21. Mc_l Length of largest metacarpal
22. Mc_w Maximum depth (width) of metacarpals 3 and 4, oriented with condyle level
23. Fe_l Length of femur, maximum
24. Fe_w Distal width of femur, condyle maximum
25. TF_l Maximum length of tibia-fibula
26. Mt_l Maximum length of mid-metatarsal
27. Mt_w Minimum width of metatarsal complex

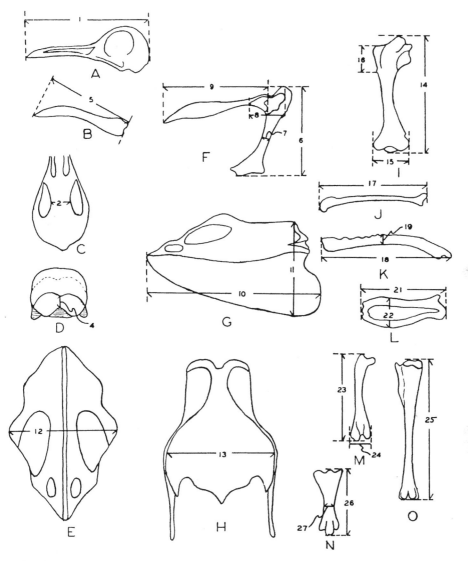

Fig. 78.—Measurements of the domestic pigeon used in studies of this species. (See p. 272 for explanation of measurements.)

Appendix D: *Knightia*

Four samples of an Eocene teleost fish, *Knightia*, were obtained from the collections of the American Museum of Natural History in New York City. Eighteen measurements, to the nearest 1 mm. or 0.1 mm., were made on each specimen by the senior author. The samples are as follows:

No.	N	Locality	Formation or Horizon
1.......	15	Fossil, Wyoming North Quarry	Green River
2.......	12	Farson Locality, Wyoming	Mid-Eocene
3.......	10	Soldier Summit, Wyoming	Mid-Eocene
4.......	11	Bayhorse, Montana	Paleocene

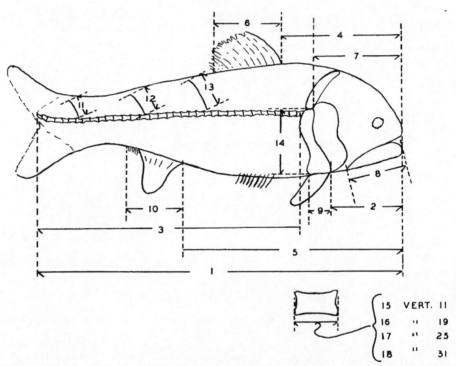

FIG. 79.—Measurements of *Knightia* used in the studies in this book. (See above for explanations.)

Specific identification of the individuals of the samples has not been made, for there is no reliable guide to assignment. It is evident, however, that all the specimens of each sample belong to one species. Although specific identity of the samples is probable, sample relationships have not been definitely established. Each sample contains a range of size that suggests that more than one age group is included, the ranges as shown in gross measurements in Tables 84–87. All specimens were laterally compressed, so that only measurements of length and depth were possible. Undoubtedly, distortion has altered dimensions somewhat from their life-conditions, but, in view of the similarity of compression, distortions appear to be comparable in all members of the samples. Only gross measurements were possible, and this fact, along with the factor of distortion, has limited the use of the samples. The following measurements were taken (see Fig. 79):

No.	Abbr.	Description
1.....	L	Distance from the anterior tip of the mandible to the posterior end of the vertebral column, measured parallel to the longitudinal axis
2.....	Mpf_l	Distance from the anterior tip of the mandible to the anterior end of the pectoral fin, measured parallel to the longitudinal axis
3.....	V_l	Length of the vertebral column from the posterior end of the opercular to the posterior end of the vertebral column, measured parallel to the longitudinal axis
4.....	Mdf_l	Distance from the anterior end of the mandible to the anterior end of the dorsal fin, measured parallel to the longitudinal axis
5...	Maf_l	Distance from the anterior end of the mandible to the anterior end of the anal fin, measured parallel to the longitudinal axis
6.....	Df_l	Length of the base of the dorsal fin
7.....	Mop_l	Distance from the anterior margin of the mandible to the most posterior point on the posterior margin of the opercular, measured parallel to the longitudinal axis
8.....	M_l	Maximum length of the mandible
9.....	Pf_l	Length of the base of the pectoral fin
10.....	Af_l	Length of the base of the anal fin
11.....	$Sp29_h$	Height of the neural spine of the 29th vertebra
12.....	$Sp21_h$	Height of the neural spine of the 21st vertebra
13.....	$Sp13_h$	Height of the neural spine of the 13th vertebra
14.....	Pg_h	Total height of the pectoral girdle
15.....	$C11_l$	Basal length of vertebral centrum of vertebra 11
16.....	$C19_l$	Basal length of vertebral centrum of vertebra 19
17.....	$C25_l$	Basal length of vertebral centrum of vertebra 25
18...	$C31_l$	Basal length of vertebral centrum of vertebra 31

Measurements are given in Tables 84–87.

Appendix E: Hyopsodus

Eight samples provide the materials of *Hyopsodus*. The specimens consist of lower jaws which include all three molars, undamaged and little worn. Measurements were made between cusps and of the lengths and widths of the individual teeth, plus the total length of the molar series (see Fig. 61). The specimens are all from the collections of the American Museum of Natural History in New York City. The numbered specimens are listed in Tables 91–98. Some, however, were unnumbered, and several specimens were entered under a single number. These have been indicated by various symbols which have been entered in the collections in New York. The specimens of each sample, with a few exceptions, were all collected from the same restricted locality, a precaution deemed necessary to insure maximum taxonomic homogeneity. This restriction, plus the necessity of excellent preservation and lack of tooth wear, has reduced the sample sizes below desirable levels in most instances (see Table 88).

The taxonomy and phylogeny of the genus *Hyopsodus* are in an unsatisfactory state and seriously in need of comprehensive revision. Such an undertaking was impossible for us, but it was necessary to undertake revision as it pertained to the species and phylogeny of the materials of our samples. The collections studied included the specimens that were the basis of the comprehensive reports of Matthew (1909) and Matthew and Granger (1915) and materials added after these studies had been completed. As far as possible, we have followed Matthew in his assignments, but, in both the taxonomy of the species and the phylogeny, departures were required. For this reason, it is necessary to present in some detail the review and revision which follow as a basis for the taxonomy and phylogeny used in chapters vii and viii of the text.

1. GENERAL, SPECIES, AND PHYLOGENY

Hyopsodus is a small condylarth that occurs in beds of Eocene age. It is abundant in the early and mid-Eocene and very reduced in numbers in the later part of the epoch. The last extensive work on the genus was carried out by Matthew in 1915. This work and his earlier study on the Bridgerian *Hyopsodus* (Matthew, 1909) have been the principal sources of published information used in the present study. It is apparent, as Matthew was well aware, that his work was by no means final and definitive. Papers since 1915 have been in large part restricted to descriptions of species from geographic areas not considered by Matthew, notation of the genus and its species in particular faunas, and discussions of the species recognized by Matthew (as Kelley and Wood, 1954).

Matthew recognized the following species in 1915:

Species	Horizon(s)
H. simplex	Sand Coulee
H. miticulus	Gray Bull
H. walcottianus	Lysite, Lost Cabin
H. mentalis	Lysite, Lost Cabin
H. wortmani	Lysite, Lost Cabin
H. powellianus	Lysite, Lost Cabin
H. paulus	Bridger B and C
H. minisculus	Bridger D
H. lepidus	Bridger C and D
H. marshi	Bridger C
H. despiciens	Bridger D

276

He considered several species, named earlier, to be variants or subspecies, as follows: *H. browni*, subspecies of *H. powellianus; H. lysitensis*, subspecies of *H. mentalis; H. minor*, subspecies of *H. wortmani;* and *H. vicarius*, subspecies of *H. paulus*. These subspecies pose an awkward taxonomic situation. The species are defined on the basis of materials referred to the named species, and the subspecies, as treated, is a variant, which, however, did not contribute to the species definition. Technically, the named species and the variant are equivalents, each a subspecies, and each should have a subspecific designation. They were not so treated, and the characteristics of the named subspecies were not considered in the definition and diagnosis of the species. We have made no effort to modify this situation except as it pertains to the samples pertinent to our studies. Two of the subspecies—*H. mentalis lysitensis* and *H. paulus vicarius*—enter into the present study. For reasons given later, it is concluded that *H. mentalis lysitensis*, on the basis of the materials assigned to it in the collections, should be considered a distinct species. *Hyopsodus lysitensis,*, as originally defined, is the name used for this material in our studies.

It is shown later, also, that the sample designated as *H. paulus vicarius* in the collections cannot be included in Matthew's definition of *H. paulus* and also cannot be included in the named species, *H. vicarius*. The specimens of this sample appear to represent an undescribed species from the Bridger B. Description of the species, without a full investigation of the genus, is unwarranted. Therefore, we designate this species, as represented by our sample, as *H.* sp. "vicarius," in which the term in quotation marks has no taxonomic significance but rather indicates that the specimens in the sample are referred either to *H. paulus vicarius* or to *H. vicarius* in the collection.

In his array of taxonomic units, Matthew lists an assemblage of specimens from the Bridger C as "intermediate mutants." These were supposedly intermediate between *H. paulus* and *H. despiciens* and mutants in the DeVriesian sense. For reference they will be called *H. paulus* "mutants."

Of the species and subspecies listed by Matthew, eight produced samples of lower dentitions suitable for the type of study that was to be carried out. Their stratigraphic positions, for the sample used, are shown in the accompanying table, with the designations as indicated in the preceding paragraphs.

SPECIES AND STRATIGRAPHIC POSITIONS OF THE
SAMPLES USED IN THE PRESENT STUDY

	H. miti-culus	*H. lysi-tensis*	*H. men-talis*	*H. powel-lianus*	*H. paulus*	*H.* sp. "vicarius"	*H. paulus* "mutants"	*H. despi-ciens*
Bridger D..........								X
Bridger C..........							X	
Bridger B..........					X	X		
Lost Cabin..........		X						
Lysite..............		X		X				
Gray Bull..........	X							

Matthew (1915) tended to agree with Osborn's earlier conclusion (Osborn, 1902) that there was a general, progressive change in the dentitions of *Hyopsodus* through the early and middle Eocene. He felt, however, that there were several phyla present and that the relationships of the species, and hence the lines of descent, were not clearly worked out. He concluded that the various species of *Hyopsodus* offered an example of evolution *in loco*. From his works of 1909 and 1915, a pattern of relationships that he seems to have considered most probable can be abstracted, as shown in Figure 80.

The eight samples treated in our work present a fair representation of what appear to be the main lines of the phyletic arrangement shown in Figure 81, but the phyla to which *H. marshi*, *H. minisculus*, and *H. lepidus* belong are not represented.

2. SPECIES CHARACTERS

Matthew summarized the species characters in a key in his 1915 publication. The four major divisions are based upon size, the development of the hypocone, and the degree of "molarization" of $P^{\underline{4}}$. A secondary character in each of the four divisions is the shape of the

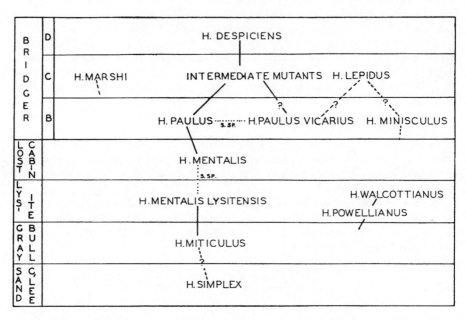

FIG. 80.—Phylogeny of species of *Hyopsodus*, reconstructed from Matthew (1909, 1915)

heel of $M_{\overline{3}}$ and involves the degree of separation of the entoconid from the hypoconulid. A tertiary character is the length of $M_{\overline{1-3}}$. Not in the key, but noted in the text, is the loss of the originally basined talonid.

Matthew mentions only in passing the small cuspule that occurs on the anterior margin of the metaconid, presumably a vestigial paraconid. This cuspule is present in the lower molars of *H. simplex* and *H. miticulus*, sporadic in occurrence in the Lysite and Lost Cabin species, and absent in the species from the Bridger. It presumably could have some taxonomic value, but its sporadic occurrence in the Lysite and Lost Cabin members makes its use difficult in precise species determination and also for measurements that involve cusps.

There are, in summary, five characters of the lower molars that may be considered of specific importance: (*a*) length of $M_{\overline{1-3}}$, (*b*) differentiation of the entoconid from the hypoconulid, (*c*) size and shape of the heel of $M_{\overline{3}}$, (*d*) presence or absence of a basined talonid, and (*e*) presence or absence or sporadic occurrence of a "paraconid."

3. DATA, ANALYSIS, AND INTERPRETATION

The measurements that were taken for our studies are shown in Figure 61. Measures are numbered from 1–10, 21–30, and 41–50 for $M_{\overline{3}}$, $M_{\overline{2}}$, and $M_{\overline{1}}$, respectively. The final digit of each number is thus the same for each comparable measure on the three molars. The digit 7 does not appear because the measure so designated was found to be unreliable. The specific characters cited by Matthew are represented by the following measures: (*a*) length of $M_{\overline{1-3}}$ (1, 21, 41); (*b*) differentiation of entoconid from hypoconulid (3, 23, 43); and (*c*) size and shape of talonid (3, 23, 43; 4, 24, 44; 5, 25, 55). In the last two instances, only the single digit numbers, those that apply to $M_{\overline{3}}$, conform strictly to his designation. No measure of the existence or degree of basining of the talonid was devised.

The basic data used in the study of taxonomy and phylogeny of the eight samples are presented in Tables 88–98.

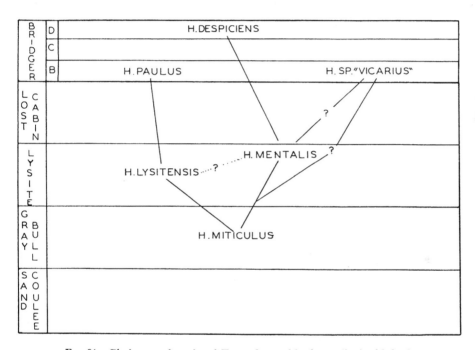

Fig. 81.—Phylogeny of species of *Hyopsodus* used in the studies in this book

The relationships of the samples to Matthew's species have been noted briefly in the explanation of the notation that we have used. The principal problems with respect to these species are (*a*) whether any of the named species, as modified in our usage, can be considered inseparable; (*b*) whether any of the assemblages are taxonomically heterogeneous; and (*c*) the status of the subspecies.

With reference to the first question, it is evident in the tables that differences between the samples do occur in the means and variances. Also, reference to other tables in the text, in chapters vii and viii, will show that differences are present in the ρ-groups, variously derived, and in the indexes. In the case of the means, based on length of lower molar series, the following do not show significant differences, with $P = .05$: *H. lysitensis* and *H. paulus;* *H. mentalis* and *H.* sp. "vicarius"; *H. mentalis* and *H. paulus* "mutants"; *H. mentalis* and

H. despiciens; H. sp. "vicarius" and *H. paulus* "mutants"; and *H. paulus* "mutants" and *H. despiciens.*

Hyopsodus paulus "mutants," so far as the sample taken is concerned, has proved to be an improper taxonomic assemblage, as discussed below and in chapter vii. Of the species that remain, *H. lysitensis* and *H. paulus* cannot be distinguished by the means, nor can *H. mentalis* be distinguished from *H.* sp. "vicarius" or *H. despiciens*. These, however, are species that can be distinguished on straightforward morphological grounds, as Matthew has shown. The "subspecies" *H. mentalis* and *H. mentalis lysitensis* differ significantly in their means, as Matthew's description implies, but they also differ in patterns of variability of molar lengths in the series $M_{\overline{3}}$, $M_{\overline{2}}$, $M_{\overline{1}}$. Confirmation of the conclusions that can be drawn from these features is found in the patterns of integration discussed in chapters vii and viii. These two species are as distinctly separated from each other as are the other recognized species in all characters studied. For this reason, they are considered distinct species, *H. mentalis* and *H. lysitensis* in the sense of the original definitions, rather than as subspecies.

The sample listed as *H.* sp. "vicarius" poses a problem. *Hyopsodus paulus vicarius* was defined by Matthew as a small variant of *H. paulus*, with a mean length of the lower molar series of 12.0. The sample used in the present study was drawn from the specimens labeled *H. paulus vicarius* in the collections of the American Museum of Natural History. There is, it is true, a significant difference between the mean length of $M_{\overline{1-3}}$ of *H. paulus*, for which our sample corresponds to Matthew's definition, and the sample drawn from the collection labeled *H. paulus vicarius*, but the latter is *larger*, not *smaller*. It is clear that the specimens of the sample do not correspond to Matthew's *H. paulus vicarius*. All the specimens in the sample, with two exceptions, are from Grizzly Buttes West, in the Bridger Basin, Wyoming.[1]

The indications of the mean length of the lower molar series, which separates it from *H. paulus*, and the patterns of variability of $M_{\overline{3}}$, $M_{\overline{2}}$, $M_{\overline{1}}$, indicate that the sample has been drawn from a species that is different from all others. Integration patterns (see chap. viii) indicate that this is not a spurious assemblage but represents a true taxonomic unit specifically distinct from the others studied. The sample appears to represent an undescribed species of *Hyopsodus*. A conventional taxonomic study is needed to determine the extent to which it is represented in the collections of the American Museum of Natural History labeled *H. paulus vicarius* and *H. vicarius*. This is not the purpose of this review. It is recognized that the specimens listed in the footnote belong to an undescribed species and, for the sake of reference, the designation *H.* sp. "vicarius" is used.

The situation of *Hyopsodus* from Bridger C, as discussed by Matthew, is somewhat confused. The sample used in the present study consists of what he called "intermediate mutants," which he considered intermediate between *H. paulus* and *H. despiciens*. The assignment of specimens as "intermediate mutants" by Matthew seems to have been based largely on size and stratigraphic position. *Hyopsodus marshi*, from the Bridger C, is noted as large, on the basis of $M_{\overline{1-3}}$ length (14.6 mm. for AMNH 11879; 15.0 for AMNH 11878). *Hyopsodus lepidus* includes the small specimens from this horizon, with the mean length of $M_{\overline{1-3}}$ of Matthew's sample = 11.7 mm. The remaining specimens were referred to "intermediate mutants." The method of assignment suggests that the association might well be taxonomically spurious. The quantitative data strongly indicate that this is the case.

The data on integration are especially pertinent. The reasoning with respect to these data is as follows. Many bonds in all the species are either between gross measures of size (1, 2, 21, 22, 41, 42) or included in large groups that involve measures of over-all size. Thus many of the bonds depend upon the factor of size. The bonds in the large ρ-groups, except those

---

[1] Collection numbers are as follows: Grizzly Buttes West, AMNH 11401, 10487, 12487, 1247*a, b, c*, 11397, 11336, 10985, 11327, 11334, 11382, 10983, 11400, 11345, 11399; Church Buttes, 11382; Cottonwood Creek, 13058.

directly between size measures, cannot be considered genetically controlled per se but only as they are dependent upon the genetic factors controlling size. Bonds not so controlled, as considered in the sections on partial correlation in the text (chap. viii), may be considered independent of the factors of size and shape, i.e., subject to other controls of genetics and to developmental influences that operate independently of size. They may be supposed to be indicative of genetic relationships not directly related to those of size, as expressed in the gross measures. The effect of size control of bonds is present in all species of *Hyopsodus* that have been studied. The removal of these effects, by partial correlation, reveals the remaining relationships. Any assemblage of specimens of the genus, regardless of specific affinities, can be expected to show arrays of bonds that are related to the size factor. In the absence of other bonds, these predominate and tend to produce a few large, size-dependent ρ-groups. A high index of integration is to be expected when the size factor is not removed. When, however, the effects of size are held constant, the value of the index should be zero, or very nearly so. This is precisely what occurs in the sample of "intermediate mutants." This phenomenon and the relatively high coefficint of variability for the length of the lower molar series (Table 88) lead to the conclusion that this is not a taxonomically coherent sample at the species level. Thus the sample is considered invalid for use in our study. It should be noted, of course, that this argument, applied conversely, adds support to the conclusion that the other units are taxonomically coherent.

In summary, the following are recognized as valid species in this study: *H. miticulus*, *H. lysitensis*, *H. mentalis*, *H. powellianus*, *H. paulus*, *H.* sp. "vicarius," and *H. despiciens*.

The status of other species has not been studied from the types of data and points of view discussed above. No comments can be made upon their standing.

4. PHYLOGENY

The phylogeny for the study of morphological integration in the evolution of *Hyopsodus* necessarily had to be established on evidence that was independent of the integrative characteristics of the species. It was hoped that Matthew's tentative phylogeny might serve, but the quantitative data, exclusive of those on integration, indicated that this was not the case. It would have been extremely fortunate, had the seven species presented an evident array of ancestor-descendant relationships in one or more phyla. The best that could be hoped for was some approximation to such relationships, one close enough that serious errors might not be introduced, as in a reasonable morphotypic series.

The evidence available for the establishment of phylogenetic relationships consists of (1) qualitative morphology, (2) stratigraphic relationships, (3) data on size, and (4) data on variability. The first and second, and in part the third, were used by Matthew. He suggested a strong general tendency toward consistent morphological change of dental patterns with time and, with some exceptions, a tendency toward increase in dental dimensions from older to younger species.

Stratigraphic considerations, of course, immediately limit the possible relationships. Size and the advanced state of the tendency toward development of a subquadrate P⁴ indicate that *H. powellianus* is not ancestral to any later species in our samples and that it probably did not arise from *H. miticulus*, although the latter cannot be definitely ruled out.

The size difference between *H. miticulus* and *H. lysitensis* is an expression of the general tendency within the genus. Cusp patterns and other dental characters provide no bar to a direct relationship but also provide no positive evidence in support of it. The same is true for the other species with permissive temporal relationships. *Hyopsodus mentalis* of the late Lysite and Lost Cabin could have come from *H. lysitensis*, and the Bridger B species, *H. paulus* and *H.* sp. "vicarius," from *H. mentalis*. The size relationships, based on the length of the lower molar series, between *H. mentalis* and *H. paulus* (Tables 88 and 90) show *H.*

paulus to be significantly smaller than *H. mentalis* ($P = .05$). An ancestor-descendant relationship would involve a reversal of general trend. This is not out of the question but does at least argue for caution in interpretation. *Hyopsodus lysitensis* and *H. paulus* and *H. mentalis* and *H.* sp. "vicarius" do not differ significantly in length of lower molar series ($P = .05$). These items are suggestive but not definitive.

The evidence of variability provides some additional clues to relationships. The data are given in Table 89. The trend of variability along the molar series, for each molar individually, is of interest. In three species, *H. miticulus*, *H. lysitensis*, and *H. paulus*, the value of V in $M_{\overline{3}}$ and $M_{\overline{1}}$ is low compared to that for $M_{\overline{2}}$. The value of V is about the same for the three molars in two species, *H. powellianus* and *H. despiciens*. Both seem to be "advanced" members of their respective phyla. In *H. mentalis*, the pattern is $M_{\overline{3}}$ high, $M_{\overline{2}}$ low, and $M_{\overline{1}}$ intermediate. This is very different from the pattern in *H. lysitensis* and *H. paulus*. The pattern in *H.* sp. "vicarius" is $M_{\overline{3}}$ low, $M_{\overline{2}}$, and $M_{\overline{1}}$ high, again different from the other species.

The data on variability strongly suggest a close relationship between *H. miticulus*, *H. lysitensis*, and *H. paulus*. The change in the series is seen in an intensification of pattern. No completely clear and reliable interpretation of the other species can be made. It seems evident that they do not belong in the *H. miticulus–H. lysitensis–H. paulus* line, although they might have come from *H. miticulus*. How they relate to each other is less certain. *H. lysistensis* and *H. mentalis* are very different, and it seems highly improbable that the latter came from the former.

If the hypothesis that an *H. mentalis–H.* sp. "vicarius"–*H. despiciens* line is entertained, the following interpretation of pattern change emerges. High variability migrated forward in the temporally successive species, for the highest V is found in $M_{\overline{3}}$ of *H. mentalis* and in $M_{\overline{2}, \overline{1}}$ of *H.* sp. "vicarius." *Hyopsodus despiciens* fails to show a high V for any molar. A stability might be thought to have arisen as, with increased molarization of premolars, the forward movement of high variability proceeded beyond the molar portion of the "molar field." Forward migration of the "molar field" does occur in *Hyopsodus*, as shown by the trend of the hypocone and the acquisition of a subquadrate fourth premolar. It is not unreasonable to assume that a shift in the pattern of variability of the individual molars might be associated with it. If this assumption is made, then the three species in question could be reasonably associated in a line. Two things should be noted. *Hyopsodus powellianius*, which is advanced in spite of its early (Lysite) age, resembles *H. despiciens* with respect to molar variability. The general tendency for forward migration of the "molar field" is present in the *H. miticulus–H. lysitensis–H. paulus* line, although rather weakly expressed, and there is no forward migration of the variability.

As is commonly the case, evidence is somewhat tenuous. The most probable interpretation seems to be as follows: *H. miticulus*, *H. lysitensis*, and *H. paulus* form a phyletic line, with no important stage missing; *H. mentalis* probably came from *H. miticulus*, but there must have been an intermediate stage. The possibility of origin of *H. mentalis* from *H. lysitensis* through an unknown intermediate stage exists, but the two are not closely related, and origin of *H. mentalis* through a different intermediate stage from *H. miticulus* seems more probable. *Hyopsodus mentalis*, *H.* sp "vicarius," and *H. despiciens* are closer to each other than to members of the *H. miticulus–H. lysitensis–H. paulus* line. They may represent a phyletic series, with stages missing, but probably they are more accurately conceived as a morphotypic series in which the members, particularly *H.* sp. "vicarius," are off the direct line of descent. This is the interpretation that has been used in the text in chapter viii. It is represented diagrammatically in Figure 81.

Appendix F: Pentremites

Six samples of the Mississippian blastoid *Pentremites* provide the basis for the studies reported in the text. Measurements (see Fig. 31 and Table 20 in text) were made by Robert Miller and by Keith Chave[1] by use of calipers and micrometer occular as appropriate. The samples are from the Renault, Paint Creek, Golconda, and Glen Dean formations of the Chester series of the Mississippian system in southern Illinois. The specimens from the Golconda were supplied by Professor J. Marvin Weller. The other samples were collected by Keith Chave and Robert Miller.

There are two basic forms of the genus *Pentremites*—godoniform, a broad-based type, and pyriform, with a slender, tapering base. These two form the basis for the species *P. godoni* (Defrance) and *P. pyriformis* Say. The two forms are readily distinguished by gross external characteristics and, although they occur in the same deposits, separation can be carried out even at relatively immature stages. For ease of reference, the two types are called "pyriform" and "godoniform" in this work, without specific reference. The oldest sample studied, from the Renault formation, is godoniform. The Paint Creek and Golconda formations have provided samples of both godoniform and pyriform types. The Glen Dean sample is godoniform. The godoniform type appears to be ancestral to both the later godoniform and the pyriform *Pentremites*. After the time at which the Renault sample existed and before the time of those from the Paint Creek, a separation into the two forms appears to have taken place. Presumably isolation was involved, but later the two came to inhabit the same environment and to exist sympatrically. The life-span of the pyriform group appears to have been short, for, to the best of our knowledge, it does not occur in post-Golconda times.

Measurements of the individuals[2] are given in Tables 100–104, and descriptions and illustrations of the measurements are shown in Figure 31 and Table 20.

[1] Dr. Keith Chave contributed importantly to the study by participating in making collections, by preparing and measuring specimens, and by contributions to the geology of the formations in which the specimens occurred. We wish to offer our thanks to him for his highly valued aid.

[2] Data for the Renault population were lost after calculation had been carried out. Thus they are not included in the tables.

Appendix G: Rattus norvegicus

Five samples of the albino rat, taken at successive postnatal stages, provide the material for study. Each sample consisted of twenty individuals. Males and females from an inbred stock were obtained and a breeding program established to assure reasonable genetic homogeneity of the samples.

Individuals were sacrificed at one, ten, twenty, forty, and approximately two hundred and fifty days after birth. Except in the case of the one-day sample, there is an error of one day in age, which is inappreciable in the type of study that was undertaken. Feeding and living conditions were kept reasonably constant for the individual during all stages of growth. Matthew Nitecki carried out the program of breeding, maintaining, and preparing the specimens under the direction of Dr. Ralph G. Johnson.

Specimens of the one-, ten-, and twenty-day samples were stained and cleared for study of the skeletons. The forty-day sample and the sample of adults were skeletonized and preserved in alcohol to minimize distortion. The stained and cleared specimens were dissected and measured with calipers, to the nearest 0.01 mm., by the senior author. The skeletonized samples were measured by Eban Matlis. A cross-check of results of measurements and techniques between the two operators was maintained, to insure that measurements were commensurate in all samples. Ossification was far from complete in the one-, ten-, and twenty-day samples. Measurements of such partially ossified elements as the limb bones, paroccipital processes, and girdles were taken to include the cartilage.

Measurements for the five samples are given in Tables 105–9. Descriptions, abbreviations, and illustrations are in Table 31 and Figure 41 in the text.

Appendix H: *Aotus trivirgatus*

The sample of *Aotus trivirgatus* was obtained from the collections of the Chicago Natural History Museum through the co-operation of Dr. Herskovitz. It was necessary to use only young adult individuals, in which the molar teeth were little worn, and only individuals in which the molar dentition was complete and the teeth were fixed in life-position. These criteria were satisfied, in spite of a collection of skulls and jaws of over a hundred individuals, only by the use of specimens from a rather wide geographic range, as shown in Table 110. Although there are undoubtedly some genetic differences in subpopulations over this range, homogeneity is certainly as high as that pertaining in the great majority of samples of fossils, upon which studies such as the one made should be of great interest. It is doubtful that the results have been seriously affected by the genetic diversity of the included individuals.

Measurements, illustrated in Figure 55, were made with calipers to the nearest 0.01 mm. by the senior author. Three upper molars, three lower molars, and the fourth lower premolar were measured. Measurements of total length of molar series, lengths and widths of molars, distances between cusps in single teeth, and distances between cusps in adjacent teeth were taken. Preliminary studies were made to determine whether bilateral asymmetry was sufficient to affect the study. Negative results were obtained, and in each case the side that had the best preservation was measured. Uppers and lowers of the same side, however, were consistently used in each individual. A check was also made for sexual dimorphism. Evident differences occur in the canine teeth, but no significant differences between males and females were detected in the molar dentitions. The two sexes were thus lumped in the sample. The measurements are listed in Table 111.

TABLE 77

MEASUREMENTS TAKEN ON SPECIMENS OF *Rana pipiens*

Measurements	SPECIMEN NUMBERS																			
	1	2	3	4	5	6	7	8	9	10	11	12	13	14	15	16	17	18	19	20
Rx1	15.4	15.0	15.5	16.4	15.4	16.3	16.5	18.0	17.0	15.9	15.9	17.5	17.3	18.1	16.4	17.6	16.4	16.6	17.5	15.9
Ocw	5.3	4.9	5.8	5.7	5.1	6.2	5.4	6.1	5.5	5.4	4.9	5.7	5.8	5.8	6.0	5.5	5.5	5.5	5.9	5.2
P2lw	4.3	4.5	4.8	4.7	5.1	5.3	4.9	5.5	4.8	4.5	4.5	5.0	4.7	5.4	5.2	5.6	5.5	5.2	4.8	4.9
Ur1	23.3	23.1	24.1	25.5	26.6	25.2	25.1	26.4	24.2	26.1	24.1	24.4	24.4	25.1	24.6	26.6	24.6	25.4	25.4	26.9
Pp1	8.7	8.7	9.2	9.3	9.3	10.9	10.2	11.4	10.4	9.4	9.3	10.2	9.9	10.3	11.2	10.8	9.4	9.5	10.8	9.4
I1	26.9	25.5	26.3	28.0	23.7	27.1	27.0	28.9	27.5	26.5	26.1	26.8	26.3	28.2	27.5	28.1	26.1	27.5	26.6	28.3
Or1	35.7	35.7	36.4	38.4	33.7	38.0	33.0	38.9	37.5	39.4	35.2	38.1	37.9	38.2	39.9	39.1	38.1	37.5	38.5	38.1
Cr_v	2.04	1.74	2.00	2.01	1.94	2.20	2.02	2.18	1.96	2.18	2.17	2.01	1.97	2.14	2.07	2.24	2.11	1.98	2.10	2.30
Ta1	17.6	17.3	17.6	17.7	19.0	19.0	18.1	20.0	18.0	18.9	17.3	19.5	19.7	19.3	20.1	19.2	18.5	18.1	18.2	18.5
Ta_v	1.11	1.10	1.30	1.25	1.29	1.22	1.24	1.36	1.35	1.24	1.33	1.39	1.32	1.44	1.39	1.38	1.57	1.32	1.32	1.29
Fe1	31.5	31.4	31.3	32.9	33.9	33.7	32.3	36.6	32.3	33.7	31.4	34.1	33.2	35.0	35.5	35.1	32.8	32.0	32.5	32.3
Fe_w	1.15	1.10	1.50	1.51	1.60	1.49	1.60	1.74	1.61	1.51	1.58	1.47	1.42	1.48	1.53	1.64	1.50	1.43	1.55	1.57
Mt1	14.1	12.9	13.1	14.1	14.6	14.6	14.0	15.6	14.2	14.5	13.6	14.4	15.3	14.5	15.2	14.7	13.8	14.0	13.8	13.7
Co1	3.9	4.0	17.5	10.5	14.6	20.3	17.6	22.3	16.8	18.0	16.8	20.5	20.8	21.1	21.0	5.3	19.8	17.8	4.2	4.5
Hu1	16.8	17.24	17.5	10.5	16.4	20.3	17.6	22.3	16.8	18.0	16.8	20.5	20.8	21.1	21.0	21.8	19.8	17.8	18.5	18.7
Hu_w	1.27	1.24	1.30	1.25	1.12	1.77	1.27	1.32	1.35	1.33	1.33	1.80	1.77	1.74	1.85	1.68	1.67	1.32	1.35	1.30
An1	9.3	10.3	10.6	11.2	11.2	11.6	9.7	12.6	10.0	9.8	9.2	11.5	11.0	11.5	11.8	12.0	11.3	10.1	9.9	10.4
An_v	1.51	1.57	1.64	1.73	1.61	2.10	1.50	2.32	1.68	1.71	1.61	1.98	1.84	2.08	2.23	2.07	1.83	1.61	1.61	1.70
Mc1	6.2	5.5	6.0	6.3	6.7	6.6	5.7	7.3	6.3	6.3	5.8	6.2	6.9	6.4	6.6	6.3	6.7	6.5	6.0	6.3
Ac_w	4.3	6.8	5.6	6.3	7.0	6.6	6.3	6.8	6.4	6.4	6.3	6.2	6.6	6.8	7.0	6.3	6.7	6.5	6.3	6.4
Sm1	3.5	3.6	4.6	4.2	5.3	4.3	4.9	4.6	4.0	4.4	4.7	4.9	4.6	4.1	4.4	5.9	4.3	5.1	4.3	4.5
FP_pt	3.1	3.3	4.6	4.5	4.1	5.0	4.9	4.9	4.1	4.4	4.7	4.3	4.3	4.1	4.4	5.0	4.8	4.9	4.3	4.1
FP_t	3.9	2.5	2.4	3.4	3.6	5.3	2.6	4.9	3.6	3.1	3.2	2.9	2.7	2.6	2.5	3.3	2.8	3.1	3.9	3.0
Sq_a	3.9	4.4	3.8	3.0	3.6	4.9	4.9	5.7	4.0	3.4	3.9	4.5	5.2	5.2	5.0	5.8	4.2	4.6	3.7	3.8
Sq_d	2.6	2.9	2.8	3.0	2.9	5.2	3.5	3.1	5.1	2.6	3.6	3.3	3.2	3.1	4.5	5.3	3.1	3.3	3.7	3.2
iaj	2.6	2.9	2.8	3.0	2.9	1.4	2.4	3.1	2.3	2.8	0.8	0.9	1.0	1.4	1.3	1.8	1.2	0.8	1.0	1.0
Cd_j	0.7	0.8	0.7	1.2	0.9	1.4	2.4	1.1	2.3	0.7	0.8	0.9	1.0	1.4	1.3	1.8	1.2	0.8	1.0	1.0
Pre1	2.0	2.3	3.3	2.1	2.2	1.4	1.8	3.7	2.9	2.8	2.9	2.2	3.1	1.8	1.7	1.7	1.7	2.1	2.1	2.8
Ja_d	2.0	1.4	1.7	3.3	3.0	3.2	2.1	2.1	1.7	1.7	3.0	3.1	3.1	3.5	3.4	3.4	3.2	3.1	3.0	2.8
FP_ld	1.3	1.4	1.8	1.5	1.3	2.0	1.8	1.6	1.7	1.7	1.6	1.8	1.6	1.8	1.7	2.2	2.5	2.1	1.9	1.5
EOint	1.3	1.3	1.7	1.5	1.3	1.3	1.5	1.6	1.7	1.3	1.3	1.5	1.6	1.7	1.7	2.2	1.5	1.5	1.9	1.5
Pro1s	3.7	3.7	1.6	3.2	3.8	4.5	3.9	4.9	4.1	3.7	3.7	4.3	4.2	4.7	4.6	4.5	4.1	3.9	4.2	3.8
Pas1c	0.55	0.72	1.14	0.35	0.67	1.42	1.01	1.51	0.83	0.59	0.93	1.16	1.17	1.60	1.31	1.05	1.05	0.90	1.22	0.73
Parc	0.55	0.72	1.14	0.35	0.67	1.42	1.01	1.51	0.83	0.59	0.93	1.16	1.17	1.60	1.31	1.05	1.05	0.90	1.22	0.73
Sca_c	1.09	1.09	1.37	1.18	1.27	1.52	1.44	1.91	1.27	1.20	1.12	1.56	1.72	1.76	1.48	1.42	1.14	1.04	1.10	1.08
Eplc_pe	4.8	4.6	4.9	4.5	5.5	5.9	4.8	6.0	5.0	4.7	4.8	5.1	5.1	5.8	5.9	5.2	5.5	4.9	5.2	5.1
St_ps	4.4	4.1	4.5	4.1	5.0	5.1	4.3	5.7	4.8	4.9	4.7	5.4	5.3	5.4	5.7	5.2	5.3	4.8	4.8	4.7
Epst_de	4.5	4.1	4.7	4.1	4.5	5.5	4.4	5.8	6.2	4.9	4.3	5.2	5.2	5.6	5.0	5.3	5.0	4.5	5.0	4.7
Hu_m	6.1	4.3	5.5	6.1	5.5	7.6	6.3	6.3	6.2	6.0	6.3	6.8	7.2	6.6	6.1	5.9	5.0	5.8	6.1	6.3
TVP3	5.2	5.8	5.6	5.3	5.6	5.8	5.1	4.2	5.3	5.6	4.8	5.7	5.7	5.8	5.6	5.8	5.7	5.8	5.5	5.4
TVP2	3.5	3.8	5.6	5.2	5.6	5.7	5.6	4.2	5.8	5.5	5.0	5.6	5.9	5.5	5.4	5.1	5.5	5.3	5.3	5.4
TVPl	5.1	3.8	5.2	5.3	3.3	5.5	5.6	5.2	5.2	5.5	4.7	5.0	5.2	3.6	3.4	3.8	2.9	4.3	5.3	5.4
TVP7	2.9	2.7	3.3	3.3	3.5	3.5	3.8	4.0	3.3	3.1	2.9	3.3	3.4	3.6	3.4	2.8	2.9	4.3	5.4	2.8
Ur_ld	12.9	11.0	13.8	12.6	15.9	12.5	10.4	15.5	12.7	14.1	12.7	13.6	15.2	13.4	14.5	12.7	12.9	14.5	15.4	13.6
P23w	4.2	4.3	4.8	4.6	4.9	5.2	4.3	5.7	4.8	4.2	4.3	5.1	5.8	5.3	5.2	5.0	4.7	5.2	5.0	4.9
P22w	4.1	4.1	5.5	4.6	5.2	5.2	4.7	4.9	2.6	4.3	4.8	4.7	4.7	4.7	5.2	5.1	4.5	5.0	5.0	4.7
I1	2.2	2.6	2.5	2.6	2.7	2.8	2.7	2.9	2.6	2.7	2.8	2.7	3.2	3.4	3.1	3.1	2.9	2.5	2.8	2.3
GrRm	4.6	4.4	4.7	4.3	5.2	4.3	5.1	5.3	4.7	4.7	4.7	4.9	4.7	5.2	5.1	5.3	4.8	4.8	5.0	4.8
Cd_d	2.1	2.1	2.0	2.4	2.6	2.3	2.2	2.3	4.4	2.1	2.4	4.9	2.1	5.2	2.2	2.4	2.1	2.8	2.3	2.2
Pz7w	4.2	4.5	4.6	4.3	4.6	5.1	4.3	5.4	4.4	4.5	4.3	4.7	1.9	4.7	4.8	4.8	4.9	4.8	4.7	42.7

TABLE 78

Measurements Taken on *Sciurus niger rufiventer*

INDIVIDUAL SPECIMENS

TABLE 79

MEASUREMENTS TAKEN ON FIVE SAMPLES OF DOMESTIC PIGEONS: SAMPLE No. 1

INDIVIDUAL SPECIMENS

Variables	1	2	3	4	5	6	7	8	9	10	11	12	13	14	15	16	17	18	19	20
Sk_l	53.3	53.0	49.0	54.2	53.1	52.6	49.5	53.3	50.7	53.2	52.8	45.9	51.6	51.9	51.2	53.6	53.2	48.9	48.6	55.7
Io_w	12.2	12.9	11.8	11.9	11.6	11.1	12.3	12.2	11.8	11.8	10.7	11.5	11.3	11.2	11.6	11.9	13.3	11.2	10.5	11.1
Sm_l	5.4	5.3	5.2	4.5	5.0	5.4	3.8	5.9	5.4	5.1	5.4	4.1	5.2	4.8	4.6	5.7	5.9	5.8	5.0	5.0
Oc_n	4.9	5.3	4.9	5.4	4.3	5.1	5.1	4.2	4.5	4.7	4.5	4.1	4.6	5.3	5.0	4.0	4.3	4.1	4.6	4.1
J_l	38.1	38.0	36.1	37.7	39.1	37.0	35.3	37.4	35.4	38.5	36.3	34.0	35.4	36.6	35.8	37.0	37.5	35.3	34.1	40.7
Co_{al}	35.4	34.8	35.3	35.8	35.0	34.0	32.7	34.7	33.2	34.2	34.8	34.1	32.6	33.4	32.4	34.7	35.9	33.3	32.1	35.9
Co_w	4.2	3.9	4.1	3.9	3.9	3.7	3.8	3.8	3.4	4.0	4.1	3.9	3.9	3.9	3.7	4.3	3.9	3.9	3.7	4.0
Gl_w	8.0	8.1	7.4	7.1	7.2	8.0	6.8	6.8	7.2	6.7	6.4	7.3	6.6	6.9	6.9	7.2	7.0	7.1	7.7	
Sc_l	45.3	45.4	45.8	43.5	44.6	41.2	43.2	44.6	42.8	45.1	44.4	42.8	43.0	44.3	42.6	45.8	46.2	44.4	42.0	49.3
Sck_l	70.6	71.8	72.8	73.3	72.1	71.5	68.5	67.8	66.2	70.1	67.4	64.9	68.9	65.1	65.1	76.4	69.2	69.5	63.9	73.6
Sck_d	28.8	30.0	28.0	30.1	29.1	28.2	27.9	28.7	29.6	28.8	27.7	28.4	28.9	28.0	31.8	30.2	29.5	27.4	29.8	
St_w	41.5	40.8	42.5	40.9	42.7	40.8	40.0	40.2	41.5	40.3	39.5	38.1	40.7	43.6	39.9	43.3	40.1	40.6	41.5	43.2
Sa_w	37.0	35.1	35.2	35.9	35.4	36.1	35.3	35.3	32.7	35.4	35.1	35.9	34.1	35.8	34.6	36.1	37.7	35.0	34.8	32.0
Hu_l	47.8	48.0	42.7	45.8	47.6	47.1	44.5	47.0	45.0	46.2	46.7	42.8	42.9	44.9	45.0	47.5	48.1	46.9	44.5	49.5
Hu_{dw}	10.8	11.0	10.9	11.2	10.8	10.7	10.2	10.9	10.3	10.8	10.9	11.0	10.3	10.5	10.3	11.0	11.4	11.1	10.9	11.3
Hu_{tl}	8.9	8.8	8.2	8.2	9.8	9.6	8.8	8.7	9.2	8.9	8.6	8.1	9.0	8.3	8.7	8.9	9.7	9.7	8.2	9.8
Ra_l	52.1	51.4	51.4	49.1	51.9	49.8	47.8	49.6	48.5	49.8	51.5	48.3	46.6	47.7	47.9	51.4	51.4	48.7	47.1	52.7
Ul_l	56.5	56.2	56.1	53.8	57.3	56.6	53.0	54.6	53.1	54.8	56.3	52.6	51.3	52.5	52.3	56.1	56.5	54.6	51.8	58.1
Ul_w	4.4	4.2	4.2	4.1	4.7	4.2	3.7	4.1	4.0	4.3	4.0	4.3	4.7	4.0	4.3	4.8	4.3	4.2	4.1	4.6
Mc_l	34.8	35.1	34.6	32.4	35.4	33.4	33.0	33.6	34.0	33.4	32.5	32.5	31.2	32.5	33.1	35.2	35.5	34.5	32.1	35.5
Mc_w	7.6	6.9	7.4	7.5	7.7	7.0	6.8	7.0	7.1	7.5	7.3	7.2	7.0	7.2	7.4	7.7	7.3	7.3	6.8	7.3
Fe_l	42.9	42.0	42.4	41.1	41.7	42.1	39.6	41.7	40.4	42.4	40.5	37.8	38.1	39.7	40.2	42.9	42.7	42.6	39.8	43.9
Fe_w	8.2	8.3	7.3	7.9	7.8	7.7	7.8	7.7	8.0	8.0	8.2	7.6	7.7	7.3	8.1	8.3	7.9	7.9	8.3	
TF_w	58.5	58.1	58.3	57.4	58.6	57.9	56.1	58.2	55.4	56.8	58.5	55.1	54.3	54.8	55.8	59.6	58.0	57.8	54.7	59.6
Mt_l	32.2	31.7	32.1	30.6	32.1	31.7	30.4	31.6	30.1	31.5	32.1	30.5	28.6	29.3	30.6	33.2	33.0	31.5	30.0	32.5
Mt_w	3.6	3.5	3.0	3.4	3.3	3.0	3.1	3.2	2.9	3.1	3.6	3.1	3.3	3.5	3.0	3.4	3.4	2.9	3.0	3.5

TABLE 80

MEASUREMENTS TAKEN ON FIVE SAMPLES OF DOMESTIC PIGEONS: SAMPLE No. 2

INDIVIDUAL SPECIMENS

VARI-ABLES	1	2	3	4	5	6	7	8	9	10	11	12	13	14	15	16	17	18	19	20
Sk_l	54.0	53.0	48.6	51.1	51.9	53.9	53.3	51.1	50.2	54.2	50.4	54.4	50.1	53.5	55.3	51.0	52.9	53.3	52.8	49.3
Io_w	12.1	11.9	10.4	10.7	10.8	11.0	11.9	10.2	10.9	11.6	10.8	11.6	10.4	10.7	12.0	12.4	11.7	11.8	11.3	11.1
Sm_l	5.2	5.1	4.7	5.0	5.9	5.3	6.0	5.2	6.6	5.6	5.1	5.7	5.1	4.7	6.5	5.1	5.4	5.8	5.8	5.9
Oc_n	4.6	5.0	4.1	5.2	4.7	4.7	5.2	4.3	4.8	4.2	4.8	4.9	4.1	3.7	6.0	4.1	4.4	4.8	4.0	5.0
J_l	38.3	37.2	34.5	36.4	36.1	38.2	37.4	36.4	35.1	38.6	34.7	39.5	34.9	35.8	37.9	36.8	38.0	37.2		
Co_{al}	37.6	34.1	32.1	32.7	35.1	35.9	36.2	32.7	35.0	35.4	33.2	36.2	33.5	36.8	34.6	34.5	34.6	34.5	34.4	35.8
Co_w	4.1	3.8	3.4	4.0	3.8	4.0	4.0	4.0	4.0	4.2	3.6	4.0	3.8	4.2	3.7	4.0	4.1	4.1	4.0	3.8
Gl_w	7.1	6.7	6.8	6.5	6.5	8.0	7.7	7.0	7.5	6.9	6.5	7.5	6.4	7.5	7.4	6.8	7.8	7.6	7.2	7.3
Sc_l	46.8	43.9	41.2	44.8	45.8	47.7	47.1	43.4	46.9	45.6	41.2	46.6	44.6	45.6	44.5	43.9	45.9	46.5	44.8	46.1
Sck_l	70.1	65.4	61.2	66.8	65.4	73.8	71.6	65.9	66.0	67.8	64.3	70.6	69.7	72.5	71.6	68.6	69.2	72.5	68.7	72.0
Sck_d	30.3	28.2	27.6	28.2	27.8	29.2	27.8	28.1	28.2	28.5	28.2	30.1	27.7	30.6	29.6	30.8	31.0	30.6	27.7	30.9
St_w	43.4	42.1	39.0	40.1	42.7	42.9	42.6	42.7	41.8	47.4	39.5	43.2	41.4	43.0	42.0	42.2	43.0	42.0	39.5	40.9
Sa_w	36.8	37.1	33.8	36.9	36.2	37.3	36.9	36.1	35.8	37.2	34.4	36.6	35.6	38.4	35.2	34.7	37.4	35.0	34.8	
Hu_l	49.1	44.6	43.8	45.5	45.6	48.3	48.2	43.7	46.4	46.7	45.0	48.1	45.8	47.4	46.4	45.6	46.9	47.4	46.3	47.3
Hu_{dw}	11.3	10.5	10.4	10.7	10.5	11.6	11.8	10.8	10.8	11.1	10.1	11.4	11.1	11.5	11.2	11.0	11.2	11.0	11.3	10.8
Hu_{tl}	9.4	9.0	7.8	8.2	9.4	9.1	10.0	9.4	10.2	8.9	8.6	9.4	8.6	8.9	8.6	9.2	9.7	8.9	9.6	
Ra_l	52.9	48.2	46.9	47.7	49.7	52.2	53.3	47.8	50.0	49.8	48.1	52.3	48.1	51.2	50.8	49.9	50.9	49.7	49.4	49.7
Ul_l	58.6	53.3	51.1	52.1	54.3	57.2	57.9	52.3	54.4	55.1	53.1	57.2	53.1	56.5	55.6	54.8	56.1	54.4	54.7	54.8
Ul_w	4.7	4.3	3.9	4.3	4.7	4.8	4.9	4.8	4.1	4.4	4.1	4.4	4.1	4.1	4.6	4.2	4.6	4.5	4.9	4.5
Mc_l	35.7	33.7	31.3	32.0	33.8	36.7	35.8	33.2	33.2	34.2	33.2	36.2	33.0	36.0	35.0	34.0	35.0	34.5	34.9	35.0
Mc_w	8.0	7.3	6.8	7.1	7.1	7.8	7.4	7.1	7.0	7.3	6.9	8.0	6.8	7.8	7.5	7.3	7.4	7.0	7.4	7.0
Fe_l	45.3	39.7	39.4	40.7	42.2	42.7	43.3	39.9	41.4	43.1	39.5	42.7	39.7	43.7	41.5	41.3	42.1	42.8	42.5	42.2
Fe_w	8.2	7.6	7.5	7.8	8.0	8.3	8.5	8.1	7.6	8.0	7.5	8.1	7.6	8.6	8.1	8.1	8.5	8.0	8.0	8.0
TF_w	61.3	58.2	54.3	55.0	57.4	58.9	58.7	54.8	57.3	57.8	55.7	59.9	56.1	59.9	58.9	68.1	57.6	58.8	58.6	58.2
Mt_l	34.3	31.8	29.0	29.1	31.9	33.7	32.4	30.8	31.2	31.6	30.3	33.7	30.1	33.1	32.0	31.2	31.6	32.0	32.6	32.2
Mt_w	3.5	2.8	3.0	3.0	3.6	3.4	3.2	3.5	3.2	3.2	3.0	3.4	3.2	3.6	3.1	3.3	3.3	3.2	3.7	3.1

TABLE 81

MEASUREMENTS TAKEN ON FIVE SAMPLES OF DOMESTIC PIGEONS: SAMPLE No. 3

INDIVIDUAL SPECIMENS

VARI-ABLES	1	2	3	4	5	6	7	8	9	10	11"	12	13	14	15	16	17	18	19	20
Sk_1	53.5	49.4	52.6	54.5	55.7	54.0	55.0	52.1	56.2	52.2	51.0	50.5	51.1	53.3	54.1	56.4	51.7	51.0	52.6	51.8
Io_w	14.8	10.6	11.3	9.9	12.8	11.9	11.3	10.9	10.9	11.7	9.6	10.4	11.5	11.8	12.7	11.9	11.1	11.7	11.3	9.9
Sn_1	5.5	4.7	4.8	4.9	5.9	5.2	4.8	4.9	6.4	5.1	5.1	4.5	5.3	4.8	5.3	5.4	4.9	4.7	4.8	5.0
Oc_n	4.9	3.8	5.5	4.6	4.7	4.6	4.6	4.0	4.3	4.9	5.4	4.6	4.8	4.2	4.6	4.7	4.4	5.6	4.6	4.3
J_1	35.5	34.6	37.3	39.7	40.1	39.1	39.0	35.5	39.2	36.3	37.8	35.6	35.4	39.0	39.0	40.6	36.8	36.8	37.6	35.5
Co_{al}	36.6	32.9	35.9	37.3	35.0	35.4	36.0	34.0	37.4	33.7	35.0	32.7	32.8	34.7	36.7	35.5	33.1	34.2	35.4	34.3
Co_w	4.0	4.0	4.1	3.9	4.0	4.2	4.2	4.4	4.0	4.2	3.5	3.5	4.1	4.0	4.4	4.4	3.6	4.1	4.1	4.0
Gl_w	7.7	6.9	7.4	7.0	7.8	7.5	8.3	7.2	7.7	7.6	6.8	7.2	7.0	8.0	8.3	8.1	7.8	7.4	7.0	7.6
Sc_1	47.0	41.9	45.8	46.1	46.8	43.9	45.2	45.0	47.8	44.6	43.8	44.1	43.0	45.1	46.2	47.9	42.3	45.1	46.8	45.6
Sck_1	76.2	65.8	68.7	71.9	72.3	66.9	72.2	66.6	74.8	67.8	68.7	66.3	67.3	67.8	73.9	70.8	65.9	68.2	70.6	68.2
Sck_d	32.2	27.7	31.1	29.6	32.0	30.4	29.3	30.1	31.0	29.8	28.1	28.3	29.5	28.6	30.8	29.7	28.6	29.7	28.3	28.2
St_w	42.4	42.3	40.8	44.5	42.4	41.2	43.1	38.6	41.5	41.0	37.1	39.4	40.2	43.2	42.7	43.3	40.5	42.1	42.5	39.6
Sa	36.3	33.7	35.9	37.2	37.5	37.2	36.7	33.5	39.2	34.1	36.7	35.0	35.4	35.9	36.4	38.5	35.5	35.2	36.0	34.9
Hu_1	50.1	44.6	46.9	49.9	48.8	46.1	47.6	44.5	48.6	45.6	46.8	44.8	43.8	47.5	48.7	47.8	44.8	45.2	46.8	47.0
Hu_{dw}	12.0	10.2	11.4	12.2	11.1	11.5	11.7	10.8	11.8	10.8	10.9	11.0	11.0	11.6	11.8	11.7	10.5	11.0	11.0	11.0
Hu_{tl}	10.0	7.4	9.5	9.3	9.8	9.9	9.7	7.8	9.8	7.6	9.0	8.9	8.4	9.4	9.8	9.5	8.5	9.0	10.0	9.3
Ra_1	52.9	46.7	51.6	53.3	53.8	50.6	50.7	48.3	52.9	49.7	49.1	47.5	47.5	51.0	51.1	51.4	48.4	47.2	51.6	49.6
Ul_1	58.8	51.1	55.7	58.5	58.9	55.3	56.1	52.6	58.1	53.7	54.5	52.1	51.7	55.8	56.4	56.7	52.8	52.3	56.8	54.3
Ul_w	4.3	4.1	5.0	4.7	5.0	4.5	4.7	4.0	4.6	4.4	3.7	4.3	4.4	4.5	4.9	4.7	4.3	4.8	4.3	4.3
Mc_1	37.0	32.0	35.4	36.5	36.0	34.6	35.0	33.0	36.5	34.5	34.0	33.1	33.1	34.7	36.3	35.1	32.8	33.2	34.8	35.1
Mc_w	7.9	6.7	7.5	7.8	8.0	7.8	7.7	7.2	7.8	7.2	7.0	7.4	7.3	7.8	8.0	7.4	7.3	7.6	7.5	7.5
Fe_1	45.5	40.4	41.7	45.3	44.4	42.5	43.8	40.1	44.8	41.2	41.0	39.8	38.8	43.3	43.7	43.7	41.2	41.0	43.0	42.3
Fe_w	9.0	7.3	7.9	8.1	8.3	8.1	8.5	7.8	8.8	8.1	7.9	7.6	8.0	8.2	8.6	8.6	7.8	7.9	8.0	8.1
TF_1	63.0	56.1	57.0	61.8	61.1	57.4	59.3	55.6	60.9	58.1	57.0	54.5	54.7	58.8	58.7	60.7	58.7	57.3	58.7	57.2
Mt_1	33.7	30.3	31.6	34.0	33.5	31.4	32.3	30.5	33.9	32.0	31.4	30.2	31.4	31.4	33.7	32.8	31.3	30.5	32.6	31.8
Mt_w	3.4	3.3	3.7	3.4	3.3	3.5	3.6	3.2	3.4	3.2	3.4	3.1	3.4	3.5	3.8	3.7	3.3	3.4	3.2	3.4

TABLE 82

MEASUREMENTS TAKEN ON FIVE SAMPLES OF DOMESTIC PIGEONS: SAMPLE No. 4

INDIVIDUAL SPECIMENS

Vari-ables	1	2	3	4	5	6	7	8	9	10	11	12	13	14	15	16	17	18	19	20
Sk_1	52.1	47.4	52.6	49.7	54.7	51.1	54.3	50.9	50.1	52.4	49.5	51.5	51.3	50.4	53.4	54.9	52.5	51.4	52.1	51.8
Io_w	11.4	11.7	11.1	11.8	11.3	9.1	11.8	11.0	11.2	12.2	11.4	8.0	10.8	10.8	10.9	13.0	11.2	11.3	9.8	9.2
Sn_1	5.1	4.6	5.4	5.5	5.2	5.0	4.8	5.1	4.4	6.5	4.8	4.9	6.0	4.8	5.7	5.5	5.8	5.6	5.5	5.0
Oc_n	4.2	4.2	4.3	4.3	4.0	3.9	5.1	3.8	5.0	5.6	4.5	5.0	4.4	3.8	5.0	5.8	4.3	4.0	5.7	4.3
J_1	37.9	33.9	37.0	32.5	39.3	35.4	36.7	35.4	34.5	37.4	34.7	36.0	38.5	38.5	38.3	39.8	36.6	38.0	36.4	37.5
Co_{al}	35.7	31.8	34.5	35.0	35.4	33.4	36.1	35.1	31.8	35.2	31.5	31.5	35.4	34.9	34.5	36.8	35.2	33.1	34.0	34.3
Co_w	4.0	4.0	4.1	4.0	4.2	3.6	4.0	4.0	3.7	4.0	3.9	3.9	4.0	3.8	3.8	4.2	4.0	3.9	4.0	4.0
Gl_w	7.6	7.1	7.7	7.5	7.2	6.0	7.0	7.4	7.1	7.5	7.2	6.7	7.8	7.6	6.8	8.1	7.0	7.5	7.3	7.0
Sc_1	46.4	42.0	44.7	43.9	46.1	42.2	46.0	46.8	42.7	44.9	42.0	42.0	46.7	44.3	44.7	46.7	42.3	44.1	43.7	45.0
Sck_1	71.9	66.9	70.1	67.7	74.7	59.0	69.4	72.4	63.3	70.6	65.3	59.0	69.9	61.3	66.8	72.4	68.0	66.0	67.3	67.8
Sck_d	30.0	28.2	31.0	29.2	30.9	27.4	29.7	30.4	27.0	30.0	27.2	26.4	31.0	28.1	30.5	30.0	31.2	27.9	29.1	
St_w	42.5	39.2	40.7	38.9	41.7	37.1	44.0	43.7	40.9	41.9	39.6	39.4	42.1	42.0	40.3	43.4	40.4	40.9	39.7	42.9
Sa	34.5	33.6	36.1	34.8	37.9	34.2	36.0	38.8	34.1	37.3	33.9	37.7	38.9	33.6	36.4	38.5	34.5	37.0	35.9	37.1
Hu_1	46.5	43.0	46.7	47.4	48.6	46.4	48.8	47.6	43.2	47.1	43.5	45.8	46.8	47.3	46.1	49.2	45.9	46.2	45.9	46.5
Hu_{dw}	11.1	10.4	11.4	11.2	11.3	10.7	11.3	10.7	10.1	11.5	10.7	11.0	11.3	11.3	10.3	11.7	10.4	10.7	10.8	11.4
Hu_{tl}	10.1	8.0	9.2	8.7	9.6	9.0	9.9	8.4	8.2	9.4	8.5	9.2	8.8	9.7	8.7	9.0	9.6	8.7	9.2	9.5
Ra_1	50.9	46.3	50.9	50.9	52.1	49.8	52.8	50.9	45.4	50.1	46.0	49.7	51.5	48.0	49.8	52.6	48.7	48.0	48.8	48.9
Ul_1	55.9	51.3	55.9	56.2	57.1	54.9	57.9	55.8	50.0	55.9	50.9	55.0	56.0	54.0	54.6	58.7	53.7	52.6	53.7	54.1
Ul_w	4.4	4.3	4.3	4.6	4.9	4.0	4.8	4.4	4.2	4.4	4.4	4.2	4.6	4.3	4.4	4.7	4.2	4.2	4.2	4.3
Mc_1	34.8	31.2	35.9	34.6	35.8	34.0	35.2	34.6	30.2	34.0	31.9	33.4	35.5	33.6	34.5	36.5	34.0	33.0	39.3	32.9
Mc_w	7.9	7.1	7.7	7.8	7.9	7.2	7.6	7.8	6.7	7.8	7.2	7.3	7.7	7.2	7.2	8.2	7.1	7.1	7.2	7.4
Fe_1	43.0	39.0	41.1	42.4	43.9	42.2	44.3	42.8	39.0	41.0	39.6	41.5	42.6	40.7	41.5	44.8	41.5	41.9	41.3	42.1
Fe_w	8.4	7.7	8.9	7.9	8.2	7.5	8.2	7.7	7.7	8.1	7.4	7.2	8.2	8.3	7.6	8.9	7.4	7.8	7.8	8.0
TF_1	60.0	53.4	55.3	58.2	60.5	58.5	61.5	59.1	53.2	58.3	53.6	56.5	59.1	57.1	58.9	61.5	57.4	57.8	57.1	68.3
Mt_1	32.4	28.3	32.5	31.8	32.8	31.3	33.1	32.7	28.9	32.0	29.0	31.4	32.7	31.4	31.2	33.0	31.5	31.0	32.3	31.2
Mt_w	3.3	3.4	3.7	3.3	3.7	3.1	3.8	3.4	3.4	3.6	3.2	3.3	3.7	3.3	3.2	3.7	3.1	3.1	3.2	3.3

TABLE 83

MEASUREMENTS TAKEN ON FIVE SAMPLES OF DOMESTIC PIGEONS: SAMPLE NO. 5

Variables	INDIVIDUAL SPECIMENS																			
	1	2	3	4	5	6	7	8	9	10	11	12	13	14	15	16	17	18	19	20
Sk_1	51.1	54.3	52.9	51.7	53.0	53.8	52.0	51.4	52.0	52.5	52.9	50.7	51.6	55.9	53.3	53.0	49.4	52.8	47.9	50.8
Io_w	12.6	11.0	10.8	11.5	12.7	12.5	12.4	10.0	11.9	10.6	11.2	9.9	10.7	10.3	10.7	12.1	10.8	11.2	11.4	7.9
Sn_1	5.1	5.5	5.9	6.1	5.2	5.0	5.9	5.0	4.9	5.3	5.3	5.1	4.9	5.8	5.0	5.6	6.1	5.1	4.8	4.9
Oc_n	4.5	4.5	3.9	5.0	4.2	5.1	4.4	4.2	4.1	5.1	4.1	3.9	3.8	3.9	5.1	5.0	4.5	4.9	4.2	4.2
J_1	35.4	39.7	38.1	39.7	37.8	39.9	37.9	39.0	37.3	39.1	37.5	35.7	36.0	41.1	37.9	37.9	37.6	37.0	34.5	35.1
Co_{al}	33.6	36.6	34.4	34.9	36.6	36.8	35.3	36.7	34.5	34.7	34.7	35.5	33.2	35.2	33.7	34.4	34.6	35.6	34.1	33.1
Co_w	4.0	4.2	4.1	3.5	4.2	4.2	4.3	4.0	4.1	4.2	4.0	3.7	3.9	4.2	3.7	3.8	3.6	3.9	4.0	3.7
Gl_w	7.9	8.1	7.8	6.8	7.1	7.9	8.1	7.8	7.5	7.1	7.6	7.8	7.3	8.2	7.0	8.0	7.0	6.9	6.7	6.3
Sc_1	44.7	48.1	44.0	45.6	45.9	46.0	46.9	45.0	43.6	44.5	43.4	45.7	43.2	47.2	43.8	44.8	44.6	46.5	43.1	38.9
Sck_1	71.3	73.9	67.2	67.5	67.3	67.5	67.5	65.9	65.4	70.5	68.8	72.4	65.8	74.9	68.3	69.9	68.2	68.0	62.9	52.5
Sck_d	29.6	31.8	29.5	28.5	28.1	31.6	30.3	29.0	30.7	30.0	30.0	30.1	27.7	29.2	29.4	29.1	30.1	30.1	28.3	22.5
St_w	41.8	46.1	41.5	39.3	42.2	44.4	41.5	41.0	41.8	41.5	39.2	38.0	36.9	44.2	40.9	41.8	39.5	43.1	34.8	29.6
Sa_w	35.9	38.5	37.2	35.4	37.3	37.8	36.1	38.0	37.1	37.5	36.8	36.4	34.8	37.2	36.3	37.1	35.7	36.7	36.2	36.4
Hu_1	45.8	48.5	45.8	46.1	47.7	48.2	48.1	47.1	45.8	48.2	45.5	47.5	44.9	48.1	46.7	46.4	46.6	48.2	43.4	44.9
Hu_{dw}	11.1	12.0	10.9	11.0	11.1	11.6	10.9	10.8	11.0	11.5	11.0	11.1	10.8	11.5	11.1	11.2	10.8	11.0	10.4	11.0
Hu_{t1}	10.1	9.4	8.5	10.0	10.5	9.3	9.9	8.6	9.1	8.6	8.9	10.1	8.4	9.5	8.9	9.5	8.9	9.4	8.1	9.7
Ra_1	48.8	52.7	48.7	50.1	51.1	53.1	51.5	50.0	50.9	49.7	49.1	50.6	48.4	50.6	49.5	49.6	49.6	51.6	46.7	46.3
Ul_1	53.8	57.5	53.9	55.8	55.8	57.8	56.3	55.1	54.9	55.8	53.9	55.7	52.9	56.1	54.1	54.8	54.1	57.0	51.0	52.9
Ul_w	4.5	4.9	4.5	4.1	4.6	4.7	4.4	4.2	4.6	4.5	4.4	4.4	4.3	4.8	4.3	4.0	4.4	4.2	4.0	4.2
Mc_1	33.6	34.2	34.3	34.5	35.2	35.1	36.0	34.5	34.5	35.1	32.6	35.0	34.0	34.3	33.7	36.1	32.5	32.9		
Mc_w	7.7	8.3	7.7	7.8	7.7	7.8	7.4	7.7	7.3	7.3	7.9	7.3	7.1	7.4	7.3	7.2	7.3	7.3	7.3	6.7
Fe_1	42.0	45.0	41.1	41.5	44.1	44.3	43.6	42.5	42.6	43.9	40.1	43.4	40.5	43.4	43.2	43.5	42.0	44.0	39.0	41.5
Fe_w	8.0	8.6	8.1	8.0	8.5	8.6	8.1	8.3	7.9	8.9	7.4	8.0	7.6	8.3	8.2	8.2	7.9	8.4	7.5	7.8
TF_1	57.6	59.8	56.9	57.5	60.1	59.6	59.0	59.6	57.7	59.1	56.6	60.0	54.4	59.9	58.1	57.6	57.0	61.0	55.6	55.0
Mt_1	31.1	32.9	32.0	32.0	33.9	32.6	31.9	32.5	32.1	32.9	30.9	32.8	29.4	32.9	31.9	31.4	32.2	34.0	30.8	30.4
Mt_w	3.3	3.8	3.5	3.5	3.8	3.6	3.2	3.4	3.5	3.6	3.2	3.0	3.1	3.2	3.3	3.1	3.2	3.0	3.4	3.7

TABLE 84

MEASUREMENTS TAKEN ON *Knightia:* SAMPLE NO. 1

Variables	INDIVIDUAL SPECIMENS														
	1	2	3	4	5	6	7	8	9	10	11	12	13	14	15
L	46	41	47	38	39	32	33	37	49	47	39	37	39	50	39
Mpf_1	13	11	14	11	11	10	9	10	14	15	12	10	11	15	11
V_1	31	28	34	27	27	21	23	25	35	33	27	26	27	35	28
Mdf_1	26	22	25	18	21	17	16	20	25	24	21	18	20	16	20
Maf_1	37	33	39	30	32	26	25	30	40	40	32	30	31	42	32
Df_1	6.0	5.7	7.1	5.0	5.5	4.0	4.9	4.0	6.8	5.6	4.9	4.3	6.5	6.6	4.3
Mop_1	15	11	15	11	11	9	9	11	14	13	11	10	12	16	13
M_1	6.0	5.2	6.6	5.1	5.2	3.9	4.2	5.0	5.6	5.5	5.3	4.8	5.5	6.5	5.7
Pf_1	1.5	1.2	1.3	1.2	1.2	0.9	0.9	1.0	1.3	1.3	1.2	1.0	0.9	1.1	0.9
Af_1	6.0	5.2	5.9	4.2	4.4	3.8	3.7	5.0	5.5	4.8	4.2	4.6	4.6	4.8	4.7
$Sp29_h$	3.8	2.9	4.1	2.8	2.3	1.7	2.0	2.0	3.2	3.1	2.7	3.0	2.6	2.9	2.8
$Sp21_h$	3.7	2.6	3.5	2.4	2.8	2.1	2.3	2.3	2.5	2.9	2.9	2.4	2.7	3.5	2.5
$Sp13_h$	2.8	2.4	3.2	2.9	3.0	2.2	2.4	2.7	3.3	3.4	2.6	2.5	2.3	3.3	2.9
Pg_h	9.7	7.0	9.1	6.5	7.4	6.4	5.8	6.5	7.9	9.3	7.6	6.5	7.0	8.1	7.9
$C11_1$	0.96	0.81	0.98	0.74	0.76	0.62	0.67	0.69	0.95	0.98	0.86	0.71	0.80	0.92	0.76
$C19_1$	1.09	0.96	1.11	0.92	0.96	0.83	0.80	0.93	1.30	1.17	0.95	0.94	0.90	1.20	0.80
$C25_1$	1.11	0.98	0.94	0.88	0.83	0.86	0.75	0.79	1.30	0.98	0.88	0.78	0.80	0.88	0.86
$C31_1$	0.73	0.67	0.67	0.57	0.57	0.53	0.53	0.56	0.83	0.64	0.58	0.42	0.66	0.84	0.48

TABLE 85

MEASUREMENTS TAKEN ON *Knightia:* SAMPLE NO. 2

Variables	1	2	3	4	5	6	7	8	9	10	11	12
L	90	80	70	71	82	80	79	73	82	75	69	83
Mpf_1	24	25	19	17	24	26	24	23	52	22	20	24
V_1	54	52	50	51	56	56	55	49	59	56	46	60
Mdf_1	39	35	31	34	39	34	37	32	35	36	34	40
Maf_1	65	65	56	59	66	64	65	61	67	59	58	69
Df_1	14.2	12.8	08.9	09.0	11.2	12.0	12.9	11.8	14.2	09.8	10.1	11.2
Mop_1	27	26	20	21	26	26	26	25	23	25	25	24
M_1	14.7	15.1	11.3	10.6	13.0	11.2	12.2	11.9	10.5	12.2	11.9	13.4
Pf_1	1.3	1.7	2.4	2.6	1.8	1.7	1.8	1.9	1.8	1.5	1.6	2.0
Af_1	13.2	8.0	10.0	9.9	11.2	8.5	12.8	9.8	11.2	11.2	8.2	11.0
$Sp29_h$	5.7	5.0	5.5	5.2	5.7	5.5	5.0	5.9	6.8	5.0	5.9	7.4
$Sp21_h$	6.5	5.8	5.6	5.5	5.1	5.6	7.0	7.1	7.0	6.2	5.3	6.8
$Sp13_h$	7.2	6.1	6.2	6.3	6.3	4.7	6.5	6.2	6.5	5.0	5.8	7.1
Pg_h	18.0	15.9	16.8	17.0	20.0	16.2	18.0	16.0	18.1	17.0	18.0	17.5
$C11_1$	1.56	1.40	1.32	1.45	1.42	1.58	1.77	1.25	1.38	1.36	1.34	1.60
$C19_1$	1.38	1.71	1.45	1.53	1.73	1.77	1.90	1.46	1.65	1.64	1.48	1.95
$C25_1$	1.86	1.46	1.68	1.72	1.85	1.91	1.85	1.63	1.80	1.86	1.31	2.06
$C31_1$	1.58	1.17	1.40	1.29	1.60	1.51	1.33	1.30	1.50	1.31	1.13	1.57

TABLE 86

MEASUREMENTS TAKEN ON *Knightia:* SAMPLE NO. 3

Variables	1	2	3	4	5	6	7	8	9	10
L	19.2	36.0	21.1	27.0	35.0	30.0	24.5	20.2	31.0	23.0
Mpf_1	4.5	11.6	6.7	8.5	9.0	9.4	6.9	5.6	7.4	5.2
V_1	12.8	23.0	14.2	17.2	24.0	21.0	16.5	13.9	20.5	14.5
Mdf_1	8.9	19.1	10.2	12.0	20.0	14.5	11.6	10.8	15.0	10.8
Maf_1	16.9	30.0	17.3	22.7	29.0	25.5	20.5	17.1	26.0	17.2
Df_1	2.8	4.7	2.7	3.8	4.0	3.1	2.9	2.2	2.7	2.6
Mop_1	5.9	12.6	6.9	9.6	11.1	10.8	7.5	6.7	10.5	8.6
M_1	2.4	5.2	3.2	4.8	5.6	4.9	4.2	3.8	4.7	3.5
Pf_1	0.55	0.85	0.65	0.72	1.30	1.10	0.85	0.56	0.99	0.55
Af_1	2.4	4.7	2.8	2.7	4.2	4.3	3.5	2.5	3.0	2.3
$Sp29_h$	1.8	2.0	1.3	2.1	2.7	2.7	2.7	1.9	2.5	2.6
$Sp21_h$										
$Sp13_h$										
Pg_h	4.0	9.0	4.1	6.7	8.6	6.7	6.0	4.0	5.5	6.1
$C11_1$										
$C19_1$	0.59	0.83	0.50	0.71	0.94	0.85	0.62	0.46	0.68	0.67
$C25_1$	0.50	0.90	0.52	0.77	1.01	0.80	0.65	0.51	0.65	0.57
$C31_1$	0.55	0.60	0.37	0.60	0.62	0.38	0.35	0.48	0.46	0.45

TABLE 87

MEASUREMENTS TAKEN ON *Knightia:* SAMPLE NO. 4

Varie-ties	I N D I V I D U A L					S P E C I M E N S					
	1	2	3	4	5	6	7	8	9	10	11
L	60	55	69	53	52	56	58	58	56	72	54
Mpf_1	20	16	21	17	14	14	17	16	15	20	18
V_1	40	36	52	39	35	38	39	41	37	51	35
Mdf_1	33	29	32	22	21	26	26	28	29	36	29
Maf_1	47	43	59	44	42	44	47	46	46	56	45
Df_1	4.6	6.7	8.3	9.1	5.9	9.4	8.4	6.7	8.4	8.1	5.4
Mop_1	21	20	21	18	17	17	21	18	20	23	19
M_1	9.3	8.6	10.0	8.4	7.2	9.3	8.9	7.5	9.1	11.3	9.8
Pf_1	1.6	1.9	2.0	2.1	1.6	1.7	1.7	1.8	1.7	2.2	1.2
Af_1	9.7	8.5	10.2	9.1	7.5	8.8	9.3	9.4	8.9	11.5	9.9
$Sp29_h$	3.9	3.1	5.7	3.3	3.2	3.5	3.6	4.2	4.1	5.4	3.2
$Sp21_h$	5.4	4.8	5.4	4.5	4.1	4.2	4.7	5.0	4.8	6.1	4.2
$Sp13_h$	3.4	3.6	4.4	3.9	3.7	3.6	3.5	4.2	4.0	4.0	3.3
Pg_h	11.6	12.5	14.7	13.4	10.5	12.4	11.2	12.0	11.8	14.1	11.0
$C11_1$	1.10	0.99	1.20	1.21	0.82	0.99	1.03	1.04	0.90	1.43	1.03
$C19_1$	1.25	1.15	1.35	1.23	1.14	1.10	1.30	1.39	1.24	1.43	1.13
$C25_1$	1.25	1.17	1.35	1.09	1.05	1.01	1.25	1.24	1.22	1.39	1.06
$C31_1$	1.18	0.71	0.98	0.94	0.79	1.02	0.85	0.90	0.88	0.95	0.70

TABLE 88

STATISTICS FOR SPECIES OF *Hyopsodus*

Species	N	$\bar{x}$	Matthew[1] Value	σ	V
H. miticulus............	19	11.18	11	0.66	5.89
H. lysitensis............	13	12.63	....	0.79	6.39
H. mentalis.............	11	13.71	12	0.84	6.13
H. powellianus..........	10	17.86	18	1.10	6.15
H. paulus..............	17	12.37	12.8*	0.33	2.59
H. sp. "vicarius"........	16	13.45	12.0*	0.90	6.69
H. p. "mutants"........	8	13.97	13.2*	1.12	8.02
H. despiciens...........	10	14.31	14.0*	0.71	4.96

[1] From Matthew (1909, unmarked; 1915, marked by *). No means were given for specimens from early Eocene. *Hyopsodus lysitensis* is Matthew's *H. mentalis lysitensis*, a subspecies for which no value was given.

TABLE 89

MEANS AND COEFFICIENTS OF VARIABILITY FOR LENGTH
OF M_1, M_2, M_3, SPECIES OF *Hyopsodus*

SPECIES	$\bar{x}$			V		
	M_3	M_2	M_1	M_3	M_2	M_1
H. miticulus.............	3.88	3.68	3.60	6.19	8.42	6.67
H. lysitensis.............	4.61	4.13	3.87	6.29	9.93	5.81
H. mentalis.............	4.96	4.47	4.28	9.07	4.25	7.00
H. powellianus...........	6.58	5.83	5.42	2.46	2.20	2.03
H. paulus...............	4.75	4.05	3.88	2.40	6.67	1.29
H. sp. "vicarius"........	4.76	4.44	4.22	6.24	9.91	9.24
H. paulus "mutants"......	4.92	4.67	4.38	9.90	8.14	6.83
H. despiciens............	5.20	4.71	4.40	5.38	5.40	6.02

TABLE 90

COMPARISONS OF MEANS OF TOOTH-ROW LENGTH IN SPECIES OF *Hyopsodus*
(Values of t and probabilities, p, of acceptance of hypothesis that $\bar{x}$'s of M_{1-3}
were drawn from the same population)

SPECIES	H. lysitensis	H. mentalis	H. powellianus	H. paulus	H. sp. "vicarius"	H. paulus "mutants"	H. despiciens
H. miticulus...... $\{$ t	3.8	7.6	17.6	5.5	7.5	6.7	9.6
$\{$ p	<.01	<.01	<.01	<.01	<.01	<.01	<.01
H. lysitensis...... $\{$ t		3.7	13.3	0.03	3.2	4.4	5.4
$\{$ p		<.01	<.01	>.05	<.01	<.01	<.01
H. mentalis....... $\{$ t			9.7	4.4	0.7	0.6	1.6
$\{$ p			<.01	<.01	>.05	>.05	>.05
H. powellianus.... $\{$ t				15.3	10.8	7.8	10.8
$\{$ p				<.01	<.01	<.01	<.01
H. paulus......... $\{$ t					3.9	6.5	6.5
$\{$ p					<.01	<.01	<.01
H. sp. "vicarius".. $\{$ t						1.2	8.3
$\{$ p						>.05	<.01
H. paulus "mutants" .... $\{$ t							0.7
$\{$ p							>.05

294

TABLE 91

MEASUREMENTS OF LOWER MOLARS OF *Hyopsodus miliculus*

INDIVIDUAL SPECIMENS

Variables	14976	14990	14994	16194	15008	14993	X_1	X_2	X_3	15004	X_4	14989	14997	15003	15000	15000	X_6	14994
1	3.81	4.18	3.78	4.07	3.88	4.33	3.39	4.13	3.38	4.22	3.64	3.74	4.12	3.93	3.69	3.91	3.78	3.80
2	2.67	3.20	2.82	3.24	2.74	3.16	2.50	3.14	2.69	2.90	2.66	2.83	2.91	2.93	2.57	3.03	3.02	2.78
3	.73	.69	.50	.72	.59	1.10	1.01	.89	.76	.75	.63	.61	.80	.80	.79	.91	.85	.60
4	1.36	1.45	1.23	1.48	1.22	1.39	1.24	1.19	1.37	1.43	1.22	1.37	1.40	1.14	1.21	1.51	1.35	1.17
5	1.21	1.28	1.21	1.33	1.17	1.03	.81	.97	1.14	1.38	1.15	1.26	1.09	1.22	1.03	1.33	1.18	1.27
6	1.15	1.12	1.06	1.30	1.12	1.25	.97	1.26	1.27	1.26	1.03	1.27	1.23	1.02	.98	1.23	1.04	1.07
8	1.41	1.57	1.37	1.35	1.30	1.47	1.38	1.58	1.31	1.51	1.30	1.66	1.68	1.53	1.34	1.61	1.42	1.40
9	1.53	1.84	1.64	1.65	1.79	1.74	1.42	1.45	1.56	1.93	1.82	1.74	1.73	1.30	1.51	1.64	1.63	1.56
10	1.65	1.63	1.48	1.86	1.41	1.69	1.51	1.74	1.61	1.58	1.58	1.83	1.97	1.47	1.50	1.82	1.59	1.67
21	3.91	4.29	3.56	3.86	3.48	4.00	3.16	3.54	3.39	4.12	3.71	3.68	3.90	3.76	3.33	3.72	3.52	3.48
22	3.33	3.39	3.17	3.59	2.95	3.40	2.82	3.24	3.04	3.22	3.36	3.05	3.36	3.21	2.94	3.34	3.18	3.00
23	.94	.98	.78	.70	.70	1.01	.84	.80	.83	1.12	.92	.76	.69	.74	.77	1.05	.85	.82
24	1.91	1.70	1.64	1.83	1.54	1.64	1.54	1.46	1.58	1.81	1.56	1.75	1.74	1.63	1.66	1.83	1.56	1.57
25	1.41	1.31	1.18	1.23	1.08	1.14	.97	1.10	1.13	1.40	1.11	1.23	1.21	1.22	1.10	1.36	1.12	1.13
26	1.57	1.47	1.23	1.29	1.24	1.38	1.23	1.32	1.13	1.46	1.20	1.28	1.36	1.35	1.27	1.43	1.24	1.27
28	1.74	1.72	1.66	1.49	1.63	1.74	1.57	1.71	1.64	1.74	1.62	1.70	1.74	1.77	1.47	1.73	1.54	1.67
29	1.80	1.80	1.56	1.80	1.65	1.75	1.61	1.53	1.66	2.03	1.89	1.68	1.93	1.56	1.56	1.86	1.86	1.58
30	1.93	1.82	1.52	1.56	1.67	1.75	1.57	1.89	1.80	1.70	1.74	1.69	1.79	1.72	1.60	1.81	1.84	1.72
41	3.71	3.74	3.47	3.80	3.33	3.86	3.25	3.42	3.28	3.93	3.58	3.90	3.73	3.80	3.36	3.91	3.43	3.43
42	2.81	2.91	2.79	2.93	2.49	2.98	2.44	2.82	2.67	3.17	2.81	2.66	3.03	2.68	2.57	2.83	2.68	2.67
43	.83	.94	.70	.57	.78	.90	.79	.56	.86	1.17	.87	.86	.76	.76	.69	.85	.94	.69
44	1.49	1.54	1.72	1.60	1.63	1.53	1.47	1.47	1.58	1.82	1.61	1.54	1.61	1.64	1.52	1.59	1.60	1.56
45	1.02	.99	1.04	1.05	1.12	1.16	.96	1.03	1.10	1.28	1.11	1.19	1.07	1.17	1.02	1.30	1.08	1.12
46	1.28	1.27	1.14	1.19	1.14	1.34	1.18	1.20	1.25	1.43	1.27	1.34	1.35	1.30	1.15	1.55	1.12	1.11
48	1.56	1.61	1.61	1.48	1.55	1.70	1.57	1.77	1.52	1.83	1.64	1.61	1.77	1.75	1.53	1.84	1.46	1.60
49	1.46	1.66	1.47	1.55	1.47	1.65	1.56	1.60	1.44	1.77	1.62	1.70	1.88	1.68	1.52	1.69	1.67	1.47
50	1.70	1.78	1.67	1.80	1.81	1.93	1.59	1.77	1.70	1.87	1.75	1.83	1.91	1.79	1.55	1.74	1.77	1.66

TABLE 92

MEASUREMENTS OF LOWER MOLARS OF *Hyopsodus lysitensis*

Variables	a	b	c	d	e	4715	14681	15621	15618	14618a	14618b	12716	4718
1	5.22	5.08	4.74	4.01	4.67	4.39	4.87	4.29	4.42	4.85	3.96	4.60	4.78
2	3.28	3.16	3.23	2.68	3.85	2.95	3.46	2.86	2.94	3.08	2.83	3.36	3.02
3	1.09	1.24	0.99	0.65	1.12	1.05	1.17	1.12	1.03	.84	.69	1.03	1.04
4	1.70	1.76	1.55	1.26	1.57	1.88	1.63	1.52	1.54	1.38	1.30	1.47	1.55
5	1.42	1.56	1.36	1.25	1.35	1.44	1.45	1.25	1.10	1.14	1.10	1.39	1.27
6	1.47	1.66	1.32	1.15	1.25	1.78	1.58	1.20	1.23	1.15	1.11	1.19	1.32
8	2.10	1.77	2.02	1.47	1.63	2.02	1.90	1.54	1.65	1.90	1.77	1.69	1.79
9	2.15	1.83	1.68	1.73	1.75	2.26	1.95	1.60	1.56	2.10	1.60	2.02	2.19
10	1.74	1.94	1.72	1.47	1.79	2.35	1.99	1.68	1.52	1.74	1.70	1.74	1.83
21	5.44	4.01	3.99	3.96	4.02	4.14	4.03	4.13	4.11	4.21	3.69	4.19	3.74
22	3.63	3.37	3.33	2.98	3.23	3.20	3.73	3.32	3.40	3.20	3.10	3.86	3.00
23	.90	1.19	.86	.77	.78	.92	1.03	1.10	.90	.93	.81	.84	.98
24	1.89	2.12	1.83	1.57	1.80	1.78	1.95	1.92	1.80	1.61	1.30	1.75	1.83
25	1.36	1.27	1.31	1.26	1.28	1.23	1.40	1.29	1.31	1.24	1.18	1.21	1.40
26	1.45	1.59	1.20	1.44	1.43	1.67	1.66	1.46	1.45	1.37	1.34	1.58	1.32
28	2.04	1.92	2.04	1.68	1.93	2.06	1.92	1.90	1.77	1.81	1.56	2.02	1.74
29	1.93	1.78	1.81	1.78	1.85	2.12	2.03	1.80	1.77	2.09	1.70	1.92	2.09
30	1.96	2.23	2.16	1.72	2.04	2.33	2.15	1.82	1.80	2.07	1.68	1.99	1.90
41	4.01	3.96	3.70	3.97	3.55	4.26	3.83	3.86	3.95	3.95	3.46	4.05	3.79
42	3.06	2.87	2.92	2.81	2.80	2.92	3.17	2.71	2.79	2.77	2.75	3.17	2.57
43	1.00	1.43	.85	.83	.83	1.02	.97	1.06	.89	1.06	.72	.95	.89
44	1.95	1.71	1.55	1.66	1.68	1.90	1.92	1.84	1.59	1.54	1.47	1.42	1.61
45	1.35	1.26	1.20	1.17	1.11	1.27	1.27	1.33	1.15	1.24	1.17	1.14	1.27
46	1.76	1.36	1.37	1.31	1.49	1.62	1.34	1.25	1.37	1.30	1.27	1.47	1.26
48	1.92	1.67	1.53	1.73	1.74	2.09	1.90	1.76	1.68	1.68	1.48	1.97	1.45
49	1.89	1.70	1.57	1.61	1.79	1.96	2.09	1.67	1.65	1.79	1.58	1.63	1.63
50	2.14	1.91	1.80	1.93	2.12	2.29	2.04	1.83	1.50	1.79	1.80	1.71	1.89

TABLE 93

MEASUREMENTS OF LOWER MOLARS OF *Hyopsodus mentalis*

	INDIVIDUAL				SPECIMENS						
Vari-ables	4704	16200	14633	12720	4102	4105	4125	14625a	14625b	16205	33
1	5.25	4.98	5.16	5.41	4.42	5.07	4.74	5.73	4.31	4.73	4.81
2	3.69	3.23	3.55	3.91	2.95	3.44	3.16	3.35	2.70	3.38	3.17
3	1.14	0.99	1.15	1.28	0.96	1.27	1.06	1.06	0.85	0.77	0.97
4	1.52	1.40	1.79	1.91	1.56	1.79	1.56	1.62	1.36	1.44	1.69
5	1.37	1.32	1.40	1.73	1.36	1.42	1.41	1.53	1.08	1.31	1.47
6	1.40	1.23	1.50	1.49	1.22	1.17	1.35	1.37	1.32	1.25	1.37
8	2.15	1.97	1.88	2.29	1.85	1.96	2.00	1.77	1.84	1.85	1.89
9	1.88	1.85	2.15	2.07	2.17	1.85	1.69	2.29	2.01	2.02	2.02
10	1.91	1.90	2.22	2.06	1.72	2.22	1.87	1.92	1.75	1.68	1.80
21	4.65	4.33	4.54	4.95	4.20	4.50	4.58	4.35	4.31	4.47	4.27
22	4.09	3.35	3.85	4.21	3.34	3.73	3.82	3.75	3.18	3.77	3.37
23	0.98	0.86	1.18	1.27	0.95	1.22	1.30	1.13	1.08	0.90	0.83
24	2.20	1.85	1.83	2.28	1.48	1.82	1.97	1.78	1.87	1.83	1.85
25	1.96	1.41	1.20	1.27	1.25	1.27	1.37	1.16	1.31	1.15	1.31
26	1.63	1.34	1.66	1.65	1.97	1.46	1.56	1.60	1.34	1.38	1.55
28	2.20	1.81	2.18	2.35	1.85	2.00	1.89	2.10	1.73	2.12	1.90
29	2.17	2.04	2.03	2.23	1.72	1.84	2.02	2.14	1.94	2.17	2.03
30	2.26	1.81	2.35	2.44	1.95	2.06	2.22	2.18	1.94	2.01	2.05
41	4.66	4.09	4.45	4.88	3.90	4.11	4.44	4.23	4.05	4.16	4.12
42	3.40	2.80	3.31	3.70	2.95	3.05	3.06	3.35	2.89	3.12	3.20
43	1.07	0.83	0.91	1.14	0.90	1.00	0.98	1.25	1.09	0.93	0.83
44	2.11	1.57	1.64	2.07	1.38	1.68	1.75	1.74	1.77	1.44	1.45
45	1.33	1.18	1.25	1.45	1.01	1.35	1.36	1.00	1.33	1.00	1.25
46	1.40	1.41	1.25	1.52	1.20	1.19	1.42	1.53	1.39	1.39	1.61
48	1.89	1.80	2.08	2.34	1.75	1.80	1.94	2.07	1.79	2.00	1.91
49	2.30	1.88	1.92	2.14	1.65	1.52	1.91	1.83	1.83	1.75	1.84
50	2.18	1.77	2.05	2.28	1.72	2.00	2.15	2.06	1.89	1.87	2.16

TABLE 94

MEASUREMENTS OF LOWER MOLARS OF *Hyopsodus paulus*

	INDIVIDUAL						SPECIMENS										
Vari-ables	10995	14564	X₁	X₂	X₃	X₄	X₅	11341	11348	11393	19217	11364	11352	10993	11387	11405	11913
1	4.87	4.77	4.93	4.71	4.68	4.52	4.77	4.85	4.80	4.65	4.95	4.95	4.76	5.07	4.90	4.72	4.49
2	3.29	3.23	3.38	3.16	3.22	3.25	3.29	3.39	3.35	3.50	3.45	3.53	3.44	3.26	3.44	3.11	3.49
3	1.07	.83	1.06	.81	.68	1.00	1.04	.90	.84	.86	.94	1.15	.91	.92	1.20	1.01	.85
4	1.42	1.38	1.73	1.30	1.59	1.34	1.65	1.53	1.41	1.30	1.62	1.67	1.69	1.47	1.67	1.68	1.57
5	1.49	1.33	1.56	1.21	1.13	1.32	1.66	1.35	1.23	1.19	1.64	1.48	1.37	1.25	1.24	1.44	1.40
6	1.41	1.36	1.49	1.35	1.51	1.30	1.45	1.43	1.36	1.42	1.44	1.53	1.42	1.40	1.53	1.63	1.38
8	1.80	1.63	1.85	1.61	1.66	1.51	1.72	1.64	1.83	1.74	1.81	1.79	1.90	1.74	1.94	1.68	1.70
9	1.82	1.82	1.92	1.85	1.99	1.89	2.08	1.75	1.63	1.79	1.97	1.92	2.17	1.93	1.88	1.98	1.90
10	1.87	1.94	1.78	1.71	1.68	1.61	1.79	1.82	1.56	1.66	2.07	1.97	1.98	2.08	1.80	1.78	1.80
21	3.98	3.93	3.95	4.10	3.77	3.83	3.99	3.78	4.24	4.19	4.35	3.86	3.95	4.28	4.15	4.25	4.30
22	3.33	3.48	3.75	3.45	3.41	3.54	3.57	3.68	3.76	3.68	3.76	3.47	3.85	3.73	3.60	3.19	3.76
23	.85	.93	.87	.93	.90	.95	1.04	.82	1.00	.97	.98	.98	1.03	.79	.99	1.09	.93
24	1.77	1.64	1.82	1.60	1.58	1.50	1.79	1.69	1.78	1.47	1.81	1.77	1.85	1.61	1.72	1.86	1.75
25	1.16	1.24	1.34	1.16	1.15	1.29	1.35	1.16	1.30	1.25	1.45	1.38	1.36	1.31	1.16	1.47	1.22
26	1.51	1.41	1.57	1.45	1.58	1.29	1.36	1.48	1.57	1.38	1.45	1.51	1.50	1.51	1.49	1.61	1.48
28	1.81	1.83	1.77	1.83	1.83	1.70	2.00	1.61	1.94	1.79	1.77	1.83	1.88	1.72	1.88	1.79	1.84
29	1.80	1.94	2.14	2.09	2.23	1.80	1.94	1.49	2.13	1.73	1.84	1.81	2.04	2.01	1.87	1.99	1.89
30	1.96	1.90	2.02	1.75	2.03	1.74	1.91	1.78	2.13	1.71	2.10	2.10	2.00	1.97	1.94	1.85	
41	3.83	3.79	3.80	3.97	3.97	3.78	3.93	3.82	4.09	3.91	3.85	3.93	3.96	3.78	3.96	3.82	3.73
42	2.95	3.08	3.15	3.00	3.09	2.95	3.04	3.15	3.21	3.15	3.18	2.93	3.29	3.10	3.23	2.75	3.12
43	.70	.85	.66	.76	.90	.80	.88	.94	.93	.85	.85	.97	1.06	.73	.88	.83	.85
44	1.54	1.51	1.52	1.51	1.49	1.47	1.47	1.54	1.68	1.45	1.68	1.53	1.90	1.46	1.66	1.85	1.55
45	1.15	1.17	1.21	1.15	1.19	1.19	1.27	1.12	1.24	1.26	1.25	1.17	1.25	1.17	1.07	1.43	1.20
46	1.29	1.29	1.29	1.30	1.43	1.21	1.27	1.28	1.54	1.27	1.32	1.40	1.57	1.24	1.34	1.38	1.31
48	1.56	1.78	1.89	1.70	1.65	1.69	1.68	1.70	1.77	1.73	1.88	1.75	1.80	1.70	1.74	1.76	1.88
49	1.70	1.85	1.85	1.86	1.82	1.66	1.63	1.46	1.90	1.68	1.69	1.76	1.87	1.84	1.67	2.14	1.73
50	1.84	1.87	1.84	1.79	1.75	1.73	1.73	1.94	2.06	1.85	1.87	2.03	2.17	1.95	1.88	2.08	1.77

TABLE 95

MEASUREMENTS OF LOWER MOLARS OF *Hyopsodus* SP. "VICARIUS"

INDIVIDUAL SPECIMENS

Vari-ables	11401	10978	12474a	12474b	12474c	11397	11336	10985	11327	11334	11382	10983	11400	11345	11399	13062
1	4.80	5.44	5.21	4.79	4.84	4.87	4.69	4.88	4.69	4.41	4.43	4.46	4.55	4.42	4.51	4.62
2	3.30	3.03	3.26	3.30	3.24	3.34	3.11	3.38	3.22	3.13	2.81	3.06	3.07	3.09	3.14	3.17
3	1.10	1.04	0.94	0.97	0.88	0.99	0.88	0.89	0.93	1.04	0.81	0.82	0.83	0.74	0.88	0.80
4	1.56	1.83	1.41	1.32	1.36	1.53	1.21	1.19	1.61	1.42	1.22	1.39	1.38	1.24	1.39	1.42
5	1.33	1.79	1.35	1.29	1.32	1.34	1.30	1.25	1.35	1.27	1.29	1.17	1.29	1.15	1.23	1.24
6	1.45	1.85	1.64	1.49	1.35	1.52	1.39	1.27	1.35	1.27	1.31	1.27	1.38	1.22	1.30	1.47
8	1.90	1.70	1.93	1.68	1.88	1.90	1.69	1.80	1.76	1.56	1.74	1.65	1.84	1.75	1.65	1.86
9	1.99	2.16	1.93	1.90	2.03	1.98	1.57	1.88	1.85	1.85	1.80	1.87	1.91	1.58	1.88	1.77
10	1.92	1.83	1.91	1.73	1.88	1.92	1.74	1.67	1.82	1.54	1.73	1.61	1.82	1.72	1.74	1.73
21	5.23	4.94	5.34	4.31	4.20	5.13	4.17	4.21	4.81	4.10	4.03	4.10	4.15	4.07	4.07	4.14
22	3.52	3.25	3.54	3.43	3.56	3.17	3.61	3.66	3.48	3.44	3.16	3.39	3.28	3.45	3.49	3.45
23	1.16	1.16	0.96	0.88	0.94	1.05	0.88	0.91	0.84	1.01	0.80	0.89	0.90	0.81	0.90	0.91
24	1.77	1.88	1.78	1.87	1.87	1.74	1.71	1.45	1.80	1.75	1.61	1.56	1.67	1.40	1.57	1.59
25	1.23	1.44	1.44	1.30	1.34	1.26	1.16	1.21	1.42	1.29	1.23	1.26	1.29	1.20	1.31	1.32
26	1.49	1.59	1.80	1.81	1.62	1.64	1.52	1.49	1.49	1.45	1.37	1.38	1.52	1.25	1.27	1.36
28	2.01	2.10	1.91	1.87	1.83	1.85	1.77	1.75	1.85	1.72	1.42	1.87	1.85	1.78	1.65	1.83
29	1.88	1.94	2.00	1.96	1.73	1.65	1.79	1.81	1.75	1.79	1.43	1.90	1.82	1.88	1.59	1.77
30	1.83	2.04	2.06	1.85	1.82	1.88	1.69	1.83	1.88	1.83	1.43	1.91	1.91	1.75	1.65	1.83
41	4.22	4.80	4.48	5.03	4.25	4.10	3.90	3.93	4.92	4.00	3.95	4.04	3.83	3.95	4.04	4.09
42	3.15	2.84	3.16	3.00	3.23	3.04	3.22	3.22	3.03	3.06	2.76	2.92	3.04	2.97	3.11	3.04
43	1.10	0.87	1.01	0.87	0.95	0.93	0.85	0.82	0.91	0.79	0.86	0.78	0.79	0.73	0.76	0.78
44	1.72	1.87	1.54	1.60	1.81	1.71	1.51	1.62	1.51	1.77	1.43	1.51	1.59	1.36	1.44	1.54
45	1.36	1.85	1.37	1.27	1.43	1.21	1.18	1.23	1.22	1.33	1.26	1.19	1.16	1.16	1.26	1.20
46	1.63	1.49	1.45	1.43	1.45	1.41	1.37	1.33	1.43	1.47	1.28	1.35	1.64	1.43	1.29	1.52
48	1.99	1.99	1.75	1.78	1.83	1.84	1.81	1.83	1.78	1.67	1.78	1.95	1.55	1.85	1.55	1.85
49	1.73	1.81	1.78	1.83	1.81	1.73	1.65	1.71	1.74	1.78	1.74	1.87	1.61	1.85	1.75	1.74
50	1.86	2.08	1.81	1.93	2.01	1.97	1.64	1.61	1.79	1.84	1.82	1.78	1.42	1.72	1.71	1.74

TABLE 96

MEASUREMENTS OF LOWER MOLARS OF *Hyopsodus despiciens*

INDIVIDUAL SPECIMENS

Vari-ables	11888	11940	11932	12486a	12486b	12483	11883	11934	11957	11892
1	4.52	4.89	5.29	5.33	5.40	5.42	5.37	5.30	5.23	5.20
2	3.01	3.25	3.47	3.43	3.68	3.57	3.36	3.42	3.54	3.64
3	1.04	0.98	1.02	1.03	0.94	1.19	1.24	1.11	1.12	1.22
4	1.37	1.44	1.34	1.59	1.66	1.39	1.57	1.65	1.76	1.70
5	1.27	1.23	1.34	1.50	1.54	1.22	1.37	1.65	1.62	1.67
6	1.28	1.37	1.36	1.51	1.38	1.42	1.56	1.45	1.72	1.44
8	1.68	2.09	1.75	2.22	1.98	2.05	1.85	2.18	2.30	2.19
9	1.76	1.85	1.89	1.97	2.18	2.09	2.01	2.00	2.07	2.07
10	1.79	1.89	1.96	2.14	2.06	1.83	1.90	1.89	2.20	2.12
21	4.18	4.49	4.59	4.85	4.83	5.03	4.70	4.85	4.81	4.82
22	3.60	3.59	3.77	3.66	3.92	4.10	3.48	3.68	3.99	3.68
23	0.98	1.04	0.95	0.85	1.02	0.97	0.85	1.03	0.90	0.81
24	1.87	1.73	1.87	1.93	1.81	1.83	1.68	1.89	1.93	1.89
25	1.29	1.26	1.54	1.50	1.49	1.28	1.44	1.39	1.50	1.39
26	1.41	1.50	1.54	1.52	1.70	1.43	1.67	1.79	1.57	1.69
28	1.80	2.01	2.11	2.22	2.20	2.12	1.85	2.15	2.15	2.20
29	1.77	1.85	2.03	2.12	2.06	2.12	1.85	2.12	2.23	2.12
30	1.98	2.06	2.24	1.97	2.39	2.14	2.10	2.12	2.22	2.21
41	4.05	4.31	4.17	4.57	4.59	4.53	4.30	4.43	4.50	4.59
42	2.82	3.16	3.19	3.25	3.44	3.63	3.02	3.32	3.44	3.21
43	1.02	0.96	0.84	0.91	0.90	1.05	0.81	1.07	0.89	0.89
44	1.80	1.70	1.50	1.82	1.76	1.94	1.62	1.92	1.64	1.81
45	1.42	1.35	1.35	1.43	1.35	1.33	1.25	1.30	1.20	1.34
46	1.43	1.57	1.42	1.43	1.65	1.67	1.48	1.50	1.60	1.47
48	1.69	1.92	1.85	1.98	2.05	2.10	1.75	1.89	2.09	2.06
49	1.72	1.94	1.89	2.12	2.17	2.08	1.82	2.00	2.06	2.17
50	1.89	2.01	2.06	1.98	2.60	2.20	2.01	2.01	1.90	1.96

TABLE 97

MEASUREMENTS OF LOWER MOLARS OF *Hyopsodus powellianus*

INDIVIDUAL   SPECIMENS

Vari-ables	12707	14628	15614	15613	12709	14976	15622	15612	14617	4
1	6.14	6.37	6.70	6.65	6.25	7.72	6.18	6.45	7.03	6
2	4.06	4.23	4.30	4.04	4.07	4.80	4.38	4.32	4.41	4
3	1.26	1.50	1.15	1.08	1.17	1.64	1.09	1.07	1.56	
4	2.03	2.02	2.05	1.87	1.63	2.25	2.12	2.15	2.45	1
5	2.24	1.70	1.80	1.74	1.60	2.14	1.83	1.75	2.05	1
6	1.67	1.74	1.80	1.64	1.72	2.05	1.77	1.73	1.82	1
8	2.56	2.63	2.76	2.21	2.38	2.95	2.71	2.31	2.55	2
9	2.41	2.37	2.72	2.68	2.44	3.20	2.44	2.60	2.74	2
10	2.21	2.30	2.46	2.04	2.31	2.90	2.28	2.10	2.34	2
21	5.27	5.44	6.10	6.09	5.40	6.28	5.71	6.14	6.16	5
22	4.76	4.65	4.74	4.71	4.50	5.29	4.55	4.87	5.00	4
23	1.38	1.39	1.30	1.05	1.18	.95	.90	1.26	1.27	1
24	2.28	2.18	2.32	2.38	2.28	2.80	2.33	2.35	2.55	2
25	1.60	1.55	1.80	1.91	1.65	2.51	1.65	1.99	2.07	2
26	1.66	1.79	1.95	1.72	1.98	2.10	1.90	2.11	2.13	2
28	2.45	2.64	2.79	2.19	2.56	3.31	2.52	2.42	2.69	2
29	2.57	2.50	2.73	2.24	2.64	2.94	2.56	2.38	2.79	2
30	2.58	2.51	2.69	2.17	2.46	3.04	2.54	2.49	2.71	2
41	4.99	5.10	5.59	5.53	5.02	5.96	5.31	5.76	5.59	5
42	3.93	3.90	4.30	3.84	3.90	4.70	4.30	4.35	3.99	4
43	1.10	1.36	1.18	0.85	1.15	.84	.84	1.31	1.20	1
44	2.03	2.17	2.15	1.74	2.25	2.55	2.25	2.33	2.38	2
45	1.45	1.45	1.55	1.38	1.50	1.91	1.75	1.91	1.94	2
46	1.90	1.96	2.23	1.90	1.80	1.90	1.81	1.84	1.93	2
48	2.48	2.72	2.67	2.24	2.60	3.21	2.57	2.43	2.61	2
49	2.42	2.38	2.59	2.34	2.55	3.13	2.19	2.47	2.50	2
50	2.47	2.55	2.68	2.38	2.67	3.30	2.66	2.70	2.67	2

TABLE 98

MEASUREMENTS OF LOWER MOLARS OF *Hyopsodus paulus* "MUTANTS"

INDIVIDUAL   SPECIMENS

Vari-ables	11903	12485a	12485b	12483	11952	12484	11917	11916
1	4.91	4.20	4.24	5.42	5.11	5.40	5.07	5.03
2	3.21	2.30	2.34	3.62	3.17	3.55	3.35	3.31
3	1.08	0.74	0.84	1.08	0.84	0.94	0.86	0.98
4	1.45	1.47	1.18	1.54	1.41	1.43	1.56	1.49
5	1.13	1.23	1.19	1.35	1.26	1.29	1.33	1.35
6	1.45	1.36	1.17	1.62	1.36	1.38	1.39	1.45
8	1.87	1.64	1.78	1.95	1.97	1.95	1.93	1.85
9	1.96	1.77	1.84	1.93	2.01	1.93	2.11	1.91
10	1.79	1.64	1.56	1.76	1.76	1.80	1.75	1.77
21	4.65	4.36	4.19	5.01	4.35	5.17	5.00	4.65
22	3.45	3.17	3.14	4.07	3.40	4.02	3.72	3.44
23	1.03	0.67	0.74	0.97	0.98	0.93	0.95	0.84
24	1.71	1.52	1.45	1.83	1.61	1.79	1.74	1.53
25	1.27	1.11	1.00	1.39	1.04	1.32	1.39	1.16
26	1.41	1.34	1.18	1.68	1.44	1.32	1.46	1.53
28	1.91	1.90	1.84	2.17	2.18	1.86	2.14	1.85
29	1.86	1.77	1.85	2.04	1.82	1.85	2.13	1.95
30	1.94	1.81	1.45	1.89	1.75	1.87	2.04	1.79
41	4.38	4.00	4.13	4.72	4.12	4.80	4.50	4.40
42	3.16	2.95	2.74	3.65	2.85	3.67	3.24	3.29
43	0.91	0.77	0.74	1.06	0.79	0.97	0.86	0.90
44	1.74	1.38	1.43	1.94	1.62	1.81	1.75	1.59
45	1.29	1.04	0.99	1.48	1.03	1.29	1.34	1.24
46	1.53	1.41	1.14	1.55	1.17	1.38	1.46	1.28
48	1.89	1.92	1.78	2.02	1.82	1.98	1.94	1.95
49	1.84	1.82	1.84	2.05	1.67	2.02	1.90	1.78
50	1.80	1.81	1.64	1.86	1.65	1.91	1.85	1.86

TABLE 99

STRATIGRAPHIC CHART OF THE CHESTER SERIES OF ILLINOIS*

SERIES	FORMATION	
	Southwest Illinois (Monroe Co.)	Southeast Illinois (Hardin Co.)
Chester Series	Glen Dean fm.†	Glen Dean fm.
	Hardinsburg ss.	Hardinsburg ss.
	Golconda fm.†	Golconda fm.
	Cypress ss.	Cypress ss.
	Paint Creek fm.†	Paint Creek fm.
		Bethel ss.
	Yankeetown ss.	Renault fm.
	Renault fm.†	
Meramec Series	Aux Vases ss.	Ste. Genevieve ls.
	Ste. Genevieve ls.	
	St. Louis ls.	St. Louis ls.

* Modified from Swann and Atherton (1948).
† Formations sampled for study.

TABLE 100

MEASUREMENTS OF SAMPLE OF PAINT CREEK GODONIFORM *Pentremites*

INDIVIDUAL SPECIMENS

Vari-ables	1	2	3	4	5	6	7	8	9	10	11	12	13	14	15	16	17	18	19	20	21	22	23	24	25	26	27	28	29	30	31	32	33	34	35	36	37	38	32	40
1	20	22	23	16	23	25	19	17	30	25	17	24	15	13	14	15	14	16	20	22	23	20	20	18	19	19	22	12	24	23	15	16	14	21	21	15	20	22	16	24
2	15	19	19	18	15	19	19	12	23	21	12	18	11	11	9	11	10	11	11	18	20	16	15	9	13	15	17	6	19	20	15	13	8	17	15	15	8	17	13	15
3	36	33	40	33	34	38	11	31	33	12	29	37	25	22	29	21	25	26	32	40	37	33	35	35	32	32	34	20	32	40	25	27	25	38	33	8	27	32	8	39
4	27	32	30	33	34	34	28	26	45	39	29	36	25	22	27	20	25	26	32	39	37	30	30	35	28	32	34	18	32	34	25	28	38	38	28	25	37	29	27	33
5	19	20	21	15	50	24	20	20	38	25	18	50	21	15	16	26	25	15	19	30	51	37	20	19	19	40	25	13	23	26	17	18	17	28	16	15	18	22	18	24
6	44	40	46	36	50	36	36	36	38	45	35	50	31	33	37	26	33	35	53	53		37		36	39	40	25	23	32	45	39	33	30	34	40	15	37	37	33	47
7		9,8	14,9	7,0	16,1	11,0	14,5	9,2	21,2	14,3	9,0	19,4	6,0	6,9	8,8	6,1	7,8	7,0	11,1	18,0	13,0	9,0	12,5	8,9	11,9	7,3	12,7	5,1	15,7	13,9	9,5	6,5	5,9	14,6	12,8	5,7	10,4	14,1	33	12,6
8	9,8	11,9	30			32	27	20	36	33	18	36	19	17	20	18	18	16	22	31	28		26	17	24	18	28	15	30	28	16	17	16	26	21	15	21	29	17	26
9	22	24	15	15	17	18					15	11	13	13	10	8	7	6						10	5		11			9							8		10	15
10	10	11	30	17	30	6	62	5	66	5	5	64	5	4	5		40	36	52	6	60	46	50	48	48	46	11	30	54	6	46	42	38	60	6	38		52	42	54
11	48	52	58	36	58	12	12	24	46	10	43	46	24	27	21	24	28	21	28	24	12	46	12	10	10	26	60	8	44	58	32	10	24	40	32	20	44	51	10	54
12	11	11	12	10	13	21	12	10	66	10	13	28	10	13	11	38	7	36	52	10	60	39	38	42	31	11	39	18	11	32	8	42	13	16	11	12	30	37	42	38
13	28	36	37	28	39	12	62	28	7	42	18	46	24	15	23	24	40	11	28	10	12	33	32	25	40	26	32	27	17	6	13	10	8	46	40	24	7	37	3	11
14	49	8	40	11	36	45	32	10	33	6	35	12	52	48	40	34	37	8	34	10	37	33	60	45	40	58	53	75	90	74	93	90	54	101	90	64	38	95	52	36
15	90	90	110	46	79	88	100	74	104	109	66	102	71	48	81	57	104	64	76	100	122	80	60	60	91	58											61			
23	5,8	6,7	8,0	4,5	9,3	7,3	8,6	6,0	11,3	7,8	3,7	10,3	1,6	4,8	5,8	1,8	4,8	4,9	8,4	10,2	7,3	5,0	2,1	5,8	7,5	5,0	7,9	5,1	8,4	8,0	1,6	3,9	1,8	5,2	6,0	2,7	6,1	8,5	3,9	7,8
24	2,5	2,1	2,5	1,9	2,2	2,5	2,3	1,7	3,0	2,8	2,6	3,5	2,6	1,6	1,8	2,0	2,3	1,9	1,6	2,6	2,7	3,0	2,0	1,7	2,8	2,0	2,1	2,1	2,6	3,2	2,1	2,1	1,8	2,6	2,3	2,4	2,0	2,7	2,1	2,9
25	4,0	3,3	3,3	2,7	3,9	4,8	2,9	2,8	4,1	3,1	3,6	5,0	3,0	2,0	2,6	3,7	2,8	3,5	2,5	3,0	3,2	2,8	4,2	2,2	2,8	4,8	4,5	4,5	4,2	5,0	3,6	2,9	3,8	3,6	3,4	5,8	2,4	5,1	2,9	4,1
26	2,5	4,5	4,2	3,9	5,9	4,8	4,4	3,9	6,9	5,4	4,9	5,8	3,8	4,0	4,2	4,2	3,0	3,5	3,8	5,1	5,1	4,9	3,1	5,0	3,0	4,2	4,5	4,5	5,2	7,1	3,6	4,9	5,3	3,8	4,9	5,8	2,6	5,4	4,9	4,1
27	3,6	3,0	3,0	2,8	6,7	3,5	3,0	2,9	5,0	3,4	3,6	4,1	2,4	2,7	2,8	4,1	3,0	3,0	2,5	3,1	3,2	2,9	3,1	3,0	3,0	3,3	2,9	3,6	3,2	5,0	2,2	4,3	4,2	2,7	3,0	4,1	2,6	3,6	3,3	2,8
28	3,4	4,6	4,9	4,0	7,6	5,5	4,5	5,2	7,9	5,2	3,1	6,4	3,3	3,3	4,1	4,1	4,5	4,5	5,6	4,7	6,9	4,6	5,1	3,9	5,1	4,3	5,6	4,3	5,6	4,3	3,6	4,7	3,6	3,6	5,5	4,7	3,7	5,9	4,7	4,8
29	3,1	6,2	7,6	0,8	4,1	7,4	6,1	3,9	11,7	8,0	2,1	10,0	1,5	0,7	3,4	0,8	2,8	1,1	1,9	4,7	6,9	4,2	5,3	2,1	5,1	1,5	5,6	0,8	7,9	6,0	1,9	1,6	2,2	7,6	4,3	1,6	4,1	7,5	2,0	5,9
30	3,6	3,5	4,1	0,8	4,1	1,7	2,7	1,7	6,4	6,4	1,8	5,1	1,5	0,6	1,5	1,5	2,8	0,6	4,7	4,7	6,9	4,2	2,9	2,8	2,9	2,9	2,9	0,8	7,9	2,9	2,5	0,8	2,5	2,2	2,3	2,3	4,1	4,0	0,8	2,5
31	12,3	13,7	16,9	9,0	18,3	15,7	17,1	10,9	23,6	16,7	10,3	22,0	8,5	9,2	11,0	8,8	10,2	9,2	14,8	20,0	15,3	11,0	15,1	11,5	11,4	10,0	15,0	7,1	16,2	16,0	11,9	8,3	8,8	16,7	11,6	8,9	12,4	16,3	8,3	14,4
32	11,0	14,1	16,1	9,0	17,7	16,9	13,9	11,0	23,2	16,1	10,3	20,0	8,0	0,7	9,1	8,0	8,8	9,6	11,1	17,5	15,3	10,5	13,1	13,0	13,0	9,4	13,9	5,9	16,9	14,1	11,1	7,7	8,7	17,0	13,0	7,5	11,3	15,4	7,7	13,9
34	6,8	7,1	7,8	5,8	9,1	8,5	6,3	6,1	11,0	8,7	5,6	10,2	5,0	5,1	5,6	5,9	5,5	5,0	6,1	8,7	8,0	6,0	7,0	6,1	6,5	6,0	8,0	4,3	8,7	7,9	5,6	4,9	5,7	8,1	7,1	5,5	6,4	8,8	4,9	7,0
35	29	38	45	23	44	43	43	27	64	46	27	51	21	20	23	23	24	24	34	49	37	28	40	28	36	25	40	14	43	38	33	21	19	44	35	17	30	42	21	36

300

TABLE 101

MEASUREMENTS OF SAMPLE OF PAINT CREEK PYRIFORM *Pentremites*

INDIVIDUAL SPECIMENS

Variables	1	2	3	4	5	6	7	8	9	10	11	12	13	14	15	16	17	18	19	20	21	22	23	24	25	26	27	28	29	30	31
1	18	20	17	21	24	20	22	21	20	15	25	24	18	18	23	18	15	20	27	27	16	18	20	18	22	16	15	22	17	16	22
2	12	17	10	16	20	12	16	15	13	10	15	16	12	12	21	10	9	16	15	20	10	15	16	12	19	12	11	11	12	12	17
3	9	10	8	10	7	8	16	9	9	8	9	16	9	11	9	9	9	8	9	11	10	7	9	10	19	7	9	10	10	8	9
4	33	35	28	30	32	33	32	34	33	28	32	36	27	30	39	30	27	32	32	33	29	28	30	27	33	28	25	35	32	24	34
5	31	30	26	33	22	27	32	28	33	21	35	34	29	29	32	25	26	29	30	30	26	23	35	25	35	23	28	24	33	26	30
6	17	17	17	22	17	17	11	19	18	11	19	22	19	18	21	25	17	22	22	13	13	15	17	17	22	15	16	20	17	16	25
7	11	40	36	40	40	36	41	40	36	29	40	45	36	33	42	36	34	39	33	38	31	33	40	34	40	36	32	32	17	32	11
8	10.9	12.0	5.9	13.5	12.6	9.0	9.6	10.1	6.0	5.9	9.0	11.7	7.6	8.5	10.0	6.5	6.2	11.2	8.9	12.1	6.2	8.0	12.2	9.0	11.4	7.9	6.1	9.9	10.6	6.9	10.9
9	21	24	14	21	24	22	20	20	19	6	19	27	18	18	21	20	18	21	21	24	14	19	22	20	23	17	16	20	22	15	22
10	10.5	9	7	8	15	5	13	10	11	6	12	10	6	11	6	12	11	16	11	12	7	11	16	5	10	10	5	12	15	9	11
11	5	11	5	5	5	5	7	7	6	6	7	9	6	11	6	6	6	6	6	7	6	6	7	7	6	5	6	6	9	9	6
12	52	50	34	46	54	46	50	50	46	38	42	58	46	50	26	48	40	48	46	52	40	44	50	44	51	42	40	48	50	40	54
13	11	11	6	12	12	10	11	11	10	10	31	13	11	12	23	12	10	12	12	12	12	11	12	10	12	10	10	10	12	10	12
14	30	27	19	32	30	27	31	31	21	20	31	36	24	27	26	36	24	26	24	31	20	22	30	26	36	21	28	29	30	23	30
15	5	6	11	9	8	10	10	17	29	23	45	12	17	10	8	36	40	15	13	13	15	6	6	16	6	9	6	15	17	10	10
20	30	50	83	28	42	33	30	34	29	35	50	60	55	37	40	40	40	30	40	42	26	50	63	37	40	38	39	67	36	40	36
21	72	64	8	90	92	60	66	34	69	50	50	60	55	80	64	82	83	70	60	68	75	66	29	29	99	70	73	51	54	66	74
22	8.4	8.1	5.0	10.4	8.0	6.8	7.8	8.1	6.9	5.5	8.0	8.7	5.5	6.8	6.6	6.0	6.1	7.6	7.3	9.0	5.6	5.6	8.9	7.3	11.1	8.0	5.8	8.0	8.0	6.0	8.6
23	2.0	2.8	2.1	2.8	3.0	2.1	2.0	2.2	2.1	1.8	2.0	2.1	1.8	2.0	3.1	2.0	2.4	2.6	7.3	2.4	2.6	5.6	2.6	2.0	1.6	2.0	1.6	2.0	2.7	2.4	2.0
24	2.3	3.1	2.7	3.1	3.8	2.5	2.6	2.9	2.5	2.3	2.5	2.8	2.5	2.7	4.0	2.5	3.0	3.0	3.1	3.1	2.7	2.9	3.1	3.0	2.5	6.1	2.5	2.5	3.0	3.0	2.7
25	5.1	6.1	5.1	4.5	4.6	5.9	4.1	4.3	5.0	4.3	5.0	5.7	4.0	4.2	7.8	4.8	5.3	3.0	3.1	3.1	3.4	5.6	5.7	3.3	3.3	4.7	3.5	3.5	5.2	5.7	4.1
26	4.1	5.0	3.6	4.5	4.6	3.9	3.3	3.0	4.4	3.5	3.6	4.3	2.9	3.2	6.4	4.3	3.8	4.6	4.7	4.0	3.5	4.1	4.0	3.1	3.1	3.5	3.4	3.5	3.8	4.1	4.0
27	2.1	5.1	1.8	6.0	6.0	3.3	3.9	3.1	3.3	3.6	4.0	3.7	3.7	3.5	5.2	2.1	4.3	4.6	4.8	4.0	3.4	4.4	5.0	4.4	3.1	2.4	4.4	3.0	4.7	4.8	3.8
28	2.1	5.1	1.8	6.3	5.9	3.3	3.9	3.0	4.4	4.6	3.0	5.0	2.5	3.5	3.8	2.1	2.7	3.8	4.8	4.0	3.5	4.1	5.1	4.4	6.2	2.4	4.4	3.0	4.7	4.8	4.0
29	2.1	2.0	1.8	2.7	2.2	1.5	1.3	3.1	1.7	1.1	3.0	2.0	2.5	3.5	4.5	1.1	1.1	2.1	1.4	4.0	1.3	3.0	5.1	1.3	6.2	2.4	1.4	3.0	3.2	1.8	3.8
30	0.8	2.0	0.9	2.7	2.2	1.5	1.3	1.5	0.6	0.5	1.1	2.0	1.1	1.2	1.5	1.1	1.1	1.4	1.4	1.8	0.3	1.2	5.7	0.5	2.9	1.1	0.4	1.1	1.9	0.8	2.0
31	16.8	18.1	10.8	19.1	17.7	15.6	11.9	11.5	11.9	10.3	14.9	18.1	11.9	12.5	14.1	12.9	12.9	16.1	11.4	16.0	10.9	12.4	18.2	11.4	19.5	13.5	10.7	15.9	15.9	11.3	16.0
32	11.0	13.1	7.9	12.8	13.5	10.3	10.6	11.3	9.0	7.6	11.0	13.3	8.8	10.3	11.7	9.6	11.8	11.4	11.8	12.9	7.5	9.9	13.1	10.6	11.0	9.3	8.0	11.0	12.0	9.2	10.9
34	9.4	10.0	6.7	9.5	9.5	9.4	8.1	11.3	6.6	6.9	9.5	10.0	7.0	8.0	8.0	7.8	7.0	8.8	10.0	9.4	6.6	7.8	9.5	8.7	9.5	8.1	6.6	8.9	12.0	7.4	8.1
35	24	28	14	29	28	26	25	22	19	16	19	29	18	18	23	17	22	27	20	28	16	19	27	20	34	19	12	19	23	17	27

TABLE 102

Measurements of Sample of Golconda Godoniform *Pentremites*

| | | | | | | I N D I V I D U A L | | | | | | | | S P E C I M E N S | | | | |
Variables	1	2	3	4	5	6	7	8	9	10	11	12	13	14	15	16	17	18
1	15	15	13	15	14	20	17	15	20	14	20	11	18	19	20	19	17	20
2	8	12	9	11	14	11	12	11	15	8	14	9	11	15	15	14	13	14
3	8	7	7	7	6	7	9	9	8	9	6	4	9	9	7	8	6	8
4	37	27	30	32	30	30	40	32	40	32	32	26	38	39	44	38	33	38
5	27	22	22	25	25	29	30	22	30	26	28	30	29	32	28	29	22	28
6	14	14	16	15	15	12	13	20	19	18	22	18	16	24	17	15	14	17
7	38	33	31	37	37	42	42	35	40	38	42	40	38	32	45	46	45	50
8	6.0	5.6	5.6	5.2	5.2	5.5	6.8	7.4	8.3	9.2	8.3	7.9	8.2	8.0	9.2	8.8	8.2	9.
9	20	17	19	17	20	16	18	20	18	24	21	18	23	20	25	24	22	22
10	10	7	8	9	11	10	8	8	9	9	10	8	8	10	10	10	10	12
11	5	5	5	5	4	4	5	5	5	4	5	4	5	5	5	5	5	5
12	2.7	2.5	2.9	2.9	3.2	1.9	2.9	3.2	3.0	3.0	2.8	2.8	3.2	3.2	3.1	3.0	2.9	3.
14	13	12	11	10	10	12	10	15	11	12	11	8	11	11	15	15	15	15
15	11	18	23	20	30	24	24	22	28	26	25	24	25	29	25	29	30	31
20	8	11	14	11	7	14	12	5	10	4	5	10	7	7	3	5	2	6
21	55	41	36	50	35	46	40	48	34	38	40	24	33	35	20	47	32	36
22	94	80	118	66	75	82	93	76	105	90	70	91	71	61	96	60	96	90
23	4.9	4.4	4.5	4.1	2.9	4.8	4.9	5.9	6.2	5.5	5.5	6.5	5.5	5.1	6.0	4.9	7.2	5.
24	1.2	1.5	1.3	1.5	1.5	1.5	2.1	2.0	1.9	2.1	2.0	1.5	2.0	1.8	1.5	1.8	2.2	2.
25	1.8	1.9	1.8	1.8	1.9	2.2	2.1	2.2	2.1	2.9	3.0	1.9	2.5	2.5	2.5	2.2	2.5	3.
26	2.4	2.8	2.5	3.2	2.5	3.1	3.0	3.3	3.5	3.9	4.1	3.0	3.1	2.8	3.8	3.2	4.0	3.
27	1.9	2.0	1.8	2.0	1.5	2.7	2.2	2.1	2.0	2.2	2.0	1.0	1.8	1.6	2.1	2.1	2.3	2.
28	2.6	2.5	2.8	3.0	2.9	3.0	2.9	2.7	3.1	4.2	4.0	3.0	4.1	3.8	3.8	4.0	4.2	5.
29	22	14	25	18	18	28	32	25	37	52	49	35	46	50	35	50	63	69
30	10	8	12	9	8	15	8	13	18	26	23	18	21	20	16	21	25	32
31	7.8	7.0	7.1	7.8	6.5	8.2	8.8	9.9	10.5	10.2	10.0	9.8	10.0	9.9	10.1	10.1	13.0	11.
32	7.1	7.0	7.2	7.5	6.9	8.0	8.2	9.9	8.8	11.0	10.3	8.0	10.8	10.3	10.3	11.1	12.2	11.
34	4.2	4.5	4.1	4.5	3.3	5.1	5.4	5.2	5.3	5.2	5.8	5.2	5.1	5.1	5.4	5.9	6.5	5.
35	21	20	21	17	17	20	22	26	27	29	29	26	29	27	30	25	33	31

TABLE 103

Measurements of Sample of Golconda Pyriform *Pentremites*

| | | | | | I N D I V I D U A L | | | | | S P E C I M E N S | | | | | | | |
Variables	1	2	3	4	5	6	7	8	9	10	11	12	13	14	15	16	17
1	25	25	10	17	25	21	15	21	15	18	16	19	20	18	22	20	14
2	28	23	10	15	21	16	15	18	13	13	12	16	17	11	15	15	13
3	11	9	9	8	10	10	7	10	8	8	10	10	18	9	10	6	8
4	58	45	34	40	48	55	35	45	37	32	30	37	50	35	39	37	28
5	50	39	24	32	35	40	23	25	30	34	30	30	37	30	35	35	30
6	33	34	15	20	20	25	20	20	17	21	23	23	23	19	22	21	20
7	50	57	36	41	45	45	33	38	43	45	37	42	57	48	50	40	46
8	15.6	15.5	5.7	6.5	11.0	13.0	5.0	12.7	6.7	8.0	8.0	8.0	10.9	10.8	11.1	10.0	6.9
9	29	29	19	20	23	35	17	26	17	19	19	21	20	20	21	21	18
10	13	13	10	10	10	11	10	11	10	10	9	11	11	11	11	13	10
11	6	6	5	4	6	6	5	6	5	5	5	5	5	5	5	5	5
12	4.1	4.9	2.8	3.0	4.0	3.6	3.0	3.2	3.9	3.0	3.0	3.2	3.4	2.9	3.5	3.1	2.8
14	20	20	12	10	12	17	10	18	14	15	12	17	15	7	10	13	12
15	26	38	22	30	40	40	20	52	19	24	22	32	23	24	42	30	19
20	2	13	13	10	5	7	14	9	5	19	2	4	22	15	5	5	6
21	35	30	23	28	34	40	61	31	53	31	38	29	34	32	30	23	47
22	110	90	80	67	48	67	95	70	96	63	60	69	80	75	110	66	48
23	14.5	15.2	5.0	5.3	10.8	11.1	9.2	11.8	7.5	8.7	6.1	9.1	11.5	8.9	9.8	9.8	7.5
24	3.1	3.5	1.5	1.6	2.9	2.5	1.9	2.4	1.9	1.9	2.0	2.1	2.5	2.0	2.9	2.0	1.5
25	3.6	4.0	1.6	1.6	3.3	2.5	2.5	3.2	2.2	2.5	2.9	2.3	3.0	2.5	3.1	2.8	2.2
26	7.2	8.2	3.0	3.7	8.0	4.9	4.2	5.5	5.0	6.0	4.0	4.8	6.2	4.9	5.8	6.2	3.2
27	5.1	6.5	2.2	2.6	6.0	3.0	3.1	2.9	3.7	4.3	2.2	3.2	4.8	3.2	3.9	4.5	4.1
28	6.2	6.2	2.5	2.7	5.2	3.8	3.2	4.3	3.7	3.0	3.9	4.0	3.8	4.0	5.0	3.9	3.1
29	95	78	18	24	55	98	13	70	25	20	47	43	44	29	65	65	21
30	32	32	7	9	21	26	5	22	14	8	16	13	14	11	29	22	7
31	25.7	28.4	9.6	10.5	23.6	20.6	12.7	20.0	14.6	17.2	13.2	16.8	20.5	15.0	17.8	18.9	13.9
32	18.2	16.6	6.9	8.1	14.0	13.8	7.8	12.9	9.2	9.3	10.1	10.8	12.1	9.9	13.1	11.8	8.2
34	15.2	15.6	5.8	6.1	14.7	11.0	9.1	10.5	9.8	10.4	6.5	10.2	12.1	9.3	9.9	11.6	9.1
35	39	36	21	21	29	31	15	33	19	22	25	21	26	22	32	27	20

TABLE 104

MEASUREMENTS OF SAMPLE OF GLEN DEAN GODONIFORM *Pentremites*

INDIVIDUAL   SPECIMENS

Variables	1	2	3	4	5	6	7	8	9	10	11	12	13	14	15	16	17	18	19	20	21	22	23	24	25	26	27	28
1	23	20	17.5	20	20	19	15	16	15	22	16.5	15	17	18	19	18	18	19	16	17	17	22	18	16	16	18	16	18
2	15	12	10	13	13	13	8	11	10	10	11	7.5	10	11	13.5	11	13	12	8	13	10	11	13	9	10	14	12	12
3	7	9.5	10	11	8	9.5	10	9.5	9	10.5	9.5	11	11	11	9	8	10	12	8	10.5	10	9	9	28	34	6	9	8
4	39	32	39	38	35	33	30	29	30	43	32	36	32	36	34	35	37	37	26	30	36	41	37	28	34	31	33	32
5	32	30	29	32	27	28	27	27	23	33	30	25	26	30	32	26	29	28	26	28	29	31	30	18	26	25	18	33
6	19	17	18	16	20	18	14	16	17	17	18	18	18	18	18	18	18	18	12	17	18	19	18	12	15	17	18	19
7	40	41	40	40	42	47	39	36	35	46	41	38	41	43	43	41	40	40	33	39	42	47	41	28	40	36	42	33
8	11.2	9.6	10.8	11.5	9.0	10.6	10.5	11.6	8.2	10.5	10.6	7.6	8.9	11.9	11.5	10.0	10.1	11.0	8.7	9.5	9.0	12.6	11.2	6.8	9.7	10.0	9.1	9.3
9	79	65	72	72	64	69	66	73	58	87	70	67	54	80	74	72	71	77	61	65	71	80	67	49	64	69	64	67
10	24	20	20	20	21	20	22	22	23	24	22	21	19	22	23	22	23	21	19	20	20	22	21	20	21	21	21	23
11	13	13	12	13	12	13	12	13	18	11	15	12	13	15	13	12	14	11	13	13	20	13	11	12	12	13	11	13
12	155	138	150	152	144	144	138	136	130	179	136	134	130	164	164	159	149	160	125	140	144	176	148	120	138	152	129	158
14	11	13	12	12	10	13	12	10	9	11	11	11	10	13	11	11	12	21	10	11	10	11	12	8	10	10	9	12
15	33	32	33	34	30	36	32	32	27	36	30	28	30	30	32	30	35	36	28	32	37	31	31	11	28	32	32	31
20	10	16	16	15	15	22	13	10	12	11	12	12	16	12	15	11	12	10	10	11	11	16	15	11	17	10	10	15
21	33	37	33	40	35	35	39	38	38	40	40	61	35	70	32	30	35	36	35	35	37	85	39	87	50	39	38	42
22	65	80	83	88	71	104	66	84	60	67	97	61	67	100	100	77	63	72	88	73	114	85	86	80	80	85	82	72
23	7.4	5.2	6.9	6.9	5.8	6.5	6.2	7.6	5.1	8.7	6.2	5.7	5.9	6.1	7.4	6.4	6.5	7.1	6.1	5.6	6.0	7.8	6.5	5.0	6.2	6.2	6.3	5.7
24	34	30	36	38	33	37	29	34	31	49	32	35	27	36	34	29	35	32	32	30	33	42	32	26	31	40	31	38
25	47	36	50	50	51	51	45	45	36	60	40	38	38	50	71	42	42	43	57	58	63	52	42	33	39	49	54	45
26	66	62	59	72	61	67	56	66	53	111	61	55	50	70	71	61	57	65	57	58	63	73	70	49	58	69	54	64
27	44	39	37	47	36	43	36	41	35	80	32	38	36	36	45	40	36	44	39	58	42	49	45	31	39	44	37	43
28	4.8	5.2	4.4	5.0	4.4	5.4	4.0	4.8	3.9	6.3	4.1	3.9	3.5	5.0	4.8	3.6	6.5	4.4	3.9	4.5	4.1	5.4	4.5	3.1	4.0	5.0	3.6	4.8
29	3.0	4.0	3.6	4.5	2.9	3.6	4.1	5.1	2.5	3.1	4.6	2.3	2.9	4.9	4.2	4.0	3.3	3.9	2.8	3.9	3.4	4.8	1.9	2.1	3.1	1.1	2.9	4.4
30	1.5	2.0	1.9	2.2	1.5	2.1	2.0	2.4	1.0	1.5	2.1	1.3	1.1	2.0	2.2	1.8	1.5	1.7	1.6	2.4	1.5	2.4	2.3	1.2	1.4	2.2	1.5	2.4
31	12.4	10.6	11.6	12.7	10.2	11.7	11.6	13.7	9.2	16.0	12.1	9.5	9.7	13.0	12.2	12.0	11.0	12.0	10.4	10.9	10.9	11.0	12.5	8.2	10.9	11.6	10.4	10.9
32	12.3	11.3	11.7	13.0	11.2	12.5	11.4	12.4	8.9	15.1	11.8	10.0	10.0	12.4	12.5	10.8	11.0	12.4	10.0	11.0	11.0	14.0	12.5	9.2	10.7	12.9	10.5	11.9
34	6.5	5.8	5.8	7.4	6.4	6.0	5.9	6.5	5.2	11.4	6.5	6.2	5.6	7.1	6.8	6.1	6.5	6.9	5.8	6.1	5.8	8.1	6.7	4.9	6.0	7.0	5.5	6.9
35	36	31	33	35	30	38	32	36	26	30	32	26	24	34	34	35	29	34	29	29	29	36	34	25	29	32	30	28

TABLE 105

MEASUREMENTS OF ONE-DAY-OLD ALBINO RATS

Variables										SPECIMENS										
	1	2	3	4	5	6	7	8	9	10	11	12	13	14	15	16	17	18	19	20
1	15.97	15.90	14.83	16.30	16.03	16.78	16.70	15.73	15.73	16.02	16.60	15.93	15.46	15.72	16.48	15.81	15.80	16.59	15.84	16.
2	7.63	8.09	7.55	7.92	7.41	6.89	6.85	7.52	7.55	7.93	7.85	6.87	7.56	7.54	7.72	7.62	7.55	7.79	7.53	7.
3	6.80	7.14	6.12	6.44	5.91	7.47	7.01	6.54	6.49	6.20	6.57	6.68	6.11	6.25	6.95	6.34	6.23	7.03	6.53	6.
4	4.36	4.58	4.26	4.69	4.63	4.74	5.00	4.77	4.43	4.57	4.81	4.46	4.37	4.56	4.83	4.72	4.48	4.77	4.74	4.
5	3.12	2.83	2.72	2.65	2.64	3.33	3.09	3.11	2.86	3.23	2.97	2.87	2.55	2.51	2.97	2.82	3.04	3.85	2.78	3.
6	6.37	6.17	5.74	6.14	5.84	6.32	5.98	5.74	5.48	5.72	5.85	5.69	5.63	5.80	6.12	5.62	5.43	5.91	5.96	6.
7	3.33	3.74	3.24	3.42	2.30	3.92	3.68	3.90	3.69	3.41	3.78	3.36	3.96	3.67	4.22	3.63	3.61	4.03	3.84	3.
8	2.98	3.33	2.84	3.23	2.84	3.34	3.67	3.02	2.93	3.12	3.19	2.57	3.22	3.08	3.39	3.34	3.15	3.55	3.32	3.
9	3.91	3.24	2.57	3.22	2.94	3.50	3.40	3.20	3.09	3.38	3.30	3.09	3.06	2.90	3.13	3.14	3.14	3.37	3.38	3.
10	5.03	5.17	4.46	4.64	4.69	5.40	4.76	5.18	4.87	4.72	4.86	4.68	4.91	4.79	5.13	4.84	4.90	5.11	5.12	5.
11	6.58	6.58	5.54	6.12	5.65	6.01	7.08	6.94	6.39	6.09	6.18	6.32	5.77	6.26	6.59	6.83	6.25	5.85	6.24	6.55
12	2.20	1.91	1.62	2.15	1.94	2.32	2.36	2.16	1.90	1.92	2.09	1.73	1.87	1.91	2.14	1.75	2.07	2.32	2.15	2.
13	5.46	5.85	4.99	6.01	5.66	6.46	6.39	4.90	5.74	5.64	6.22	5.87	5.66	5.23	5.90	5.52	5.68	6.11	5.95	6.
14	2.51	2.25	2.06	2.24	2.24	2.71	2.74	2.28	2.37	2.31	2.53	2.53	2.20	2.20	2.53	2.24	2.48	2.77	2.42	2.
15	4.75	4.97	4.25	4.81	4.49	5.00	5.15	4.81	4.80	4.53	5.10	4.83	4.60	4.77	4.82	4.80	4.68	5.03	4.88	4.
16	2.07	1.97	1.68	1.93	2.29	2.10	2.23	1.87	2.04	1.96	1.87	2.14	1.92	1.91	2.13	1.96	2.02	2.22	2.15	1.
17	1.74	1.85	1.64	1.94	1.90	1.73	2.12	1.92	1.89	1.87	1.82	1.84	1.81	1.78	1.91	1.85	1.84	1.98	2.00	1.
18	3.34	2.92	2.88	2.45	2.81	3.38	3.37	3.18	2.75	2.93	3.27	2.92	2.86	3.22	3.33	3.02	3.25	3.49	3.48	3.
19	3.42	3.61	2.93	3.00	3.26	3.77	3.95	3.80	3.54	3.68	3.77	3.50	3.37	3.45	3.70	3.68	3.96	4.04	3.78	3.
20	6.52	6.47	5.84	6.21	6.28	7.93	7.83	7.56	6.42	7.03	7.23	7.26	7.10	6.76	7.45	7.10	7.41	7.62	7.10	7.
21	1.64	1.93	1.65	1.42	1.53	1.71	1.99	1.74	1.66	1.62	1.47	1.60	1.67	1.78	1.70	1.79	1.70	1.76	1.71	1.
22	6.34	7.09	6.18	6.81	6.48	7.01	6.51	7.20	6.50	6.77	6.86	6.89	6.77	6.40	7.27	6.99	7.14	7.13	7.00	7.
23	5.27	5.41	4.85	5.11	5.12	5.46	5.82	5.61	5.00	5.48	5.65	5.57	5.32	5.15	5.68	5.46	5.32	5.86	5.65	5.
24	6.32	6.93	6.14	6.78	6.82	7.57	6.50	6.95	6.74	6.81	7.04	6.73	6.72	6.04	7.07	6.73	6.88	7.36	6.98	7.
25	2.38	2.83	2.10	2.50	2.72	3.02	3.09	2.76	2.73	2.62	2.76	2.75	2.88	2.40	3.13	2.78	2.81	3.05	2.86	2.
26	6.97	6.68	5.75	6.78	6.26	7.92	7.68	6.68	6.53	7.43	6.78	6.41	6.87	6.91	7.37	6.90	6.98	6.91	7.03	7.
27	1.00	0.93	0.87	0.98	1.11	1.09	1.08	1.06	0.99	0.90	1.07	0.98	1.01	0.95	1.05	1.04	0.99	0.98	0.93	1.
28	4.86	5.09	4.29	4.84	5.24	5.87	5.76	5.41	5.06	5.20	5.17	4.85	4.84	4.77	5.21	4.93	4.97	5.47	5.25	5.
29	1.75	2.16	0.86	1.62	1.75	2.29	2.16	2.20	1.82	1.75	1.93	1.79	2.15	1.80	2.16	1.97	1.83	2.10	1.93	2.
30	5.59	5.33	4.81	5.16	5.48	6.94	6.75	5.97	5.42	5.66	6.24	5.27	5.18	5.54	6.20	5.85	5.98	6.21	5.98	6.
31	1.14	1.11	1.12	1.11	1.10	1.24	1.25	1.16	1.10	1.09	1.20	1.16	0.97	1.06	0.89	1.14	1.12	1.19	1.10	1.
32	5.45	5.72	5.23	5.52	5.94	6.76	6.78	6.45	5.68	6.34	6.05	6.06	5.74	5.76	6.26	5.93	6.14	6.88	6.40	6.
33	5.02	5.21	4.75	5.13	5.42	5.53	6.11	5.98	5.25	5.81	5.78	5.40	5.17	5.21	5.84	5.35	5.86	6.30	5.43	2.
34	2.44	2.02	1.74	2.24	1.85	2.55	2.71	2.50	2.45	2.83	2.50	2.40	2.19	2.14	2.67	2.36	2.28	2.46	2.27	

TABLE 106

MEASUREMENTS OF TEN-DAY-OLD ALBINO RATS

Variables										SPECIMENS										
	1	2	3	4	5	6	7	8	9	10	11	12	13	14	15	16	17	18	19	20
1	25.80	26.00	26.20	26.10	26.10	26.30	26.30	26.40	25.40	25.50	25.50	24.50	25.80	26.10	25.50	24.10	26.40	25.30	26.00	25
2	10.10	10.53	9.98	10.22	10.24	10.46	10.11	10.55	10.45	10.27	10.55	9.83	10.01	10.48	10.48	10.56	9.91	9.93	10.20	10
3	9.74	10.81	10.21	10.77	10.14	10.11	10.46	10.48	10.09	9.64	9.69	9.14	10.00	9.08	9.93	10.05	9.62	9.91	10.15	10
4	6.42	6.39	5.89	5.83	6.01	6.04	5.93	6.19	6.13	5.82	6.05	5.55	5.76	6.21	5.98	5.92	5.78	5.88	5.85	6
5	4.58	4.75	4.72	5.53	5.63	5.37	5.40	5.19	5.32	5.02	5.49	5.30	5.08	5.45	5.58	5.27	5.53	4.90	4.93	4
6	7.79	7.96	7.72	7.62	7.68	7.92	7.52	8.26	7.77	7.49	7.86	7.30	7.57	7.91	7.89	7.41	7.81	7.60	7.78	7
7	6.93	6.63	6.96	6.24	6.42	7.32	6.86	6.96	6.96	6.86	6.74	6.33	6.45	6.81	6.92	6.28	6.73	6.50	7.15	6
8	5.54	5.69	6.09	5.97	5.61	5.72	5.76	5.96	5.63	5.30	5.75	5.52	5.81	5.76	5.63	5.57	5.89	5.70	5.87	5
9	5.29	5.43	5.52	5.44	5.70	5.53	6.20	5.80	5.58	5.21	5.83	5.47	5.24	6.14	6.08	5.30	5.79	5.40	5.80	5
10	7.48	7.76	7.57	7.79	7.70	7.77	8.14	8.07	7.83	7.48	7.82	7.64	7.74	7.65	7.54	7.63	7.81	7.73		7
11	8.90	9.34	8.83	9.05	8.71	9.05	8.82	9.17	8.28	8.75	8.76	8.64	8.54	9.21	8.91	8.48	9.64	9.01	8.92	8
12	3.84	3.68	4.12	4.25	3.55	3.74	4.14	3.82	3.84	3.88	3.86	3.95	3.87	3.63	3.66	3.63	4.01	3.61	3.99	3
13	8.49	9.23	8.78	8.32	8.59	8.85	9.15	8.88	9.04	8.62	8.87	8.30	8.50	8.73	8.61	8.03	8.57	8.73	8.70	8
14	3.90	3.80	3.51	3.62	3.66	3.82	3.79	3.76	3.84	3.81	3.73	3.39	3.71	3.68	3.86	3.69	3.94	3.64	3.86	3
15	7.54	7.53	7.45	7.64	7.26	7.27	7.60	8.14	7.38	7.20	7.63	7.35	7.46	7.60	7.56	6.89	7.59	7.31	7.36	7
16	2.85	3.43	3.42	3.65	3.48	3.40	3.42	3.59	3.28	3.27	3.35	3.24	3.20	3.24	3.30	3.05	3.33	3.12	3.28	3
17	2.92	3.31	3.07	3.20	3.16	3.21	3.19	3.38	3.24	3.10	3.12	3.02	2.99	3.19	3.23	2.91	3.30	2.71	3.27	
18	5.18	5.30	5.02	5.60	4.94	4.95	4.85	5.08	4.88	4.49	4.87	4.19	4.46	4.98	4.70	4.22	4.90	4.62	4.70	
19	5.88	6.17	5.57	5.81	5.84	5.71	6.12	6.07	5.71	5.60	5.57	5.20	5.76	5.86	5.74	5.26	5.56	5.46	5.76	
20	13.06	13.43	14.26	13.80	13.35	13.33	14.13	14.19	14.16	13.13	13.56	12.91	12.85	13.10	13.11	12.62	13.91	13.39	13.61	13
21	2.52	2.74	2.59	2.55	2.43	2.45	2.65	2.77	2.46	2.40	2.56	2.10	2.35	2.60	2.44	2.04	2.77	2.23	2.49	2
22	11.23	11.12	11.18	11.24	10.84	11.37	11.39	11.73	11.00	10.42	11.09	10.69	10.34	11.18	10.95	10.15	11.33	10.98	11.41	10
23	8.57	8.41	8.92	8.38	8.23	8.78	8.83	8.86	8.78	8.10	8.56	7.96	8.30	8.41	8.62	7.73	8.85	8.49	8.74	
24	11.43	11.70	11.85	12.21	11.52	11.91	11.87	12.09	11.57	11.60	11.65	11.06	11.49	11.45	11.39	10.84	11.88	11.17	11.90	10
25	5.53	5.00	4.83	4.81	4.82	4.85	4.80	5.00	4.63	4.48	4.82	4.26	4.42	4.85	4.42	4.40	4.69	4.51	4.75	4
26	13.78	13.33	13.61	13.45	13.73	14.27	13.96	13.74	13.51	13.22	13.64	13.36	12.88	13.32	13.38	12.98	14.05	13.29	13.99	12
27	1.47	1.60	1.63	1.72	1.69	1.73	1.56	1.73	1.48	1.51	1.54	1.43	1.46	1.51	1.55	1.58	1.68	1.62	1.71	1
28	10.04	9.75	10.53	10.25	10.32	11.08	10.73	10.02	9.96	10.15	10.06	9.90	10.06	9.70	10.11	9.64	10.37	9.93	10.42	10
29	4.43	4.93	4.43	4.29	4.31	4.25	4.31	4.54	3.80	4.37	3.83	4.28	4.34	4.37	4.30	4.02	4.68	4.15	4.60	3
30	11.59	11.48	11.72	11.88	11.60	12.23	11.85	11.87	11.67	11.63	11.60	11.21	11.39	11.54	11.67	11.31	12.06	11.50	12.15	11
31	1.80	1.85	1.83	1.74	1.63	1.72	1.79	1.87	1.83	1.72	1.67	1.64	1.50	1.69	1.65	1.65	1.83	1.74	1.65	1
32	13.06	13.55	14.15	13.78	14.04	14.25	13.94	13.89	13.57	13.50	13.90	13.23	13.60	13.50	13.71	12.93	14.21	13.15	13.65	13
33	12.98	12.78	13.05	11.75	12.21	12.57	12.38	12.61	11.84	12.13	12.01	11.60	11.80	11.92	12.10	12.03	11.95	11.71	12.73	11
34	5.75	5.64	5.96	5.82	5.86	6.02	6.21	5.91	5.74	5.77	5.53	5.68	5.37	5.76	5.80	5.94	6.14	5.60	6.01	5

TABLE 107

MEASUREMENTS OF TWENTY-DAY-OLD ALBINO RATS

Variables									S P E C I M E N S											
	1	2	3	4	5	6	7	8	9	10	11	12	13	14	15	16	17	18	19	20
1.	31.50	31.70	30.50	30.90	30.50	29.10	32.30	31.80	32.70	30.00	32.50	32.70	32.40	29.70	31.80	30.70	31.70	29.80	30.10	29.10
2	10.74	11.18	11.02	10.37	10.42	10.99	10.92	10.70	11.53	10.68	11.46	11.11	11.17	11.09	11.07	11.12	10.72	10.42	11.14	10.53
3	10.83	10.97	11.06	11.12	10.67	10.69	11.38	11.17	10.85	10.78	11.08	11.44	11.64	11.37	10.77	11.10	10.71	10.49	11.34	10.09
4	6.58	6.21	5.93	5.75	5.77	5.90	6.45	6.30	6.25	6.20	6.14	6.30	6.04	6.32	6.10	6.05	6.34	5.84	6.29	5.82
5	6.40	6.10	5.79	5.69	5.46	5.80	5.92	6.31	6.94	6.41	6.76	6.45	6.32	5.42	5.66	5.95	6.47	6.34	5.40	5.67
6	8.47	9.11	8.40	8.62	8.44	8.30	9.20	9.24	9.07	9.13	8.73	8.72	9.01	8.23	8.65	8.94	9.10	8.46	8.38	8.52
7	7.62	8.36	7.45	7.95	7.75	7.12	8.61	8.17	8.42	8.22	7.32	8.37	8.48	7.61	8.23	7.92	8.28	7.54	7.55	7.42
8	6.81	6.71	6.89	6.87	6.42	6.24	6.98	6.96	7.21	6.91	7.02	7.16	7.08	6.36	7.11	6.87	6.88	6.70	6.84	6.54
9	7.43	7.15	6.58	6.68	6.75	6.74	7.78	7.09	7.38	7.17	8.38	7.55	7.40	6.74	7.43	7.20	7.28	6.24	6.27	6.37
10	9.44	9.35	9.27	8.96	8.96	8.05	9.22	9.18	9.28	9.06	9.20	9.28	9.31	8.84	9.35	8.87	9.21	8.66	8.51	8.49
11	10.38	10.28	10.56	10.26	10.13	9.42	10.66	10.45	10.47	10.12	10.55	11.07	10.61	10.36	10.27	10.30	10.73	10.15	10.28	9.25
12	4.87	5.31	4.93	4.93	4.67	4.38	5.35	5.05	5.08	4.54	5.13	5.13	5.20	4.87	4.89	4.84	5.15	4.53	4.98	4.50
13	11.82	11.94	11.18	11.08	10.93	10.17	11.64	11.28	12.29	11.52	11.73	12.15	11.54	10.96	11.73	11.70	11.08	10.80	11.38	10.22
14	4.74	4.52	3.90	4.10	4.38	3.94	4.33	4.39	4.65	4.37	4.54	4.72	4.33	4.20	4.28	4.47	4.35	4.34	4.32	4.02
15	8.78	8.79	8.68	8.43	8.52	7.97	8.86	8.81	9.02	8.99	8.97	8.72	8.95	8.49	8.74	8.55	8.97	8.37	8.38	8.33
16	4.14	4.04	3.98	3.98	3.92	3.63	4.07	4.19	4.49	4.25	4.14	3.98	4.04	3.87	4.08	3.91	4.13	3.77	3.75	3.74
17	4.18	3.93	3.98	3.85	3.84	3.59	3.99	3.98	4.23	4.14	4.04	4.15	3.90	3.75	4.08	3.94	4.12	3.78	3.65	3.77
18	7.17	6.85	6.14	6.11	5.94	6.02	6.65	7.15	7.21	6.83	6.74	6.80	6.17	6.35	6.84	6.36	6.42	5.97	6.18	6.38
19	7.17	7.41	7.14	7.02	7.03	6.45	6.96	7.82	8.01	7.60	7.48	7.86	7.17	7.45	7.27	7.21	7.36	6.95	6.91	6.62
20	19.25	19.26	18.30	17.94	17.95	15.17	19.75	19.15	20.24	19.53	20.10	20.93	19.74	17.72	18.86	18.90	19.20	16.87	17.06	16.90
21	3.70	3.90	3.61	3.71	3.56	3.26	3.89	3.38	3.62	3.87	3.74	3.91	3.79	3.31	3.37	3.68	3.76	3.53	3.68	3.17
22	14.55	14.67	14.30	13.79	13.26	12.03	14.98	14.95	14.77	14.73	14.79	14.98	14.39	13.57	14.09	13.66	14.84	12.74	13.28	12.54
23	10.50	10.98	10.56	10.66	10.05	9.20	11.27	11.13	11.02	10.96	11.07	11.12	10.19	10.94	10.99	11.18	10.03	10.33	10.21	
24	14.95	15.32	14.25	14.46	14.49	13.10	15.28	14.89	15.93	15.13	15.53	15.75	15.59	14.11	15.30	14.71	14.99	13.93	13.85	13.68
25	5.28	5.30	5.31	5.23	5.33	5.08	5.38	5.34	5.76	5.47	5.39	5.51	5.42	5.14	5.43	5.34	5.45	5.03	5.14	5.00
26	18.70	18.33	17.54	17.55	16.98	15.38	18.33	18.35	18.69	18.31	18.90	18.41	18.50	16.53	18.45	17.69	18.42	16.79	17.07	16.56
27	1.83	2.13	2.01	2.12	1.91	1.88	1.97	1.88	2.08	2.04	2.15	1.96	2.11	1.77	2.21	2.11	1.97	1.76	2.11	1.86
28	14.11	13.95	13.00	13.16	12.87	11.55	13.81	13.48	14.07	13.47	13.97	14.22	14.12	12.57	13.57	13.08	13.71	12.62	12.52	12.53
29	5.74	5.49	4.88	5.27	4.97	4.84	5.84	5.41	5.44	5.55	5.83	6.21	5.57	5.12	5.79	5.17	5.55	5.27	5.10	5.08
30	16.60	16.95	15.18	15.90	14.91	13.18	16.49	16.21	16.28	16.66	17.30	17.08	15.09	16.34	15.93	16.34	14.78	14.54	11.48	
31	2.45	2.23	2.11	1.94	1.83	1.81	2.29	2.49	2.32	2.19	2.32	2.43	2.19	2.15	2.25	2.30	2.31	1.82	1.85	1.79
32	20.08	19.53	18.24	18.35	17.64	15.65	20.12	19.21	20.27	19.25	19.87	20.48	20.23	17.07	19.61	18.30	19.30	17.31	17.20	16.92
33	18.75	17.81	16.31	16.97	16.41	14.15	19.67	17.41	15.25	18.08	18.37	18.67	18.51	15.30	17.49	17.00	17.44	15.23	15.47	15.13
34	8.20	8.31	7.43	7.65	7.21	6.71	8.28	7.79	8.71	8.07	8.84	7.96	8.46	6.75	8.22	7.55	8.34	6.83	7.09	7.03

TABLE 108

MEASUREMENTS OF FORTY-DAY-OLD ALBINO RATS

Variables									S P E C I M E N S											
	1	2	3	4	5	6	7	8	9	10	11	12	13	14	15	16	17	18	19	20
1.	34.40	36.00	35.90	37.30	38.00	34.60	36.50	37.60	35.60	35.40	38.20	36.20	38.50	34.10	38.10	38.00	36.00	36.60	35.50	35.40
2	10.98	10.98	10.79	10.82	11.06	10.62	10.47	11.22	10.82	10.91	11.14	10.54	11.07	10.80	11.01	10.46	10.43	10.64	10.38	10.87
3	11.24	10.87	11.43	11.87	11.04	11.21	11.72	11.36	10.90	10.81	11.11	10.98	11.51	11.34	11.69	11.78	11.50	11.73	11.60	11.46
4	6.32	6.14	6.31	6.34	6.31	6.26	6.26	6.44	6.06	6.08	6.35	6.09	6.48	6.29	6.76	6.42	6.34	6.25	6.51	6.21
5	5.80	6.90	6.05	6.30	6.22	6.29	5.97	6.00	6.06	6.22	6.36	6.27	6.84	5.94	5.49	5.85	5.61	6.22	6.35	6.15
6	8.79	9.08	8.92	9.24	9.21	9.19	9.38	9.92	9.57	9.82	10.33	9.71	10.31	8.75	10.17	9.20	9.47	9.40	9.86	9.30
7	10.09	10.14	9.86	10.34	10.26	9.90	10.07	10.56	9.84	9.96	10.41	10.17	10.56	9.51	10.22	10.55	10.08	10.07	10.15	9.69
8	6.85	7.17	7.13	7.10	6.86	6.48	6.71	6.85	7.12	6.79	7.13	6.89	7.40	6.61	7.02	7.28	7.11	6.93	6.86	6.80
9	7.90	8.08	8.13	8.38	8.56	7.45	8.33	8.48	7.92	8.34	8.34	8.30	8.38	7.71	8.77	8.52	8.03	8.32	8.10	7.90
10	9.37	9.24	9.28	9.78	9.70	9.13	9.75	9.46	9.32	9.34	9.75	9.22	10.08	9.14	9.67	9.90	9.39	8.90	9.83	9.10
11	11.23	11.86	11.01	11.98	11.62	10.96	10.91	11.21	10.76	11.39	11.72	11.30	11.92	10.98	11.85	12.07	11.64	11.50	11.71	11.07
12	6.27	6.60	6.07	6.51	6.01	5.87	6.19	6.23	6.16	6.25	6.57	5.80	6.72	5.91	6.39	6.41	6.37	6.28	6.34	5.78
13	13.47	13.83	13.29	13.97	14.44	12.62	14.35	14.34	13.02	13.25	14.64	14.03	14.82	12.70	14.59	14.34	13.49	14.16	13.75	13.66
14	5.55	5.48	5.40	6.05	5.92	5.27	5.58	5.77	5.34	5.51	5.91	5.78	6.08	5.28	5.81	5.59	5.53	5.74	5.69	5.40
15	8.93	9.57	9.12	9.23	9.71	8.88	9.27	9.48	9.51	9.01	9.84	9.41	10.02	9.07	9.99	10.34	9.45	9.45	9.14	9.28
16	3.97	4.38	4.43	4.47	5.94	3.93	5.37	5.15	4.31	4.37	4.76	4.15	4.98	4.01	4.84	4.42	4.23	4.45	4.72	4.09
17	5.55	5.53	5.53	5.68	5.77	4.98	5.82	5.75	5.55	5.39	5.78	5.42	6.02	5.16	6.35	6.00	5.63	5.57	6.09	5.18
18	7.61	8.18	7.40	8.02	8.18	7.70	8.03	8.05	7.38	7.83	8.47	8.05	8.63	7.54	8.90	8.86	7.96	8.29	8.48	7.67
19	9.02	9.89	8.19	8.93	8.57	7.93	8.38	9.38	8.52	8.90	9.16	9.99	9.30	8.13	10.29	10.36	9.34	9.47	10.08	8.57
20	27.20	29.80	28.20	29.20	29.60	26.20	29.30	30.30	25.40	26.70	30.00	28.70	31.20	25.20	30.50	31.00	29.10	29.10	30.30	26.60
21	6.16	6.41	5.92	5.94	6.27	5.57	6.61	6.91	5.69	5.72	6.79	6.32	6.70	5.47	6.95	6.34	5.97	6.23	6.57	5.87
22	14.09	15.71	15.56	15.76	16.34	14.59	16.11	16.36	14.37	14.41	16.42	15.70	16.96	13.69	16.71	16.87	15.72	15.35	16.18	14.74
23	15.94	17.31	17.20	17.72	18.15	15.74	18.26	18.38	15.14	16.10	18.25	17.65	18.87	15.64	18.17	18.72	17.57	17.68	18.54	16.43
24	18.73	19.11	18.42	19.45	19.44	17.64	19.60	20.20	18.11	18.05	19.53	19.22	20.20	17.10	20.40	20.59	19.03	19.21	19.90	17.97
25	4.59	5.73	5.52	5.81	5.99	4.84	5.69	5.93	5.54	5.49	5.92	5.65	6.23	5.19	6.18	5.86	5.75	5.83	5.63	5.47
26	19.23	21.22	20.62	21.20	21.20	18.80	20.90	20.80	18.71	18.85	20.80	19.97	20.70	18.35	20.90	21.30	20.10	20.10	20.60	18.81
27	1.73	1.93	1.72	1.75	2.05	2.04	1.81	2.12	1.97	1.79	2.02	1.88	1.93	1.76	2.12	1.90	1.95	2.12	1.72	
28	17.27	16.28	15.49	17.94	17.91	15.20	16.37	16.65	14.97	15.26	16.49	15.67	16.83	14.97	16.80	16.79	16.16	16.12	16.61	15.06
29	6.84	5.71	5.69	5.88	5.94	5.13	5.71	5.98	5.28	5.49	6.08	5.61	6.23	5.25	5.76	5.89	5.52	5.85	5.36	5.50
30	23.50	24.50	23.40	24.20	24.70	22.00	24.50	25.40	23.90	22.80	24.90	24.00	26.00	22.60	25.90	26.10	24.30	24.50	25.20	22.70
31	2.97	2.85	3.00	3.18	3.29	2.89	3.05	2.85	2.93	2.78	3.23	3.07	3.31	2.29	3.29	3.27	3.12	3.22	3.22	2.95
32	27.50	27.40	26.70	28.80	28.80	26.20	28.80	29.60	25.90	26.00	27.90	27.10	28.90	25.30	30.00	29.80	28.60	28.20	29.60	26.40
33	23.50	24.60	23.90	25.10	25.00	22.60	25.00	25.70	22.80	23.50	25.40	24.10	26.50	21.60	26.10	25.80	24.50	24.30	25.50	23.00
34	10.43	10.44	10.51	10.91	11.44	9.69	10.80	11.35	9.81	10.13	11.11	10.39	11.33	9.48	11.08	10.98	10.73	10.46	11.38	9.93

TABLE 109

MEASUREMENTS OF ADULT ALBINO RATS

Variables	1	2	3	4	5	6	7	8	9	10	11	12	13	14	15	16	17	18	19	20
1	47.50	45.60	48.90	46.70	47.60	46.50	47.10	48.60	46.60	50.70	46.70	47.40	51.00	45.30	48.40	50.00	50.60	48.20	51.50	45.80
2	9.28	9.13	9.09	9.28	9.32	10.22	10.55	9.70	9.68	10.19	9.56	9.75	10.56	9.54	9.31	10.19	10.27	9.29	11.07	10.25
3	11.83	11.95	12.26	12.22	11.72	11.89	12.56	12.42	12.00	12.85	12.34	11.89	13.13	12.14	13.07	12.69	13.55	12.75	13.48	12.38
4	6.85	6.81	7.10	6.85	7.09	7.44	7.65	7.35	6.64	7.77	6.87	7.40	7.65	6.95	7.05	7.44	8.00	7.19	7.96	6.83
5	8.09	6.95	7.37	7.69	7.52	6.14	7.22	7.51	7.97	6.95	6.82	7.45	6.77	5.92	7.53	7.03	8.04	7.17	7.52	6.73
6	12.14	12.46	12.39	11.34	12.76	12.25	12.71	11.92	12.00	13.17	11.63	12.88	13.51	11.31	12.35	13.59	14.22	12.79	14.11	12.25
7	14.05	13.37	14.03	13.64	13.78	13.14	13.79	14.36	13.25	14.38	13.62	13.77	14.64	13.07	14.23	15.28	14.52	14.37	14.86	13.44
8	9.08	8.84	8.48	8.53	8.90	8.52	8.58	9.37	9.07	9.62	9.17	9.11	9.64	8.40	9.15	10.19	9.89	8.56	9.46	8.75
9	11.57	11.09	11.77	11.22	11.64	10.82	11.55	11.58	11.20	12.57	11.21	11.60	11.75	10.97	11.88	12.10	11.56	11.32	12.54	11.09
10	10.50	10.43	10.43	11.45	10.57	11.08	10.08	10.30	11.20	11.48	11.48	9.82	10.61	10.21	10.28	10.36	10.50	11.20	15.70	9.87
11	13.72	12.93	14.85	13.56	14.48	13.00	12.98	14.53	13.13	14.95	14.06	13.17	15.01	14.03	15.03	15.03	14.69	14.20	15.70	13.64
12	9.57	9.59	10.50	9.76	10.03	9.62	9.92	10.47	10.04	10.30	9.49	9.86	10.27	9.30	9.93	10.77	10.34	9.91	10.87	9.94
13	19.64	19.53	21.10	20.00	20.10	20.20	20.10	20.00	19.70	21.60	19.40	20.50	20.90	19.60	20.60	21.60	21.30	20.20	22.20	19.70
14	8.07	7.52	7.87	7.75	7.87	7.36	8.00	7.85	7.66	7.90	7.61	7.71	8.20	7.19	7.48	8.37	8.10	7.62	8.34	7.07
15	14.10	13.19	14.79	13.05	14.31	13.18	12.97	14.59	13.41	14.12	14.12	13.29	14.90	14.59	14.65	15.62	14.59	13.99	15.32	13.54
16	5.40	5.58	5.79	6.19	5.70	4.98	5.30	5.33	4.85	5.87	5.06	4.96	5.83	4.83	5.46	5.45	5.74	5.31	6.09	4.82
17	8.50	7.34	8.70	7.73	7.42	7.31	7.06	8.11	7.22	8.84	7.05	7.22	8.70	6.75	8.45	8.91	8.51	7.97	10.25	7.45
18	11.80	10.80	11.65	11.30	11.90	11.60	11.52	12.88	10.55	12.72	10.32	10.45	12.80	10.57	11.60	12.01	12.11	11.39	12.84	10.66
19																				
20	46.50	46.90	53.10	48.08	50.00	45.00	47.20	50.90	47.10	52.60	46.50	46.70	51.50	45.50	51.20	52.80	51.20	49.10	53.20	46.40
21	10.04	9.87	10.14	10.10	9.57	9.76	9.44	10.34	9.01	10.96	8.84	9.84	10.56	9.84	10.82	9.84	10.98	10.45	11.90	9.38
22	24.90	24.60	27.90	26.20	25.80	23.10	25.40	27.60	23.70	28.10	24.90	25.00	28.60	23.10	27.30	28.00	27.80	26.30	28.40	24.00
23	28.10	26.80	31.50	27.30	27.90	25.50	27.80	32.10	27.00	31.20	26.80	28.00	31.70	25.90	30.00	30.80	31.40	29.60	31.40	26.30
24	28.20	28.00	31.70	27.80	29.60	27.70	28.60	32.10	28.20	31.80	28.00	28.00	31.70	27.50	30.90	32.20	31.80	29.90	32.30	28.10
25	7.00	7.23	7.71	6.91	7.85	7.35	7.53	7.66	7.11	8.08	7.07	7.35	8.02	7.19	7.66	8.10	7.57	7.22	7.63	7.05
26	33.10	29.80	33.60	30.00	32.40	30.00	30.90	33.20	30.40	32.08	30.50	30.50	33.70	30.30	33.00	34.50	32.70	31.90	33.70	31.10
27	2.11	2.07	2.14	2.14	2.00	1.97	2.11	2.35	1.99	2.36	1.99	1.80	2.14	2.02	2.26	2.25	2.26	2.24	2.36	1.41
28	26.50	24.24	27.00	24.00	25.90	23.90	24.40	26.00	24.10	26.00	24.00	24.30	26.80	24.00	26.10	27.10	25.80	25.20	26.10	24.50
29	7.15	6.49	7.32	7.00	7.05	6.25	7.12	7.02	6.85	6.70	6.39	6.85	7.70	6.96	7.02	7.11	7.24	7.11	7.69	6.57
30	38.10	37.20	42.40	37.10	40.80	36.40	37.80	40.80	37.20	41.40	37.50	37.40	41.90	36.60	41.20	42.70	41.20	39.90	42.80	36.60
31	3.20	3.80	4.27	3.97	4.09	3.96	4.25	4.51	3.98	4.82	3.70	4.11	4.65	3.65	4.79	4.68	4.71	4.35	5.01	4.11
32	42.50	40.00	45.30	40.80	42.80	40.50	44.10	44.70	40.90	45.40	41.50	41.00	45.00	40.50	44.90	45.70	45.70	43.70	46.20	40.40
33	38.70	37.30	45.90	38.10	39.60	37.20	38.80	41.80	38.10	41.50	38.00	38.30	42.40	38.10	41.70	42.50	41.40	40.60	42.60	37.40
34	14.72	13.74	14.92	13.79	14.76	14.19	14.34	14.48	14.68	14.80	14.54	13.56	14.94	13.64	14.57	14.49	14.95	14.55	15.36	14.07

TABLE 110

The Sample of *Aotus trivirgatus*

CNHM No.	Sex	Locality
25344	F	Peru: Tingo Maria
25354	M	Peru: Tingo Maria
41492	M	Ecuador: Rio Babonazo
43219	M	Ecuador: Rio Pindo Yaco
43221	M	Ecuador: Rio Pindo Yaco
54332	F	Paraguay: Chaco
55412	M	Peru: Loreto
55414	F	Peru: Huanaco
62074	F	Peru: Loreto
63864	F	Paraguay: Chaco
68324	F	Paraguay: Chaco
66430	F	Peru: Cuzco
68851	F	Columbia: Bolivar
68852	F	Columbia: Bolivar
68855	F	Columbia: Bolivar
70675	M	Columbia: Hulla
75131	F	Peru: Cuzco
78677	F	Peru: Cuzco

TABLE 111

Measurements of Teeth of *Aotus trivirgatus*

INDIVIDUALS

Variables	66430	25344	70675	68324	43219	41492	54332	62074	43221	55414	68852	68855	63864	78677	75131	68851	55412	25345
1	8.95	8.64	8.87	9.16	8.95	8.62	9.15	9.22	8.86	9.27	7.90	8.36	9.28	8.46	9.00	8.93	9.49	9.19
2	4.06	3.75	3.53	4.20	3.87	4.00	4.11	4.24	3.92	3.94	3.97	3.82	4.13	3.85	4.28	3.76	4.38	4.07
3	1.68	1.58	1.74	1.82	1.70	1.82	1.98	1.72	1.82	1.81	1.67	1.87	1.66	1.98	1.72	2.04	1.86	1.86
4	2.06	1.85	2.24	2.02	1.99	2.02	2.34	2.40	1.89	2.13	2.20	2.04	2.10	2.05	2.28	2.20	2.30	2.47
5	2.14	2.05	2.00	2.32	2.20	1.90	2.55	2.61	2.05	2.38	2.17	2.14	2.22	2.12	2.49	2.06	2.35	2.34
6	1.52	1.53	1.48	1.49	1.57	1.50	1.53	1.66	1.56	1.75	1.29	1.32	1.75	1.26	1.47	1.29	1.60	1.53
7	2.00	2.76	2.28	2.39	2.42	2.33	2.56	2.65	2.25	2.42	2.25	2.43	2.52	1.98	2.34	3.02	2.50	2.36
8	1.28	1.23	1.16	1.24	1.33	1.23	1.36	1.12	1.14	1.27	1.25	1.21	1.28	1.00	1.27	1.32	1.35	1.26
9	1.69	2.04	1.93	1.90	1.95	1.92	1.84	1.86	1.82	1.93	1.76	1.72	1.91	1.76	2.08	1.88	1.94	1.81
10	3.30	3.35	3.27	3.52	3.37	3.39	3.56	3.37	3.21	3.50	3.35	3.33	3.19	3.58	3.53	3.68	3.43	3.43
11	1.31	1.41	1.37	1.37	1.36	1.43	1.55	1.51	1.10	1.36	1.41	1.52	1.37	1.25	1.30	0.97	1.40	1.39
12	2.95	3.75	3.55	3.53	3.53	3.87	3.83	4.06	3.77	3.83	3.62	3.53	4.14	3.65	4.15	3.42	4.08	3.83
13	1.57	1.49	1.58	1.55	1.71	1.45	1.43	1.71	1.55	1.51	1.67	1.74	1.73	1.59	1.69	1.78	1.78	1.59
14	1.75	1.43	2.24	1.81	1.92	1.83	2.02	2.71	1.82	1.95	1.99	2.02	2.09	1.84	1.97	2.01	2.22	1.92
15	1.85	1.67	2.21	2.28	2.36	2.19	2.44	2.72	2.15	2.41	2.26	2.15	2.43	2.20	2.55	2.31	2.75	2.17
16	1.69	1.53	1.44	1.40	1.57	1.39	1.50	1.66	1.40	1.83	1.32	1.42	1.86	1.41	1.64	1.47	1.61	1.42
17	2.23	2.24	2.08	2.20	2.34	2.23	2.40	2.53	2.22	2.45	2.28	2.17	2.52	2.05	2.48	2.94	2.75	2.20
18	1.12	0.87	1.14	1.17	1.24	1.06	1.22	0.87	0.89	1.22	0.85	1.04	0.96	1.04	1.02	1.11	1.05	1.11
19	1.71	1.76	1.73	1.96	1.86	1.72	1.70	1.82	1.93	1.86	1.84	1.73	1.87	1.95	1.91	1.87	2.17	1.78
20	3.07	2.96	3.07	3.25	3.35	3.14	3.00	3.12	3.02	3.30	2.32	2.26	3.04	3.04	3.47	3.41	3.28	3.06
21	1.28	0.97	1.34	1.24	1.40	1.25	1.10	1.25	1.12	1.33	1.15	1.23	1.05	1.27	1.20	1.30	1.27	1.31
22	3.43	3.20	3.05	3.28	3.23	3.35	3.22	3.25	3.17	3.07	2.82	2.95	3.47	3.18	3.41	2.95	3.35	3.51
23	0.88	0.74	0.89	1.01	0.92	0.85	0.70	0.71	1.07	1.17	0.75	0.82	0.80	1.32	1.28	0.70	0.95	1.06
24	2.06	1.63	1.46	1.34	1.96	1.83	2.05	1.81	1.79	1.28	1.91	2.05	2.13	1.74	1.22	2.14	2.22	1.85
25	1.66	1.82	1.92	2.16	2.18	1.94	1.83	2.27	2.16	1.99	1.39	1.75	2.27	2.10	2.44	1.98	2.38	2.06
26	1.12	1.24	1.32	1.38	1.24	1.15	1.13	1.15	1.35	1.43	1.69	1.19	1.45	1.02	1.10	0.81	1.30	1.35
27	1.49	1.25	1.03	1.53	1.51	1.35	1.11	1.84	1.59	1.51	0.99	1.34	1.40	1.38	1.41	1.18	1.67	1.52
28	2.25	2.10	2.26	2.48	2.33	2.34	2.02	2.27	2.36	2.38	1.71	2.16	2.53	2.12	2.16	2.18	2.41	2.43
29	2.19	1.72	1.79	1.90	1.67	1.86	1.79	1.47	1.63	1.88	1.73	1.40	1.35	1.61	1.64	1.85	1.95	1.67
30	2.04	1.91	1.82	1.67	1.47	1.72	1.56	1.62	1.46	1.75	1.19	1.36	1.42	1.60	1.41	1.90	1.51	1.44
31	1.72	1.92	1.77	1.68	1.40	1.33	1.49	1.62	1.38	1.43	1.12	1.27	1.35	1.38	1.52	1.81	1.53	1.43
32	1.75	1.46	1.41	1.76	1.78	1.50	1.51	1.52	1.63	1.64	1.55	1.43	1.40	1.63	1.66	1.82	1.81	1.59
33	9.98	9.35	9.41	10.37	9.47	9.77	9.50	10.16	9.69	9.22	10.04	10.42	10.15	9.67	10.17	9.22	10.89	9.60
34	2.98	2.62	2.68	2.90	2.92	2.65	2.82	2.92	2.95	2.60	2.53	2.53	2.81	2.59	2.79	2.78	3.05	2.77
35	2.06	1.97	1.79	2.00	1.78	1.71	1.89	2.20	1.68	1.45	1.77	1.77	1.81	1.47	1.73	1.87	1.96	1.97
36	2.01	1.98	2.14	2.18	2.29	2.21	2.35	2.44	2.05	1.97	1.76	1.92	2.30	1.65	1.97	1.98	1.97	2.27
37	1.66	1.64	1.73	1.74	1.95	1.63	1.65	1.65	1.76	1.75	1.51	1.67	1.75	1.40	1.27	1.71	1.72	1.90
38	1.87	1.81	1.55	1.99	1.79	1.57	1.86	2.06	1.83	1.60	1.69	1.61	1.85	1.65	1.66	1.48	1.96	1.95
39	2.89	2.76	2.80	2.97	2.81	2.70	2.86	2.93	2.72	2.52	2.40	2.51	2.83	2.52	2.34	2.59	2.63	2.03
40	1.06	0.97	1.04	1.26	1.08	1.00	1.00	1.09	0.99	0.91	0.88	0.92	0.84	0.93	0.99	0.99	1.07	0.95
41	1.13	2.17	2.13	2.26	2.28	2.11	2.05	2.32	1.95	1.87	2.07	1.91	1.93	1.95	2.09	2.05	2.31	2.10
42	3.36	3.18	3.17	3.53	3.39	3.33	3.13	3.45	3.35	3.03	3.15	2.84	3.04	3.15	3.18	3.06	4.60	3.24
43	0.96	1.21	1.26	1.45	1.24	1.08	1.25	1.10	1.23	1.17	0.99	1.16	1.26	1.30	1.26	1.18	1.15	1.15
44	1.18	0.97	1.37	1.27	1.45	1.29	1.12	0.88	0.96	0.93	1.09	0.95	1.07	1.01	1.21	0.89	1.25	1.25
45	3.00	2.68	2.82	2.92	2.97	2.78	2.85	2.87	2.99	2.76	2.59	2.61	2.76	2.77	2.87	2.79	3.16	2.96
46	1.19	1.73	1.56	1.81	1.71	1.55	1.76	2.11	1.75	1.69	1.67	1.65	1.68	1.56	1.54	1.83	2.14	1.84
47	1.79	2.07	1.94	2.08	1.96	1.87	2.18	2.17	2.01	1.88	1.97	2.02	2.12	1.95	1.86	2.06	2.15	2.13
48	1.82	1.94	1.90	1.89	2.03	1.77	1.96	2.05	1.77	1.68	1.65	1.70	1.87	1.69	1.63	1.67	1.99	2.02
49	1.96	1.72	1.50	1.90	1.70	1.55	1.80	2.06	1.83	1.73	1.42	1.56	1.73	1.60	1.58	1.48	2.15	1.85
50	2.82	2.59	2.42	2.78	2.77	2.70	2.83	2.91	2.59	2.70	2.54	2.48	2.85	2.51	2.45	2.52	2.94	2.77
51	1.01	0.99	0.92	1.24	1.17	0.97	0.98	1.04	0.95	0.94	0.87	0.96	0.84	0.88	1.01	1.02	1.04	0.96
52	2.19	1.92	1.85	2.17	2.24	1.76	2.02	2.13	2.06	2.02	1.53	1.81	1.91	1.95	2.06	1.82	2.39	1.94
53	3.31	3.26	3.01	3.43	3.42	3.04	3.17	3.44	3.27	3.17	3.17	2.88	3.04	3.07	2.98	2.97	3.53	3.07
54	1.10	0.98	0.87	1.20	1.56	0.99	1.13	1.12	1.34	1.19	1.40	1.30	1.11	1.31	1.26	1.05	1.35	1.04
55	0.95	0.99	1.21	1.34	1.17	1.07	1.24	1.02	1.23	0.89	1.22	0.90	1.41	1.04	1.24	1.18	0.92	1.10
56	2.65	2.43	2.51	2.80	2.43	2.43	2.48	2.56	2.51	2.52	2.32	1.88	2.43	2.40	2.46	2.46	2.81	1.72
57	1.67	1.35	1.36	1.50	1.32	1.32	1.39	1.73	1.43	1.47	1.31	1.39	1.51	1.31	1.27	1.55	1.46	1.64
58	1.39	1.45	1.62	1.83	1.63	1.62	1.54	1.73	1.53	1.42	1.69	1.54	1.87	1.51	1.56	1.51	1.71	1.59
59	1.78	1.65	1.61	1.99	1.79	1.61	1.48	1.87	1.69	1.76	1.50	1.44	1.76	1.80	1.48	1.66	1.79	1.91
60	1.74	1.61	1.61	1.80	1.41	1.59	1.54	1.73	1.54	1.81	1.12	1.37	1.74	1.34	1.45	1.47	1.98	1.85
61	2.30	2.12	2.17	2.45	2.14	2.26	2.31	2.47	2.27	2.36	2.08	2.23	1.99	2.13	2.21	2.26	2.45	2.42
62	0.98	0.87	0.91	1.09	1.22	1.10	1.00	1.10	0.99	1.03	0.83	1.12	0.76	0.86	1.20	0.83	1.14	0.90
63	1.66	1.55	1.56	1.74	1.39	1.69	1.76	1.77	1.83	1.63	1.31	1.58	1.89	1.60	1.65	1.65	1.92	1.87
64	3.22	3.01	3.09	3.38	2.72	3.35	3.15	3.27	3.05	2.98	2.92	3.18	3.36	3.14	3.27	3.06	3.50	3.20
65	0.82	0.77	0.80	1.09	0.90	1.09	1.08	0.85	1.19	0.85	1.03	0.91	1.25	0.81	1.02	0.86	1.21	0.83
66	1.01	0.74	0.76	0.93	0.89	1.04	0.99	0.86	1.06	0.79	1.04	0.89	1.12	0.96	0.78	0.99	0.80	0.82
67	1.40	1.48	1.54	1.57	1.73	1.80	1.50	1.44	1.51	1.45	1.43	1.46	1.37	1.41	1.54	1.80	1.37	1.38
68	1.18	1.49	1.26	1.72	1.82	1.27	1.43	1.42	1.66	1.64	1.24	1.58	1.61	1.66	1.64	0.98	1.83	1.41
69	1.20	1.13	1.10	1.25	1.51	1.48	1.27	1.25	1.46	1.35	1.46	1.53	1.49	1.41	1.42	0.96	1.42	1.20
70	1.23	1.39	1.48	1.15	1.48	1.49	1.09	1.22	1.40	1.38	1.37	1.22	1.03	1.43	1.61	1.44	1.30	1.22
71	0.96	1.45	1.19	1.02	1.33	1.16	0.90	0.88	0.88	0.96	0.89	0.97	1.01	0.95	0.99	1.31	1.09	0.90
72	1.49	1.63	1.27	1.52	1.38	1.26	1.37	1.10	1.15	0.93	0.85	0.90	1.07	1.00	1.11	1.14	1.05	0.65
73	2.48	2.38	2.34	2.64	2.52	2.36	2.64	2.44	2.38	2.40	2.22	2.32	2.46	2.34	2.68	2.09	1.66	2.47
74	1.43	0.88	1.12	1.30	1.00	1.22	1.08	1.30	1.03	1.12	0.98	1.01	1.19	0.82	0.94	0.86	1.23	0.96
75	2.00	1.93	1.47	1.96	1.82	1.63	1.48	1.84	1.38	1.60	1.60	1.60	1.87	1.58	1.59	1.42	1.65	1.53
76	1.62	1.52	1.49	1.69	1.44	1.55	1.60	1.75	1.25	1.51	1.30	1.38	1.47	1.39	1.27	1.46	1.58	1.75
77	1.35	1.32	1.10	1.35	1.07	1.27	1.37	1.37	1.23	1.29	1.31	1.18	1.32	1.06	1.16	1.21	1.47	1.45
78	2.44	2.19	2.00	2.45	2.32	2.15	2.05	2.41	1.96	2.17	2.11	1.95	2.26	2.05	1.96	1.98	2.21	2.26
79	1.15	1.04	0.78	1.27	1.31	1.04	0.89	1.13	1.04	0.86	0.85	1.07	1.06	0.96	1.03	0.93	1.05	1.14
80	1.67	1.46	1.39	1.45	1.05	1.47	1.35	1.42	1.45	1.34	1.37	1.27	1.45	1.25	1.47	1.28	1.49	1.22
81	2.94	2.51	2.53	2.87	2.44	2.59	2.60	2.67	2.72	2.58	2.35	2.37	2.78	2.43	2.63	2.69	2.84	2.47
82	1.22	1.09	0.91	0.72	0.90	0.68	0.77	0.76	0.84	0.80	0.82	0.74	0.75	1.16	0.84	0.77	0.89	0.72
83	1.14	0.98	0.75	1.45	0.80	0.82	0.69	1.06	0.85	0.78	0.64	0.86	0.93	0.83	1.02	0.69	0.73	0.82

Literature Cited and Selected References

ANDERSON, E. 1949. Introgressive hybridization. New York: John Wiley & Sons.

BADER, R. S. 1955. Variability and evolutionary rate in oreodonts, Evolution, 9:119–40.

BARNARD, M. 1935. The secular variations of skull characters in four series of Egyptian skulls, Ann. Eng., London, 6:352.

BERKSON, J. 1929. Growth changes in physical correlation—height, weight, and chest circumference—males, Human Biol., 1:463–502.

BERTALANFFY, L. 1938. A quantitative theory of organ growth. II. Inquiries in growth laws, Human Biol., 10:181–213.

BURMA, B. H. 1953. Studies in quantitative paleontology: an application of sequential analysis to the comparison of growth stages and growth series, J. Geol., 61:533–43.

BUTLER, P. M. 1937. Studies of the mammalian dentition. I. The teeth of *Centetes ecaudatus* and its allies, Proc. Zoöl. Soc. London, s.B., 107:103–32.

———. 1939a. Studies of the mammalian dentition: differentiation of the post-canine dentition, *ibid.*, 109:1–36.

———. 1939b. The teeth of Jurassic mammals, *ibid.*, pp. 329–56.

———. 1941. A theory of the evolution of mammalian molar teeth, Am. J. Sc., 239:421–50.

CATTELL, R. B. 1952. Factor analysis. New York: Harper & Bros.

CRONEIS, C., and GEISS, H. L. 1940. Microscopic Pelmatozoa. I. Ontogeny of the Blastoidea, J. Paleontol., 14:345–55.

DANFORD, C. H. 1939. Genetic and hormonal factors in biological processes, Harvey Lect., 1938–39, pp. 246–64.

DAVID, F. N. 1938. Tables of the ordinates and probability integral of the distribution of the correlation coefficient in small samples. London: University College, Biometrika Office.

DE BEER, G. R. 1930. Embryology and evolution. Oxford: Clarendon Press.

———. 1954. *Archaeopterix* and evolution. Advancement Sc., 42:1–11.

DIXON, W. J., and MASSEY, F. J. 1951. Introduction to statistical analysis. New York: McGraw-Hill Book Co.

EAKIN, R. M. 1949. The nature of the organizer, Science, 109:195–97.

ECKER, A., WIEDERSCHEIM, R., and GAUPP, E. 1896. Anatomie des Frosches. Vol. 1. 3d ed. Brunswick: F. Viewig & Sohn.

FELLER, W. 1950. An introduction to probability theory and its applications. New York: John Wiley & Sons.

FISHER, R. A. 1946. Statistical methods for research workers. (Biol. Mon. Manuals, Vol. 5.) Edinburgh and London: Oliver & Boyd.

GLASER, O. 1938. Growth, time and forms, Biol. Rev., Cambridge Phil. Soc., 13:20–58.

GRAY, J. 1929. The kinetics of growth, Brit. J. Exper. Biol., 6:284 ff.

GREEN, C. V., and FEKETE, E. 1933. Differential growth in the mouse, J. Exper. Zoöl., 66:351–70.

GREGORY, P. W., and CASTLE, W. E. 1931. Further studies on the embryological basis of size inheritance in the rabbit, J. Exper. Zoöl., 59:199–211.

309

GREGORY, W. K. 1934. A half century of trituberculy, the Cope-Osborn theory of dental evolution, with a revised summary of molar evolution from fish to man, Proc. Am. Phil. Soc., 63:169–317.

HOEL, P. G. 1947. Introduction to mathematical statistics. New York: John Wiley & Sons.

HOLMES, B. E. 1927. Vocal thermometers, Scient. Month., 25:261–64.

HUXLEY, J. S. 1932. Problems of relative growth. London: Methuen & Co.

JACKSON, C. M., and LOWREY, L. G. 1912. Of the relative growth of the component parts (head, trunk, and extremities) and systems (skin, skeleton, musculature, and viscera) of the albino rat, Anat. Rec., 6:449–74.

JENKISON, J. W. 1912. Growth, variability, and correlation in young trout, Biometrika, 8: 444–55.

JOHNSON, R. G. 1955. The adaptive and phylogenetic significance of vertebral form in snakes, Evolution, 9:367–88.

KELLEY, D. R., and WOOD, A. E. 1954. The Eocene mammals of the Lysite member, Wind River formation of Wyoming, J. Paleontol., 28:337–66.

KERMAK, K. A., and HALDANE, J. B. S. 1950. Organic correlation and allometry, Biometrika, 37:30–41.

KRUMBEIN, W. C., and MILLER, R. L. 1953. Design of experiments for statistical analysis of geological data, J. Geol., 61:510–32.

KRUSKAL, W. 1953. On the uniqueness of the line of organic correlation, Biometrics, 9:47–58.

KURTÉN, B. 1953. On the variation and population dynamics of fossils and recent mammal population, Acta zool. Finn., Vol. 76.

———. 1955. Contribution to the history of a mutation during 1,000,000 years, Evolution, 9:107–18.

LERNER, J. M., and GUNNS, C. A. 1938. Temperature and relative growth of chick embryo leg bones, Growth, 2:261–66.

LEWENTZ, M. A., and WHITELEY, M. A. 1902. A second study on the variability and correlations of the hand, Biometrika, 1:345–60.

LINSDALE, J. M., and TOMICH, P. Q. 1953. A herd of mule deer. Berkeley, Calif.: University of California Press.

LUNDELIUS, E. 1957. Skeletal adaptations in two species of *Sceloporus*, Evolution, 11:65–83.

MATTHEW, W. D. 1897. A revision of the Perco fauna, Bull. Am. Mus. Nat. Hist., 9:259–323.

———. 1909. Carnivora and Insectivora of the Bridger Basin, Middle Eocene, Mem. Am. Mus. Nat. Hist., 9:508–22.

MATTHEW, W. D., and GRANGER, W. 1915. A revision of the Lower Eocene Wasatch and Wind River faunas. II. Order Condylarthra, family Hyopsodontidae, Bull. Am. Mus. Nat. Hist., 34:311–28.

MERRIL, M. 1931. The relationship of individual growth to average growth, Human Biol., 3:37–70.

MICHENER, C. C., and SOKAL, R. R. 1957. A quantitative approach to a problem in classification, Evolution, 11, 130–62.

MILLER, R. L. 1950. Biometrical analysis of skull morphology. Unpublished doctoral thesis, available in Culver Biological Library, University of Chicago.

———. 1954. Analysis of the interaction of quantitative variables in a modern environment of sedimentation (abstr.), Bull. Geol. Soc. America, 56:1285.

MILLER, R. L., and OLSON, E. C. 1955. The statistical stability of quantitative properties as a fundamental criterion for the study of environments, J. Geol., 4:376–87.

MILLER, R. L., and WELLER, J. M. 1952. Significant comparisons in paleontology, J. Paleontol., 26:993–96.

MOMENT, G. B. 1933. The effects of rate of growth on the postnatal development of the white rat. Philadelphia: Wistar Institute Press.

OLSON, E. C. 1951. *Diplocaulus*, a study in growth and variation, Fieldiana: Geol., **11**:57–154.

———. 1952a. The evolution of a Permian vertebrate chronofauna, Evolution, **6**:181–96.

———. 1952b. Fauna of the Upper Vale and Choza. No. 6. *Diplocaulus*, Fieldiana: Geol., **10**:147–66.

———. 1953. Integrating factors in amphibian skulls, J. Geol., **61**:557–68.

———. 1957. Size-frequency distributions in samples of extinct organisms, J. Geol., **65**:309–33.

OLSON, E. C., and MILLER, R. L., 1951a. Relative growth in paleontological studies, J. Paleontol., **25**:212–23.

———. 1951b. A mathematical model applied to a study of the evolution of species, Evolution, **5**:256–338.

ORTON, G. 1955. The role of ontogeny in systematics and evolution, Evolution, **9**:75–83.

OSBORN, H. F. 1902. American Eocene primates and the supposed rodent family Mixocetidae, Bull. Am. Mus. Nat. Hist., **16**:169–214.

OUTHOUSE, J. 1933. The rate of growth. I. Its influence on the skeletal development of the white rats. Philadelphia: Wistar Institute Press.

PARPART, A. K. 1949. The chemistry and physiology of growth. Princeton, N.J.: Princeton University Press.

PATTERSON, B. 1949. Rates of evolution in taeniodonts. In: G. L. JEPSEN, E. MAYR, and G. G. SIMPSON (eds.), Genetics, paleontology, and evolution, pp. 243–78. Princeton, N.J.: Princeton University Press.

PEARSON, K. 1899. Mathematical contributions to the theory of evolution. V. On the reconstruction of the stature of pre-historic races, Phil. Tr. Roy. Soc. London, s.A., **192**:169–244.

———. 1907. Mathematical contributions to the theory of evolution. XI. On the influence of natural selection on the variability and correlation of organs, *ibid.*, **200**:1–66.

RAO, C. R. 1948. The utilization of multiple measurements in problems of biological classification, J. Roy. Statist. Soc., Suppl., **10**:159.

———. 1952. Advanced statistical methods in biometric research. New York: John Wiley & Sons.

REEVE, E. C. R. 1940. Relative growth in the snout of the anteaters, Proc. Roy. Zoöl. Soc. London, s.A., **3**:279–302.

ROBB, R. C. 1929. On the nature of hereditary size-limitations. II. The growth of parts in relationship to the whole, Brit. J. Exper. Biol., **6**:311–24.

SCHAEFFER, B. 1956. Evolution in subholostean fishes, Evolution, **10**, 201–11.

SCHMALHAUSEN, I. 1949. Factors of evolution: the theory of stabilizing selection. New York: McGraw-Hill Book Co.

SIMPSON, G. G. 1944. Tempo and mode in evolution. New York: Columbia University Press.

———. 1949. Meaning of evolution. New Haven: Yale University Press.

———. 1951. Horses. New York: Oxford University Press.

———. 1953. The major features of evolution. New York: Columbia University Press.

SIMPSON, G. G., and ROE, A. 1939. Quantitative zoölogy. New York: McGraw-Hill Book Co.

SINNOTT, E. W., and BAILEY, I. W. 1914. Investigations on the phylogeny of the angiosperms. III. Nodal anatomy and the morphology of stipules. Am. J. Bot., **1**:441–53.

SPORNE, K. R. 1945. A new approach to the problem of the primitive flower, New Phytologist, **48**:259–75.

———. 1948. Correlation and classification in dicotyledons, Proc. Linn. Soc., London, **160**:40–47.

Sporne, K. R. 1954. Statistics and evolution of dicotyledons, Evolution, 8:55–64.

Stebbins, G. L., Jr. 1951. Natural selection and the differentiations of angiosperms, Evolution, 5:299–334.

Suner, A. P. 1955. Classics of biology, chap. xvi, pp. 300–323. English translation by Charles M. Stern. New York: Philosophical Library.

Swann, D. H., and Atherton, E. 1948. Subsurface correlations of the Lower Chester strata of the Eastern Interior Basin, J. Geol., 56:269–87.

Tessier, G. 1948. La relation d'ilométrie, sa signification statistique et biologique, Biometrics, 4:14–52.

Thoday, J. M. 1953. Components of fitness. *In:* Symposia of the Society for Experimental Biology. VII. Evolution, pp. 96–111. New York: Academic Press, Inc.

Thompson, D. W. 1942. Growth and form. Cambridge: Cambridge University Press.

Waddington, C. H. 1940. Organizers and genes. Cambridge: Cambridge University Press.

Weatherburn, C. E. 1947. A first course in mathematical statistics. Cambridge: Cambridge University Press.

Westoll, T. S. 1944. The Haplolepidae, a new family of late Carboniferous fishes: a study in taxonomy and evolution, Bull. Am. Mus. Nat. Hist., 83:1–122.

Whiteley, M. A., and Pearson, K. 1900. Data for the problem of evolution in man: a first study of the variability and correlation in the hand, Proc. Roy. Soc. London, Vol. 65.

Wright, S. 1932. General, group, and special size factors, Genetics, 17:603–19.

———. 1934. The method of path coefficients, Ann. Math. Statist., 5:161–215.

Yule, G. U. 1924. A mathematical theory of evolution, Phil. Tr. Roy. Soc. London, s.B., pp. 21–87.

Index

313

Morphological Integration: Forty Years Later

BARRY CHERNOFF AND PAUL M. MAGWENE [1]

I. INTRODUCTION

The vision set forth by Olson and Miller expanded the critical thinking of Thompson (1917) and Huxley (1932) about the evolutionary transformation of organismal form. Olson and Miller sought to invigorate and stimulate studies into the evolution of organismal phenotypes from a perspective of the covariation of parts. At the core of Olson and Miller's concept of morphological integration is an emphasis on a holistic view of organismal structure and evolution. While emphasizing the contributions that studies of a single trait or a few traits have made to our understanding of evolution, Olson and Miller argue that a multivariate approach provides a more sensitive and powerful tool for addressing evolutionary questions. Ontogenetic, developmental, functional, and temporal consequences were identified as critical hypotheses of constraint on the expression, covariation, and evolution of the phenotype. Although individuals are viewed as units of evolution, morphological integration is a population-level phenomenon meant to inform us about variation and covariation taken in descriptive or evolutionary perspectives. Interestingly, we come to understand the meaning of morphological integration not from an explicit definition, which Olson and Miller never provided, but rather from the ideas they presented.

Defining Morphological Integration

We define morphological integration as the correspondence of patterns of covariation among traits to *a priori* or *a posteriori* biological hypotheses. Estimates of covariation are derived from samples of individuals taken to represent biological populations. Morphological integration is a measure of effect, whereas the biological hypotheses are a statement of cause. Quantitative traits covary as a result of underlying developmental, functional, or other biological factors. Morphological integration is useful both to estimate those factors as sources of covariation as well as to test the effects of known or speculated factors. We see the scope of investigations of morphological integration as encompassing both phenomenological studies of

---

[1] The names of the authors have been listed in alphabetical order; no seniority of authorship is implied.

effect as well as experimental tests to validate statements of cause. Even though morphological integration is observed as a pattern, both cause and effect fall within the scope of study.

Our definition corresponds to the population-level integration of Cheverud (1996), who subdivided the concept into genetic and evolutionary integration. Genetic integration is the genetic modularization and co-inheritance of morphological traits; modules and sets of morphological traits being more or less independent of one another. Genetic integration can result from pleiotropy (Cheverud 1996) or linkage disequilibrium (Falconer 1989). Evolutionary integration describes the coordinated evolution of morphological traits. Cheverud (1996) notes that this may occur because either traits are inherited together (i.e., the result of genetic integration) or because they are selected together, though inherited separately (Felsenstein 1988).

Cheverud (1996) recognized integration at the level of the individual with functional integration and developmental integration as components. The former term refers to the interactions among morphological elements within an individual organism that affect their joint performance. The latter term refers to interactions, direct or indirect, among traits during morphogenesis. We agree with Cheverud's (1996) concepts and note that individual organisms are intrinsically integrated entities. We, as did Olson and Miller, focus upon the population level, where morphological integration can be viewed and estimated as a pattern that has underlying mechanistic causes and evolutionary consequences. The level of the studies chosen and the nature of experiments undertaken will determine whether one can decompose morphological integration into component parts.

Perspectives on the Measurement of Covariation

Estimation of covariation among traits may be approached from separate analytical perspectives in quantitative morphology. The perspectives include analyses of linear measurements, landmark configurations, meristic or serially repeated elements, as well as aspects of cellular structure and histomorphometry. Olson and Miller measured covariation as correlation in a population of individuals using sets of linear measurements taken across biological structures. Whether the interpretation of covariation is derived from simple analyses of structural or regional adjacency (e.g., Olson and Miller) or from explicit models of development or function (Wright 1932; Cheverud 1982; Bookstein et al. 1985; Zelditch 1987), hypotheses of morphological integration can be tested or imputed.

Another perspective estimates morphological integration based upon the reports of shape changes derived from deformations of landmarks. The resulting deformations, whether as vectors of landmark displacements (Walker 1996) or as derivatives of thin-plate splines (Bookstein 1991; Zelditch et al. 1992; Rohlf 1993), estimate covariation from regional and spatial patterns of homogeneity. Again, underlying causes due to ontogeny or function may be explored or tested from analyses of landmark deformation.

Zelditch et al. (1992) eschew analyses of correlations derived from measurements among landmarks and construe Thompson's (1917) text as somehow inconsistent with the analyses and interpretations of Olson and Miller. Thompson (1917) noted that integration was due to underlying causes and that in patterns of covariation we should search for fewer causes than the parts we are studying—an overall philosophy identical to that of Olson and Miller. Perhaps rescinding an extreme position, Fink and Zelditch (1996: 62) later stated that: "As he [Thompson (1917)] saw it changes in shape due to growth gradients would be expressed as a modification of a large number of organs and regions; allometric change would be global and graded across the organism." A notion that could have been paraphrased from Olson and Miller. While extremely promising, we do not believe that current landmark-based techniques are sufficiently devoid of limitations (Rohlf 1998) to warrant the dismissal of elegant studies from a multivariate morphometric tradition (e.g., Zelditch 1987, 1988). Nor are landmark methods sufficiently universal to be applied to the diversity of biological forms (e.g., Lohmann and Malmgren 1983).

Roadmap

In this Afterword, we discuss a variety of topics to bring the reader up to date on methods and issues that have arisen since the first publication of *Morphological Integration*. In many cases we extend beyond the originally published material to include new topics or those underrepresented in the original text. The sections that follow include: a discussion of heuristic approaches and phylogenetic frameworks, a brief discourse on the use of genetic versus phenotypic correlations, a survey of methods used to analyze integration, the quantitative genetic approach to morphological integration, a review of some theories on the evolution of integration, and finally a prospectus outlining future avenues for investigation.

Given our limitations in space, we do not intend this to be a complete treatment of morphological integration but rather a perspective derived from Olson and Miller. A perspective that seeks to explain morphological evolution and morphological diversity as a result of changes to the underlying causes of morphological integration within populations or species.

II. FRAMING HYPOTHESES

In this section we consider two conceptual frameworks essential for studying morphological integration. The first deals with *a priori* versus *a posteriori* formation of hypotheses. The second considers the phylogenetic context necessary for the study of evolutionary morphological integration.

The Case for Heurism

Olson and Miller's methodology for analyzing morphological integration, the ρF-model (described below), was initially designed as a test of *a priori* hypotheses of integration (Olson and Miller 1951). However, in their reformulation of the model (chapter 3), Olson and Miller abandoned *a priori* construction of F-sets to

derive hypotheses *a posteriori*. These hypotheses appear formally similar to their
ρF-set hypotheses. Olson and Miller demonstrated through a series of sampling experiments (chapter 6) that they discovered identical patterns of integration whether
from ρF-sets defined *a priori*, or from heuristic evaluations of ρ-sets. They reasoned
that because underlying causes affect the correlation structure of the sampled
traits, such structure will be inducible.

The debates concerning hypothetico-deductive approaches in relation to induc-
tive approaches (heurism) abound in the scientific literature (e.g., Hull 1988) and
are beyond the scope of this Afterword. Both deductive and heuristic approaches
have an important role in studies of morphological integration. Heuristic approaches derive from data exploration and provide the bases for many interesting
hypotheses of function and development that can be subsequently tested in more
mechanistic studies.

Heurisitic approaches can also involve models but the form and specific components of the models may be determined through exploration. An example of
this would be the path model of allometric coherence (Bookstein et al. 1985) which
was an elaboration of the method by Wright (1932). Ontogenetic covariances are
partitioned into an allometric size factor and secondary or specific factors that explain positive residual covariances. The number and composition of the secondary
factors is left unspecified but the residuals from an allometric factor are explored
to see if patterns of correlation explain covariation at spatial scales or for functional
units. The median fin factor of the fish genus *Atherinella* is such a factor (Bookstein et al. 1985), as were the spatio-functional secondary factors discovered by
Wright (1932). Once models of ontogenetic covariance are hypothesized they can be
written as path diagrams and tested with maximum likelihood methods (Zelditch
1987; Bollen 1989; Marcus 1990). Models of morphological integration can also
be derived from landmark configurations without specific *a priori* hypotheses. Because knowledge of underlying factors of development and function are far from
complete, heuristic approaches allow us great insights into potential underlying
mechanisms in advance of mechanisitic studies which are not always feasible. However, we must recognize that hypotheses derived from heurism as well as *a priori*
hypotheses are usually limited by our preconceptions about development and
function.

Nevertheless, when *a priori* hypotheses are available we gain degrees of freedom
in the study of morphological integration. Hypotheses can be written as path models or statistical models and covariation among traits can be analyzed to see if the
models provide adequate and unambiguous explanation for the covariances observed (e.g., Cheverud 1982; Zelditch 1987, 1988).

Phylogenetic Frameworks

From an evolutionary perspective, the proper framework for interpreting morphological integration is a phylogenetic framework of species or population relationships. This modern perspective was present in the philosophy espoused by

Olson and Miller (chapter 10). As with other studies of their time, they used the fossil history of a group as an hypothesis of phylogenetic history (cf. Hennig's rule of paleontologic precedence). Though they relied heavily on the fossil record, they also included character-based information; their spirit was definitely one of comparing patterns of morphological integration in a phylogenetic context.

The role of phylogenies in historic and evolutionary analysis has long been discussed (Eldredge and Cracraft 1980; Brooks and McLennan 1991; Martins 1996). Identifying where patterns of trait covariation have changed in the evolutionary history of a group is critical to understanding historical patterns of developmental stability or lability (Fink and Zelditch 1996). Using a phylogenetic hypothesis, Fink and Zelditch (1996) identified congruent patterns of evolutionary transitions for sets of ontogenetic characters. The assessment of ancestral conditions of covariance has been addressed recently by Steppan (1997a, 1997b), who used common principal component structures to assess the portions of phenotypic covariance matrices that were shared by sister species and sister groups.

III. HIERARCHICAL HYPOTHESES OF INTEGRATION

Ontogenetic studies of morphological integration have a distinct advantage in that hypotheses of integration are logically hierarchical: nested sets of traits covarying due to developmental or functional causes. The ontogenetic perspective that we address here differs from that expressed by Olson and Miller. Their ontogenetic analyses compared integrative patterns among delimitable ontogenetic stages. Olson and Miller's approach is temporally ordered but not hierarchical. The patterns they discovered were not nested in relation to the ontogenetic stages.

If one studies a major portion of an ontogeny without subdividing into stages, artificial or natural, then one can address different sets of questions regarding morphological integration. For example, questions about rates and timing, allometry, heterochrony, and heterotopy fall into a natural hierarchic framework. Elements of this framework can be found in Thompson (1917), Wright (1932), Bookstein et al. (1985), and Zelditch and Fink (1996).

Within the hierarchical framework we recognize the potential for genetical, developmental, and functional processes to operate at different spatial scales. The levels of the hierarchy are necessarily nested but from a computational perspective the nested levels do not necessarily imply independence (orthogonality) among levels or among units within a level. An hierarchichal investigation begins with an analysis of size and allometry and proceeds to less inclusive levels of spatial scale or adjacency defined by other criteria (Fig. 1; see section on adjacency below). A key point here is that hypotheses of adjacent or functional units are nested within each of the levels including that of size.

Allometric Integration

One of the primary features associated with the ontogeny of many organisms is size change. We here follow the conventions of Bookstein et al. (1985) and Bookstein (1991) and distinguish *size* from *scale*. Scale is an aspect of the physical dimension of an organism, the most robust measure for a set of landmarks being "centroid size" (Bookstein 1991). Size, on the other hand, is construed as a factor containing information about dimension as well as information about allometric change; allometry is size dependent shape-change (Gould 1977; Bookstein et al. 1985). An allometric or size factor can be expressed from either a traditional multivariate or landmark-based geometric perspective. In the case of the former it is often the largest principal component or general factor within a single population (Bookstein et al. 1985; Zelditch et al. 1995). In the case of the latter, there may be more than one shape component (warps or other measures) that covaries with centroid size (i.e., shape deformation as a function of scale). In both cases, the expression of allometry across the organism is viewed as a reasonable hypothesis of underlying developmental programs that govern the changes in shape associated with changes in scale. Variable coefficients or landmark displacement vectors indicate the rates of change of the parts over the range of scale of the sampled population.

So how then do we examine hypotheses about integration, heterochrony, and heterotopy in relation to size change? Heterochrony is the difference in rates of growth or timing of appearance among parts of the body. Heterochrony and allometry are thus related concepts. Huxley (1932) presented a framework on differential growth as the differential equations dx/dt and dy/dt, where x and y are measured traits and t is time. The heterochronic relationship of $x{:}y$ becomes $(dx/dt)/(dy/dt) = dx/dy$, the rate of change of x relative to the rate of change of y. The rates of change of x and y are well estimated by the allometric coefficients on a size factor. The relative allometries on a size factor or landmark vector displacements on an allometric landmark shape component describe the heterochronic relationships among traits estimated at the level of the population.

Within an overall measure of size change, one can examine the specific rates of change or displacements of suites of traits or landmarks corresponding to functional, developmental, or spatial criteria. For example, one can investigate whether or not all traits or landmarks associated with particular functional groups scale in a similar manner (i.e., at the same rate).

The degree of morphological integration associated with a measure of allometry or heterochrony can be measured by the fit of individuals to the overall heterochronic description. That is, the width or scatter of individuals around the size factor or allometric landmark shape component. This is measured by correlation or some function of the trace of the covariance matrix. One could further test the heterochronic covariances due to size change under certain hypotheses of growth (e.g., biomechanical models predict specific allometries for limb scaling, Biewener 1990).

Subsequent Levels of Integration

After covariation due to size, subsequent analytical levels of the hierarchy describe components of integration corresponding to heterotopic patterns at decreas-

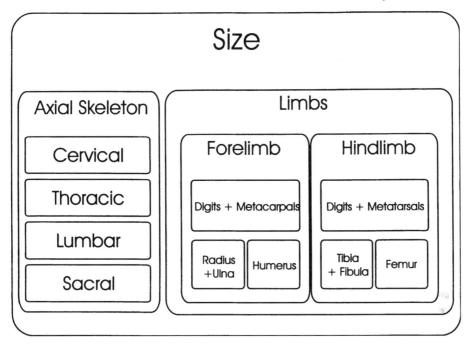

FIG. 1—A representation of the hierarchical nature of integrative hypotheses, based on a hypothetical example. This figure represents a number of morphological traits which are integrated at the broadest level due to covariances with size. At a less inclusive level traits covary due to developmental/functional associations (axial skeleton, limbs). At yet lower levels, covariation is due to anatomical (spatial) association.

ing spatial scales (Fig. 1; Zelditch and Fink 1996). At the least inclusive level, suites of traits or landmarks covary only within suite. Hypotheses of development and function may be associated with these integrated sets and are measured or tested by analyses of correlations or covariances, whether from inter-landmark distances or directly from their coordinates.

At more inclusive levels, heterotopic sets may join with each other in ways that suggest further hypotheses about adjacency or function. For example, Olson and Miller demonstrated (Table 24, p.114) that traits were integrated within functional groups (e.g., limb locomotor) at relatively high levels of correlation, and at slightly lower levels of correlation among functional groups. Similarly, to the extent that biological information is described by partial warps (Rohlf 1998; Adams and Rosenberg 1998), Fink and Zelditch (1996) depicted how distinct patterns of form change associated with anterior and posterior body regions were integrated at more inclusive spatial scales during the ontogeny of several species of piranhas.

Hierarchical investigations into morphological integration can lead to crucial hypotheses about development and function in relation to the entire phenotype. Hierarchical analyses should not constrain traits from participating in more than one functional or adjacency group. Furthermore, there is no *a priori* reason to constrain the hierarchical levels to be orthogonal to one another. The style of analyses should be driven by the nature of the data and by the hypotheses being tested or sought.

IV. GENETIC AND PHENOTYPIC CORRELATIONS

Since its inception, the study of morphological integration has focused on the analysis of correlations between phenotypic traits. In the modern formulation of the concept there have been two schools of thought regarding the type of correlations to analyze: phenotypic correlations (**P** matrices) or genetic correlations (**G** matrices).

Interest in genetic correlations in relation to integration stems from Lande's (1979) model for multivariate selection (Cheverud 1982, 1984; see below). From the perspective of a quantitative genetics framework, genetic correlations are more important than phenotypic correlations for predicting responses to selection. However, genetic correlations are rather difficult to estimate; they require large breeding designs to compute with sufficient accuracy (Falconer 1989). If one focuses solely on genetic correlations one ignores a wide class of questions and approaches of evolutionary interest which can only be addressed via the use of phenotypic correlations: studies of fossil organisms, hard-to-breed or rare species, and comparative studies involving large numbers of taxa.

Cheverud (1988) recognized this problem, and undertook a comparison of genetic and phenotypic correlations in order to assess under what circumstances phenotypic correlations can be substituted for genetic correlations.

Similarity of G and P Matrices

Cheverud (1988) recognized two circumstances under which one might expect similarity of genetic and phenotypic correlations. The first circumstance is when trait heritabilities approach unity. Empirical evidence suggests that this is rare. The second circumstance is the case of similar patterns of genetic and environmental correlations. Because phenotypic correlations are the sum of genetic and environmental correlations, under these conditions the pattern of phenotypic correlation will mirror that of genetic correlation no matter the level of heritability.

In a survey of forty-one studies in which data on both genetic and phenotypic correlations of morphological traits were available, Cheverud (1988) found that in cases where effective sample sizes were large, genetic and phenotypic correlations were similar across a range of heritabilities. He interpretted this result as indicating a broad congruence between genetic and environmental effects on development. While recognizing that their use undoubtedly introduces error, Cheverud (1988) concluded that the substitution of phenotypic correlations for genetic correlations is acceptable in many cases. Willis et al. (1991) objected to Cheverud's (1988) conclusions on a number of theoretical grounds, but recent empirical studies

(Roff 1995, 1996; Koots and Gibson 1996) have, with similar caveats, supported Cheverud's conclusions regarding the similarity of phenotypic and genetic correlations of morphological traits when genetic correlations are well estimated.

Long-term Consequences of G and P

Lande's (1979) model for multivariate selection response is not without its faults. There are the usual criticisms of population genetic models; for example, the assumption of infinite population size. A more telling critique relates to the assumption required in order to make predictions about long-term responses to selection: constancy of the genetic correlation matrix. The assumption of constancy has been widely criticized on theoretical grounds (e.g., Turelli 1984, 1985), and empirical studies have only resulted in weak support for constancy (e.g., Lofsvold 1986; Kohn and Atchley 1988).

A more fundamental issue relates to the action of selection. It is the phenotype, summarized by the phenotypic correlation matrix, which is the target of selection. Genetic correlations may or may not be useful for predicting long-term responses to selection, but it is phenotypic correlations which are the filter through which genotype and environment interact. As such, phenotypic correlations are of great interest, regardless of their predictive value, in studies which attempt to understand the functional, developmental, or other underpinnings of morphological integration.

V. ANALYZING INTEGRATION

Much progress has been made in the last four decades towards establishing a body of statistical and analytical methods appropriate to analyses of morphological integration. However, as with any evolutionary study dealing with character states across taxonomic groups, homology of traits is a fundamental issue. We briefly discuss homological concepts, and then provide an overview of the methods which have been developed in order to study morphological integration.

Homology

A critical issue inherent in the study of morphological integration is that of homology. In chapter 10, Olson and Miller mentioned that they had used both analogous and homologous characters in their study of fossil mammal teeth. Olson and Miller did not, however, discuss what constitutes analogy and homology for quantitative traits. The question is an important one because it governs the types of traits or data that can be used in studies of morphological integration. Homology remains a contentious issue in systematic and developmental biology not only conceptually but also with regard to the use of quantitative traits in evolutionary studies.

The issue of homology would not constrain a descriptive inquiry into the morphological integration expressed by a single population, species, or taxon. When dealing with comparative evolutionary morphological integration, however, a clear statement of the homology concept should be provided by the researcher. The particular concept guides the selection of traits and the limits of conclusions that can be drawn. In the following paragraphs we will try to clarify briefly the concept of homology as it will affect evolutionary morphological integration and the types of traits that can be used.

From an evolutionary or a phylogenetic perspective, homology is the ancestor-descendant relationship among parts or systems of organisms (Hennig 1966; Mayr 1969; Nelson 1994). Homology forms an information continuum across the phylogeny of life (Van Valen 1982) and, therefore, is the counterpart to phylogeny. Phylogeny is the relationship among taxa and homology is the relationship among parts of organisms (Nelson 1994; Rieppel 1994). As such, both phylogeny and homology are unknowable. There is a single, true phylogeny of life and a single, true homology among parts of organisms. Our evolutionary and phylogenetic studies provide estimates of homology first and phylogeny second.

Smith (1990) pointed out the inherent circularity in the evolutionary and phylogenetic homology concepts. He noted that estimation of homology is a two-step process. The first step corresponds to an application of classical or phenetic mappings made across organisms in which parts are compared or measured based upon the similarity of their structures, embryologies, compositions, or positions. These statements are then subject to tests of congruence among other such characters. From the patterns of congruence among derived character states we estimate both the homological relationships among the traits (i.e., synapomorphy) and phylogenetic relationships among taxa. Thus, synapomorphy is not synonymous with homology (cf. Patterson 1982; Nelson 1994). Synapomorphy is an hypothesis of homology. If we add taxa or traits to our studies, our notions of synapomorphies and hypotheses of phylogenetic relationships may change. However, the true phylogeny and true statements of homology remain unaffected and unknowable.

HOMOLOGY AND MORPHOMETRIC TRAITS

Bookstein (1994) claims that morphometric characters are fundamentally flawed as purveyors of homological information, stating that (p. 224): "Morphometrics cannot supply homologous shape characters, but must be informed about them in advance, . . ." If we take Bookstein's (1994) arguments seriously, then no trait, quantitative or otherwise, could be used in evolutionary studies. In so arguing, he fails to recognize that homology is an estimated relationship and that all traits must suffer the same, necessary fate of the test of congruence.

Bookstein's (1994) other arguments involve ambiguity of character description. We agree that ambiguous trait descriptions are to be avoided, but again, this problem is neither unique to, nor incorrectable in, quantitative traits. For example, while the similarity of cylinders is ambiguous if we use volume as the descriptor of shape, their similarity is not ambiguous if we compare the heights and diameters of cylinders. Thus, in studies of evolutionary morphological integration care must be taken to include quantitative traits for which an hypothesis of homology can be formed (with or without landmarks), and one must test for the potential ambiguity in our measures, descriptors, and analyses.

Zelditch et al. (1995) have criticized the use of traditional morphometric characters (i.e., linear measurements across biological structures) on two bases: i) failure of homology; and ii) loss of local information about shape change. They ascribe the failure of homology because the "actual" homologues are the endpoints of measurements; the intervening regions have no particular homology function and interpretation may be difficult. They recommend instead that one should analyze directly the changes in shape due to changes in geometric configuration of homologous landmarks.

Rieppel's (1994) discussion of homology and topological similarity help one navigate through the arguments of Zelditch et al. (1995). When one measures the length between landmarks one makes an assumption that there is a homological correspondence among the structures being measured. For example, in measuring the length of the supraoccipital bone from its base on the parietal to its tip, one equates the intervening lengths among the sample of supraoccipitals. True enough, it is not possible to determine if the tip or the base is growing (receding) but from a homological point of view one assumes that the bony tissue between the tip and the base is congruent; that the similarity in structure and length is due to patterns of common ancestry. This is the very same process involved in the description of qualitative character states in systematics. For example, the descriptions—supraoccipital long or short—presume the same homological correspondence among the sections of the bone between its base and its tip.

Zelditch et al.'s (1995) criticism about loss of local information in traditional measures is often true. A limitation of using measured lengths in studies of morphological integration involves rates of change (e.g., allometries). Long measurements may cross regions that express different growth biologies. The reported rates of change will average the different rates of growth in proportion to their lengths (Bookstein et al. 1985). Shorter measures are to be preferred. However, the landmark-based method used by Bookstein (1991), Zelditch et al. (1992, 1995), and Fink and Zelditch (1996), partial warp analysis, does not escape inter-landmark effects upon the resulting spline and warps. This is apparent immediately from the smoothing algorithm employed, in which the thin-plate spline is calculated from a partial differential equation integrated over the x- and y-coordinates (see Bookstein 1991: 29). The effect is greatest when landmarks are close together. Interpretation of the displacement vectors of the partial warps as changes in landmark configuration must be couched in the context of the entire configuration and *not* as a set of independent displacements. The reported displacements at each of the landmarks for any partial warp is a function of the following: i) the distances among the landmarks in the reference configuration; and ii) the heterogeneity in the direction and magnitude of the vector deformations required to fit the reference configuration onto that of the target specimen. Therefore, both traditional and landmark-based techniques can be affected by heterogeneity and local information can be lost. If one desires information about biological shape deformation in the small, then finite element scaling may be preferable (Cheverud, pers. comm.). The interpretation of such analyses must take into account the assumptions and limitations of methods and data.

Zelditch and Fink (1996) also criticized the comparability of within-group multivariate estimates of size factors. They state that although these within-sample estimates describe

... ontogenetic changes of homologous features, the within sample PC1's [multivariate estimates] cannot be regarded as a homologous feature. When taxa diverge in shape correlates of size there is no common ontogenetic shape axis on which to compare them. (p. 243)

We disagree with their assessment for two reasons. The first reason is that their point about shape space for comparison is irrelevant to the main hypothesis of underlying developmental programs and homology. One could plot the multivariate vectors in the original character space and graph the values for individuals in the same space in order to compare forms at a variety of sizes. The second and principal reason is that the multivariate estimates describe the rates of change of the homologous traits. This provides a Huxley-like (1932) interpretation of growth gradients over the measured regions and is an hypothesis of underlying developmental cause—a developmental program that governs the relative rates of growth of the parts during ontogeny. The comparison of the multivariate estimates among populations or species, preferentially in a phylogenetic context, provides clues to the divergence of such developmental programs. From the comparison we can postulate functional or environmental causes, ancestral conditions, etc. But the homology of the multivariate estimates derives from the assumption that the developmental causes for growth of homologous structures are also homologous. This assumption can be verified from the test of congruence.

Methodological Approaches

Olson and Miller established the precedent of using character correlations to search for and to test patterns of morphological integration. Their model for analyzing integrative patterns developed over a period of several years. As first presented (Olson and Miller 1951), their approach, called the ρF-model, involved testing empirically derived clusters of traits (ρ-groups) against trait sets derived by qualitative assessment of function or development (F-groups). By the time that *Morphological Integration* was published (1958), they were sufficiently convinced of the validity of ρ-groups as indicators of real biological factors to argue that the comparison with qualitatively formed F-groups was no longer a necessary component of their "formal model." In this model, described in chapter 3, ρ-groups are defined as maximally inclusive sets of traits such that within each set, all pairwise correlations exceed a predetermined level. Often ρ-groups showed considerable overlap, and Olson and Miller designed the "basic pairs" method to minimize overlaps. The basic pairs approach is rather unconventional by today's standards, but given the lack of readily available computing power in the early to mid 1950's, it was not an unreasonable approach to cluster analysis. Despite its flaws, the modern reader will find that their emphasis on a quantitative, multivariate methodology distinguished their work from the typically qualitative methodology prevalent particularly in the paleontological literature.

Most modern studies have followed the lead of Olson and Miller in assuming that correlation structure is indicative of underlying biological factors. However, the majority of these studies of integration adhere to the spirit of the original ρF-model, in that they attempt to test the validity of specific hypotheses of integration against empirical data. Statistical advances have made it easier to test alternative hypotheses of integration, thereby eliminating some of the "difficulties" Olson and Miller refer to with regard to comparing the fit of quantitative data against qualitative models.

A renewed interest in the concept of morphological integration was generated by a number of papers published in the early and mid 1980s by James Cheverud (Cheverud 1982, 1984; Cheverud et al. 1983; Cheverud et al. 1989). Cheverud's contributions included the formulation of statistically sound and easily comprehensible tests of integration and, perhaps more importantly, he placed the concept of morphological integration firmly in the framework of quantitative genetics. These influential papers led to a number of investigations of morphological integration in a variety of taxa (e.g, Cheverud 1982, 1989, 1995; Leamy and Atchley 1984; Zelditch 1987, 1988). Along with new studies came the development of new techniques suitable for investigations of morphological integration. Methodological issues of interest include: i) testing for integration, ii) assessing overall levels of integration, and iii) comparing integrative patterns. Below, we review the statistical methods which have been proposed to address these questions.

TESTING FOR INTEGRATION

Studies attempting to demonstrate integrative patterns fall into two major camps. On one hand are investigations which test the fit of specific hypotheses of morphological integration against empirical data. On the other hand are those studies which delimit potentially integrated sets of characters in an *a posteriori* fashion following statistical analysis.

Attempts to delimit integrated sets of traits in a post-hoc fashion (e.g., Leamy and Atchley 1984; Zelditch et al. 1992; Zelditch and Fink 1995; Cane 1993) have utilized a variety of multivariate statistical techniques, typically based on eigenanalyses of covariance or correlation matrices. This type of analysis has resulted in varying degrees of success, with the delimited integrative trait sets subject to more or less biological interpretability.

Recent papers by Zelditch and co-authors (e.g., Zelditch et al. 1992; Zelditch et al. 1993; Zelditch and Fink 1995) exemplify a trend to try to discern integrative patterns using landmark morphometric techniques. The hypothesis of integration inherent in these investigations is one based on spatial criteria. While these studies are laudatory in their goals, we believe the approach they use to detect integration is potentially problematic. Their approach is based on Bookstein's thin-plate spline (TPS) family of methods (Bookstein 1989, 1991). These methods are designed to reveal shape changes at successively smaller spatial scales. However, the notion of spatial scale inherent in these techniques is a consequence of the configuration of landmarks in the "reference" specimen used in the analysis. There is no *a priori* reason to believe that spatially integrated patterns should be manifested at these particular scales (Rohlf 1998). Interpretations of integration based on such analyses should be viewed with caution.

Studies attempting to test specific hypotheses of integration include those of Cheverud (1982, 1989, 1995), Zelditch (1987, 1988), and Wagner (1990) among others. The approach taken by Cheverud (1989, 1995) and Wagner (1990), and advocated in Cheverud et al. (1989), is based on testing pattern similarity between an empirically derived correlation matrix and a theoretical matrix which represents the hypothesis of integration. Similarity of the two matrices is determined by calculating a matrix correlation (Sneath and Sokal 1973; Cheverud et al. 1989). The significance of matrix correlations can be assessed using a variety of distributions (e.g., Mantel 1967) or by the use of Monte Carlo methods (Cheverud et al. 1989). This matrix correlation approach is relatively straightforward and simple. Additionally it allows one to compare the goodness of fit of alternative hypotheses of integration (Dow and Cheverud 1985; Cheverud et al. 1989).

A more powerful approach to testing hypotheses of integration, as exemplified by Zelditch (1987, 1988), is to use path models and confirmatory factor analysis (Bollen 1989). Hypotheses of integration are stated in the form of path diagrams (Wright 1932) and the fit of the models to the empirical data can be evaluated using the χ^2 statistic. This approach also facilitates tests of alternative hypotheses but the models to be compared must be hierarchically nested. The critique that Cheverud et al. (1989) raise, that the method must be based on covariance rather than correlation matrices, is no longer valid (Bollen 1989).

INDICES OF INTEGRATION

Related to the subject of testing for integration among traits is the issue of quantifying overall levels of integration. Olson and Miller provided an index for overall integration among a set of traits (chapter 6, pp. 153—158). Their index, which ranges from 0 to 1, was based on a ratio of the number of bonded pairs to non-contained ρ-groups, corrected for the total possible number of bonds. Zelditch (1988), Cheverud et al. (1989), and Cane (1993) also provide indices of overall integration. Zelditch's (1988) measure for the intensity of overall integration is based on the standardized χ^2 value for the fit of the model of no integration (i.e., complete independence of traits). A high χ^2 value relative to degrees of freedom indicates interdependence among the traits (Zelditch 1988). Cheverud et al. (1989) prefer Wagner's (1984) use of the variance of the correlation matrix's eigenvalues as a measure of overall integration. The higher the variance the narrower the ellipsoid represented by the correlation matrix. Cane's (1993) index is simply the average of the absolute values of pairwise correlations.

COMPARING PATTERNS OF INTEGRATION

A common endeavor has been to compare patterns of morphological integration among ecophenotypes, populations, or species (e.g, Olson and Miller 1951, 1958; Berg 1960; Cheverud 1982) or between ontogenetic stages (Zelditch 1988; Cane 1993). Both matrix correlations and factor analytic approaches (described above) have been used to test such associations. Recently, other techniques related to common matrix structure have been receiving increased attention.

Cheverud et al. (1989) highlight two issues related to the similarity of correlation stucture: i) levels and patterns of variance, and ii) levels and patterns of covariance. These topics are further elucidated and expanded by Steppan (1997a). Degrees of matrix similarity can be arranged in a hierarchical fashion, from most to least similar: matrix equality, matrix proportionality, and whole and partial common structure (Flury 1987, 1988; Steppan 1997a). A technique designed to characterize the various levels of this hierarchy is the common principal component (CPC) method of Flury (1987, 1988). CPC is a generalization of principal components analysis from single samples to multiple samples (Steppan 1997a). CPC analysis works by testing the various levels in the hierarchy of matrix similarity. The significance of a given level of similarity may be tested by calculating likelihoods. Biological applications of this method may be found in Airoldi and Flury (1988), Klingenberg and Zimmermann (1992), and Steppan (1997a, 1997b). Illustrative because of their bearing on the evolution of integrative patterns are Steppan's (1997b) CPC analyses of phenotypic covariance structure in leaf-eared mice, *Phyllotis*. Steppan demonstrated that only a small portion of the subspecies he examined share common principal components, and that divergence in covariance structure did not have a strong phylogenetic signal. The CPC method has been shown to be generally useful and we expect this approach will become increasingly popular in comparative studies of integration.

An Exploratory Approach to Uncovering Integrative Patterns

As argued above, we consider heuristic and exploratory methods to be useful and necessary tools for studying integration. In this spirit we present a brief overview of a new method which, given reasonable, broad criteria, allows the researcher to generate a smaller set of specific hypotheses to be explored in further studies. This method is described in greater detail in another paper (Magwene, in prep.).

ADJACENCY MAPS

The concepts of "adjacency," "connectedness," and "contiguity" are useful tools for generating and testing hypotheses of integration. Our definitions of these terms differ from those presented by Gabriel and Sokal (1969), who were concerned primarily with geographic criteria. Adjacency, as we use it, refers to whether two traits can be considered "neighbors" with respect to a specific criterion. The criteria for specifying adjacency may be functional, developmental, spatial, or based on other biological models. Given a criterion, one can typically define an "adjacency map"—a graph, matrix, or other representation of adjacency for all the traits in question. An example of an adjacency map is the Gabriel graph (Gabriel and Sokal 1969; Fig. 2). The edges of a Gabriel graph specify which points are considered adjacent under the criterion of spatial proximity.

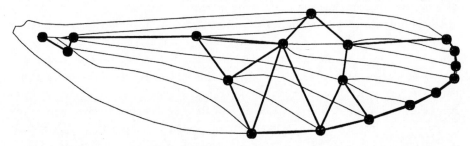

FIG. 2—A Gabriel graph based on the mean landmark configuration of a sample of mosquito wings (Rohlf 1993). Edges between landmarks indicate spatial adjacency. See Gabriel and Sokal (1969) for a discussion of spatial criteria for adjacency.

Related to adjacency is the concept of "connectedness." Connectedness is also specified in an adjacency mapping. Any two traits are connected if they are adjacent, or if they may be linked via a chain of continuous adjacency comprising any number of traits (Fig. 3). The notion of "contiguity" refers to the relationships of connectivity for sets of two or more traits. A contiguous set is one in which all included traits are connected (Fig. 3). Contiguity relationships also can be inferred from an adjacency map. When based on biological criteria, adjacency maps can be viewed as specifying the universe of allowable hypotheses under the rules specified by those criteria.

Adjacency, connectedness, and contiguity can be used in exploratory analyses of integration. We advocate the following methodology: 1) specify criteria for adjacency; 2) define adjacency mappings based on those criteria; 3) reduce the family of models implied by the adjacency map to one or a few most likely hypotheses by testing the fit of each model against the empirical data. In the case of similar fits, those hypotheses containing the largest contiguous sets are to be preferred.

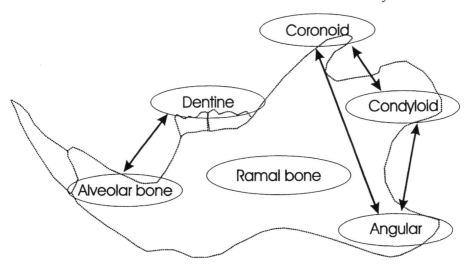

FIG. 3—One of several possible adjacency maps implicit in Atchley and Hall's (1991) model for the development of the rodent mandible. We use this map to illustrate notions of connectivity and contiguity. Double-ended arrows denote adjacency. The coronoid and condyloid traits are a connected pair. Angular, coronoid, and condyloid traits form a contiguous set. The three largest contiguous sets are {alveolar bone, dentine}, {ramal bone}, and {coronoid, condyloid, angular}. The set {alveolar, dentine, ramal bone} would not form a contiguous set because the ramal bone trait is not connected to either the alveolar or dentine traits.

Specifying an adjacency mapping greatly reduces the allowable trait combinations (i.e., potential hypotheses). For example, given six traits there are 203 possible models.[2] However, if the traits have the simple adjacency mapping specified in Fig. 4a, there are only 31 possible hypotheses congruent with the mapping.

Once an adjacency mapping has been used to generate hypotheses based on the specified criteria, models in the reduced set can be tested for their fit to the empirical data using the investigator's method of choice (e.g., matrix correlations or confirmatory factor analysis). We suggest using matrix correlations (which are less computationally intensive than confirmatory factor analyses) to eliminate all models which don't fit at least as well as a single factor model. Matrix correlations provide a useful way to further reduce the number of hypotheses before preceeding with more detailed tests of significant fit. The preference for maximally contiguous sets further constrains the results towards those solutions requiring the fewest underlying factors.

In the absence of alternative hypotheses, or simply as a matter of course, we suggest that spatial relationships are a logical null-hypothesis for defining adjacency.

[2] The number of models in an unconstrained model of n traits can be calculated as a "Bell Number": the number of ways a set of n elements can be partitioned into non-empty sets whose union equals the original set (Bell 1934; Rota 1964). For example, the set {1,2,3} has five such partitions: {{1},{2},{3}}, {{1,2},{3}}, {{1,3},{2}}, {{1},{2,3}}, and {{1,2,3}}.

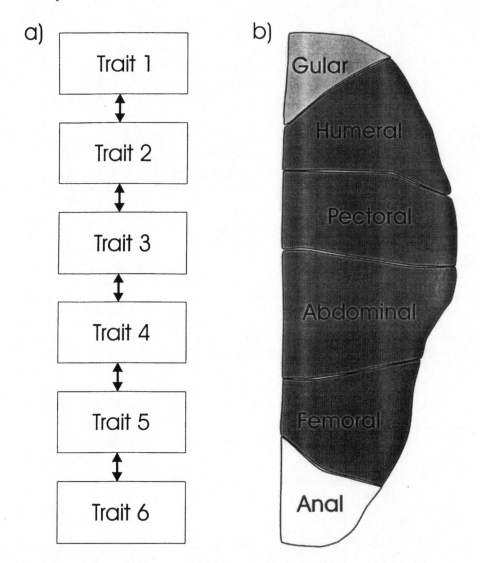

FIG. 4—a) A simple adjacency map of six traits. Double-ended arrows indicate adjacency. b) Results of an analysis of plastral scute integration in hatchling painted turtles (*Chrysemys picta*). The adjacency map used was identical to that in 4a. Analysis suggests two or three modules of integration (denoted by shading).

This criterion is especially applicable to traits in the form of landmark data (e.g., Fink and Zelditch 1996).

In Fig. 4b we show the results of the adjacency map approach applied to data on the plastral scutes of the painted turtle, *Chrysemys picta*. The analysis is based on correlations of scute areas estimated from a sample of fifty specimens aged two weeks. The adjacency criterion used was spatial proximity. The results suggest the existence of two or three modules of integration (Magwene, in prep.).

Adjacency maps are powerful representations of families of hypotheses based on particular criteria. This approach is distinct from other clustering techniques because it enables one to explore data under specific biological hypotheses. This places the emphasis on formulating and considering underlying biological cause while retaining the flexibility engendered by the ability of the method to handle multiple alternatives. This approach allows one to test the support for various integrative schemes with greater flexibility and with fewer assumptions than is required by standard hypothesis-testing techniques (e.g., Cheverud et al. 1989; Zelditch 1987). Of course, with greater flexibility comes less power to differentiate between hypotheses. In view of this, the technique should be viewed as a heuristic tool, and "answers" generated by such an approach should be considered hypotheses for further study.

VI. QUANTITATIVE GENETICS AND INTEGRATION

Olson and Miller recognized that integrative patterns are themselves subject to selection and evolution. While Olson and Miller presented the reader with a variety of interesting questions about the evolution of integration, they provided no specific theoretical or quantitative infrastructure upon which to build theory or expectation. Modern quantitative genetic theory (e.g., Lande 1979, 1980; Cheverud 1982) provides the necessary framework for exploring this topic.

Multivariate Selection and Integration

Current quantitative genetic models related to morphological integration (e.g., Cheverud 1982, 1984; Atchley et al. 1994) stem from Lande's (1979) model for evolutionary responses to multivariate selection. Lande's (1979) model indicates that the existence of phenotypic correlations between traits is not sufficient to ensure that traits will evolve in a coordinated fashion (Lande 1979; Cheverud 1982, 1984). Instead, it is the pattern of genetic correlations which is critical for determining the evolutionary responses of correlated characters (but see Gromko 1995 regarding the effects of negative pleiotropy).

Genetic correlations are estimated from phenotypic correlations. The two types of correlation are related by the following formula:

$$r_{\mathrm{P}} = h_{\mathrm{X}} h_{\mathrm{Y}} r_{\mathrm{G}} + e_{\mathrm{X}} e_{\mathrm{Y}} r_{\mathrm{E}}$$

Where X and Y are the two traits of interest, h represents the square root of the heritabilities, e is the square root of the phenotypic variance due to environmental factors, and the r's are phenotypic (P), genetic (G), and environmental (E) correlations (Falconer 1989).

A good deal of theoretical consideration has been directed at modeling the evolution of genetic correlations and morphological integration. Cheverud (1984, 1988,

1989b) argues that "functional integration causes genetical integration which in turn leads to evolutionary integration of morphological elements." Cheverud's argument is simple—genetic and evolutionary integration are linked by Lande's (1979) equation for the response to selection:

$$\Delta z = G\beta$$

Because the elements of β, the directional selection gradient, represent the partial regressions of each trait on fitness with all other traits held constant, the correlated response to selection depends only on G, the genetic variance-covariance matrix. Hence, evolutionary integration results from genetic integration. The connection between genetic and functional or developmental integration is due to the relationship between the genetic covariance matrix and the fitness surface relative to the additive genetic values. This fitness surface reflects patterns of stabilizing selection. Given uniform mutation, stabilizing selection is non-uniform when a subset of the traits are involved in some functional role or developmental process. It is patterns of stabilizing selection reflecting functional or developmental requirements which results in genetic integration (Cheverud 1984, 1989b).

Wagner (1988a, 1988b, 1996) argues that stabilizing selection alone isn't strong enough to lead to integration among traits. The scenario he points to is one in which integrative patterns evolve due to a combination of stabilizing and directional selection. This argument is based on simulations using his preferred model for adaptive landscapes—"malignant" corridor models. Corridor models can be visualized as rising ridges in a fitness landscape. Under these models, periods of strong directional selection on functionally or developmentally integrated traits with simultaneous stabilizing selection on other traits will lead to maintenance or increase of pleiotropy among the functionally or developmentally related characters. At the same time it results in a decrease in pleiotropy between unrelated traits.

Developmental Constraints and Integration

The quantitative genetic framework has also been used to draw a link between the phenomena of developmental constraint and morphological integration. Maynard Smith et al. (1985) defined developmental constraints as "biases on the production of variant phenotypes or limitations on phenotypic variability caused by the structure, character, composition, or dynamics of the developmental system." Both Cheverud (1984, 1988, 1989b) and Wagner (1988a,b) have argued that modern quantitative genetic theory incorporates the phenomenon of developmental constraint and that constraints are causally related to integration. This argument is based on the interpretation of developmental constraints as a form of "internal" stabilizing selection. "Internal" factors may modify or filter the amount of phenotypic variation that is presented to the external environment and available for "external" selection. Cheverud (1982, 1984) suggests that genetic correlations will evolve to mimic patterns of stabilizing selection, with developmental constraints represented in the genetic covariance matrix. If one accepts that genetic correlations are the proximal cause of morphological integration, this formulation implies a causative link between developmental constraints and integrative patterns.

Wagner (1988a, 1988b) has argued that developmental constraints as expressed in patterns of integration may facilitate the evolution of multi-trait phenotypes. Wagner (1988a) defines unconstrained phenotypes as those in which the variances and heritabilities of all traits are identical and there are no phenotypic or genetic correlations between traits. Using corridor models of fitness functions, Wagner (1988b) has shown that rates of evolution are very slow in unconstrained multi-character (>3 traits) systems. Correlations between traits imply constraints given the above definition of "unconstrained." Trait coupling, expressed in correlations, facilitates evolution by decreasing the variance perpendicular to the fitness corridor during bouts of selection for particular functions.

VII. THE EVOLUTION OF MORPHOLOGICAL INTEGRATION

Phenotypic Perspective

The evolution of morphological integration is ascertained from comparisons of morphological integration within populations or species. The study of morphological evolution can be approached at two levels. The first examines changes in the specific patterns of morphological integration at various levels of the integrative hierarchy. The second evaluates how an overall measure of integration changes over the evolutionary history of a group. The two approaches are complimentary. We argue, however, that calculating only overall integration statistics is in and of itself not sufficient to understand evolutionary patterns of morphological integration.

A fundamental issue in the study of evolutionary morphological integration concerns the stability of patterns through time. Fink and Zelditch (1996) reasoned that from the perspective of growth gradients there should be a conservatism governing evolution of developmental integration. That is, it becomes impossible to conceive of developmental associations constraining trait evolution if the covariances (i.e., patterns of integration) constantly fragment and reassociate (Bookstein et al. 1985; Fink and Zelditch 1996).

Despite some localized lability, patterns of integration found within small groups of piranhas (Fink and Zelditch 1996; Zelditch and Fink 1996), poeciliid fishes (Strauss 1992), and other characiform fishes (Chernoff et al. unpublished) are stable over larger portions of phylogenetic history. This phylogenetic stability contrasts with other studies in which ontogenetic covariances were conserved among only a few sister species or subspecies (e.g., Steppan 1997a). Fink and Zelditch (1996: 67) concluded that "evolutionary coordination arises from changes in integration, not from inherited patterns." Nonetheless, the lack of stability in inherited patterns (Fink and Zelditch 1996; Raff 1996 and references therein) conflicts with our *a priori* expectations and those that might be drawn reasonably from Olson and Miller. We will return to this point below.

The phenomena of integration can be classified as follows: coupling, decoupling, dissociation, reassociation, reorganization, and strengthening. We distinguish *decoupling* from *dissociation*. Some authors (e.g., Fink and Zelditch 1996) refer to both without definition while the majority refer only to dissociation. We think that it is useful to distinguish between them, as well as between *reorganization* and *reassociation*, in order to separate pattern from process. We add strengthening and coupling to these four in order to complete the description of patterns.

Dissociation refers to the separation of previously linked developmental processes (Needham 1933; Strauss 1992; Raff 1996); whereas reassociation is the linking of developmental processes. We define decoupling as a pattern that describes a decrease in the level of covariation between two or more traits in the phenotype; whereas reorganization is a pattern of increased covariation. The proximate cause(s) of decoupling and reorganization is left unspecified.

All four terms require a phylogenetic framework in order to identify the polarity of transformations. Decoupling and reorganization are comparative descriptions of differences among patterns of morphological integration. The phylogenetic transformations of covariation, viewed as gains or losses to integrative systems (Strauss 1992; Fink and Zelditch 1996), can serve as hypotheses about evolutionary processes and also address issues about the stability of integrative systems. Correlated changes of patterns of morphological integration with historic changes in ecology or biogeography would potentially provide fascinating insights into organismal evolution.

Heterochrony is thought to result in large part from dissociation of developmental processes (Raff 1996) leading to decoupling of phenotypic traits. The repartitioning of covariance or patterns of covariation deserves careful scrutiny in order to estimate evolutionary processes. Decoupling or reorganization of traits are only two possible outcomes. One should also consider whether the covariation among traits is *strengthening*. This strengthening could be due to the actions of selection or developmental programs that canalize unique variance into existing integrative systems. This may correspond to mechanisms of "transitionism" postulated by Duboule and Wilkinson (1998). Evidence for strengthening would be relative constancy of the trace of the covariance matrix synchronous with an increase in off-diagonal covariances. The increase in covariation would be at the expense of the unique trait variances. Developmental processes that result in strengthening should result in stability of morphological integration over that portion of the phylogeny for which such developmental processes were active.

Another possible repartitioning of phenotypic variance would be to increase the degree of integration by way of the addition of trait(s) into an existing suite. If the trait was previously associated with a different module or suite, then this would be an example of *reorganization*. If the trait was not associated historically (i.e., having no apparent covariance with a suite identified on functional, spatial, or developmental grounds), then this would be an example of a *coupling*. How would coupling be viewed with respect to the "stability" of the plesiomorphic integrative system? Should one recognize an integrative suite at a particular spatial scale as basically the same or fundamentally different by the addition of one or more traits to an existing suite? We suggest that the suite or module be recognized as the same if the suite implies identical underlying hypotheses of development, adjacency, function, etc.

These are but a few of the possible changes among descriptions of integration that have important consequences for evaluating evolutionary theories of morphological evolution. Though not universally true, many of the changes in form and integration seem to be associated with dissociative processes that result in heterochronic patterns (Gould 1977; Strauss 1992; Raff 1996). This would result in the apparent instability of morphological integration during the evolutionary history of a group. Yet within clades, many groups of organisms appear very similar in shape or bauplan. We reconcile the apparent discrepancy with the fact that organisms reported upon (*Peromyscus*: Lofsvold 1986; *Rattus* and *Mus*, Atchley et al. 1992; poeciliid fishes: Strauss 1992; piranhas: Fink and Zelditch 1996; sigmadontine rodents: Steppan 1997a,b) by and large do not manifest major reorganizations in shape or conformation. Thus, the changing patterns of integration that are phylogenetically unstable produce relatively small-scale shape changes in relation to the shape diversity of larger clades. This signifies that the measures of trait covariance in the published accounts may not be carrying information about integrative patterns associated with the larger, "gestalt," shape differences that we see among clades.

Olson and Miller looked at patterns of evolution of morphological integration across ontogenetic stages and over phylogenetic time. To better summarize their results, Olson and Miller constructed an index of integration (see discussion above). In their studies of changes in integration over ontogeny, Olson and Miller examined the integration index for each of several age-groups of rats. The index of integration did not necessarily increase between age-groups. In fact, there seemed to be a relaxation of integration in middle ontogenetic stages before traits were reassociated at latter ages. Assuming that this is not an artifact of sampling, their data suggest that growth is not a continuous function during ontogeny. Rather, various elements experience onsets and cessations of growth even during active growth phases of the entire organism. This pattern of variable growth is well documented in the literature (Raff 1996) and has even been found for fishes with indeterminate growth (Martin 1949; Chernoff et al. 1991; Chernoff and Machado 1999).

In a careful examination of allometric integration, Strauss (1992) discovered that corresponding larval and adult growth patterns were evolutionarily decoupled. But more importantly, he discovered that larval patterns of integration were more highly conserved among species and phylogenetically congruent than were adult patterns of integration. The generality of Strauss's (1992) findings need to be verified for many other taxa and across groups of organisms. Nonetheless, the work of Olson and Miller, Strauss (1992), and Raff (1996) demonstrates clearly the potential effects of dissociation. These studies highlight the fact that growth processes may be altered during the entire ontogeny, and we should not have the *a priori* expectation that as ontogeny proceeds so should morphological integration increase.

Olson and Miller investigated whether integration would increase towards the Recent in lineages known only from fossils. Their study of molariform teeth belonging to species of the extinct mammal genus *Hyopsodus* illustrates how changes in integration relate to their phylogenetic hypothesis. The overall results were mixed (Fig. 70, p. 244). Dramatic increases in integration towards the Recent were discovered for the M1 molar in each of two lineages—that leading to *H. paulus* and that leading to *H.* sp. There was a more modest increase in the integration of the M2 molar for the same two lineages. However, the M3 molar displayed either relative stasis in integration or a dramatic decrease in the level of integration in the lineage leading to *H. despiciens*.

A more complex pattern of change is found in the blastoid echinoderm *Pentremites*. Because of the quality of the fossil record, Olson and Miller were able to separate the effects of gradual phyletic and anagenetic change from those consequences of lineage splitting (speciation). They found an overall orderly increase of integration with evolution towards the Recent. This was especially true during periods of phyletic change due to strengthening, coupling, and reorganization. The overall index of integration fluctuated over time periods due to diminished integration (decoupling and increased variability) that occured after crossing the boundaries of time horizons, particularly the Paint Creek Horizon.

Whether the results of Olson and Miller are found to be general or just circumstancial is less crucial at this time than is the necessity to undertake studies of morphological integration across a variety of groups. Analyzed within a phylogenetic framework, studies of morphological integration will produce the necessary patterns to develop hypotheses about evolutionary processes. We must determine whether there is a commonality to the underlying developmental structures that permit morphological diversification. In the prospectus section we pose a number of key questions that require answers in order to use morphological integration as a critical paradigm to investigate and develop a cohesive theory about morphological evolution.

Genetic Models

An important issue related to the evolution of integration is the topic of null-models for trait correlations. Two alternate theories for the evolution of integrative patterns are fusion and fission models (Riska 1986; integration and parcellation *sensu* Wagner 1996). The fusion model hypothesizes that patterns of integration arise via the establishment of pleiotropic effects among initially independent traits. The expectation under fusion models is that the "unintegrated" state is one with little or no pleiotropy between traits. Fission models, on the other hand, suggest that patterns of integration evolve via the dissolution of pleiotropic effects among unrelated traits with simultaneous maintenance or strengthening of pleiotropy among traits serving common functional or developmental roles. Fission hypotheses predict that distinctive integration patterns evolve from a state of initially high levels of pleiotropy (Wagner 1996; Wagner and Altenberg 1996). While little empirical data exists to allow us to judge the relative frequencies at which the two models operate, a number of lines of reasoning lead us to argue for fission as the more fundamental process. Riska (1986) argues that "high genetic correlation is inherently simpler than and antecedent to zero genetic correlation . . ." In support of his argument he cites the hierarchical nature of developmental processes. Simple models show that shared ontogenetic history is likely to impose genetic covariance on traits developing from the same developmental precursors (Riska 1986). Also from an ontogenetic perspective, we argue that the simplest model for the coordination of growth is one based on a single factor with systemic effects (e.g., a growth hormone affecting all tissues equally). An agglomerative scheme would require differential responses for each trait (under a single factor growth model) or different factors affecting each trait. Also in support of fission models, Wagner (1996) and Wagner and Altenberg (1996) cite the tendency for repeated elements (e.g., mammal teeth or insect segments) to differentiate through evolutionary time. Initially uniform correlations between units may be differentially broken down into integrated sets according to functional or developmental criteria (e.g., a uniform dentition into shearing and chewing sets).

Besides being of theoretical interest, the preference of one model over the other has methodological implications with regard to statistical tests of morphological integration. A fission model implies that a relatively large amount of non-zero correlation is to be expected between traits. From a statistical point of view this suggests that a single-factor (overall integration) is the appropriate null hypothesis to be tested against. Only in the face of convincing evidence to the contrary should this be rejected in favor of more complex (i.e., requiring more factors) integrative schemes. In this respect we agree wholeheartedly with Zelditch's (1987, 1988) confirmatory factor analysis approach in which a general size factor was taken as the benchmark against which to accept or reject the fit of other models. In our view, a complete lack of integration (i.e., complete independence among traits) requires the most complex developmental mechanism and should be rejected when simpler hypotheses suffice.

VIII. A PROSPECTUS FOR THE STUDY OF INTEGRATION

In the preceeding pages we have outlined some of the many conceptual and methodological advances which have been made since the publication of *Morphological Integration*. In closing it seems appropriate to provide a prospectus detailing some of the fields from which new contributions are likely to come. We focus on five major areas: genetics, manipulations of development, analytical techniques for generating and testing integrative hypotheses, phylogenetic frameworks, and macroevolutionary issues.

Genetic Approaches to Integration

Both quantitative and population genetical models continue to yield interesting insights and predictions about the expected behavior of correlated traits under a variety of evolutionary scenarios.

Quantitative genetic models for the multivariate evolution of correlated characters have become increasingly sophisticated and incorporate an ever wider variety of effects. For example, some of the models presented by Atchley et al. (1994) include not only the standard direct effects (additive genetic + environmental variance), but a variety of epigenetic effects as well, including maternal effects and cascades of interactions between traits (e.g., Atchley et al. 1994). Unfortunately our ability to accurately estimate the parameters of these models has only increased incrementally. We hope to see new techniques and approaches to estimating model parameters.

Population geneticists also continue to probe topics bearing on the concept of morphological integration. Of particular interest are models for exploring evolution under pleiotropy (e.g., Wagner 1989; Price and Langen 1992; Gromko 1995; Waxman and Peck 1998). One of the prime results suggested by these models is that correlated responses to selection might depend on the details of pleiotropic effects, rather than simply being a consequence of genetic correlations (Gromko 1995). This could, then, represent a significant challenge to the importance of genetic correlations for determining evolutionary dynamics. Advances in modeling pleiotropic effects and integrating the results of these models into a coherent theory of correlated responses to selection must be addressed by evolutionary geneticists.

On the interface between quantitative and molecular genetics is the technique of quantitative trait loci (QTL) mapping (e.g., Cheverud and Routman 1993; Cheverud et al. 1997). QTL mapping uses a combination of molecular and statistical techniques to estimate the chromosomal locations of loci affecting quantitative characters. Such studies are of obvious interest to any research program dealing with trait variation and covariation. In a study bearing on the genetic bases of integration, Cheverud et al. (1997) showed that the effects of pleiotropic loci tended to be restricted to developmentally or functionally related suites of traits.

Molecular developmental genetics continues to provide fascinating insights into the molecular underpinnings of morphology and morphogenesis. While so-called "master control" genes (Halder et al. 1995) represent exciting discoveries bearing on these topics, regulatory genes and mechanisms involved in subtler variations deserve greater attention. An example of research exploring the type and scale of variation relevant to the morphological integration concept is the work of Tsukuya and colleagues (Tsukuya 1995; Tsuge et al. 1996). Their studies of leaf morphogenesis have revealed the existence of families of mutant alleles which independently affect leaf width (*angustifolia* mutants) or length (*rotundifolia* mutants) in *Arabidopsis* (Tsuge et al. 1995). Tsuge et al. (1995) postulate the existence of one or more factors which are suppressed by the products of the *angustifolia* and *rotundifolia* genes. It seems reasonable to suggest that the degree to which these underlying factors coordinate growth in width and length is the proximal cause of integration (or lack thereof) between these two traits.

Manipulations of Development

By and large, studies of morphological integration have been phenomenological in nature. While valuable, this type of approach can, at best, only tell us how well or poorly a specific hypothesis of integration fits the observed data. An alternative approach for exploring integrative patterns is via experimentation designed to alter patterns of integration. A straightforward implementation of this approach is to cause disruptions of development, either by tissue manipulation or modification of gene products, in a manner suitable to testing specific hypotheses. This may help not only to test the validity of integrative hypotheses but may also provide valuable information about compensatory mechanisms which operate when the modular nature of integration is disrupted. The study of allocation tradeoffs during development exemplifies one such approach (Nijhout and Emlen 1998). Nijhout and Emlen showed via experimental manipulation of developmental resources that certain traits are "competing" for limited developmental substrates. In artificial selection experiments, they showed that this competition for developmental resources induced a negative genetic correlation between traits. This finding coincides with Riska's (1986) model for the evolution of correlated characters, suggesting that developmental processes can cause genetic constraints.

Hypothesis Building and Testing

Along with the many new techniques and approaches outlined above, we argue that considerable advances remain to be made in what may be the most basic of requirements for studies of integration: generating and testing hypotheses.

The study of morphological integration has, to a large degree, been formulated in terms of trait association. However, many functional and developmental models make predictions not just about the degree of association among traits but about the scaling of traits as well. We suggest that studies such as that by Westneat (1991) on the biomechanics of jaw protrusion in the slingjaw wrasse, *Epibulus insidiator*, provide an opportunity to test integration over development in relation to functional constraints. The four-bar linkage system that explains jaw protrustion in these fish makes an exact prediction about how morphological elements should covary over development in this species. Incorporating expectations such as these into studies of integration strengthens the argument for integration when it is found, and broadens the appeal of the concept to workers in peripheral fields (e.g., functional morphology, biomechanics, physiology).

Along with new approaches to model building, new ways to test these models will be important as well. In our discussion of the adjacency map technique we have presented one approach to testing broad families of models. New approaches, incorporating a variety of types of data (qualitative data, landmark data) will also be needed as we develop new models and expectations.

Phylogenetic Perspectives

The phylogenetic issues and concerns relevant to comparative studies (e.g., Harvey and Pagel 1991; Martins 1996) of one or a few traits are equally applicable to evolutionary studies of morphological integration. Many interesting macroevolutionary issues (see below) require, or are best addressed in the context of, specific phylogenetic hypotheses. Unfortunately many of the problems inherent in comparative studies, such as the estimation of ancestral states, are compounded in integrative studies involving numerous traits (e.g., Steppan 1997a,b). One of the challenges will be to generate theory and methodology applicable to addressing these issues.

Additionally, while phylogenetic methods have advanced by leaps and bounds with regard to the analysis of discrete data, the analytical techniques available for the analysis of quantitative data are still rudimentary (e.g., Archie 1985), and no coherent theory or methodology has been presented that deals with quantitative trait covariation in a phylogenetic context. One potential application of integrative studies is to code patterns of integration as qualitative characters to be used in phylogenetic studies. While the few studies attempting to do this have been contentious (e.g., Rohlf 1998, Zelditch et. al. 1998), we feel that such an approach merits further investigation.

Macroevolutionary Issues

In chapters 2 and 9, Olson and Miller explored the potential role of morphological integration in the context of various macroevolutionary phenomena. Among the phenomena they considered were directional evolution (evolutionary trends), rates of evolution, speciation and adaptation, convergence and parallelism, and hybridization. The theoretical role of morphological integration in these subject areas remains largely unexplored. We encourage the reader to take note of Olson and Miller's recommendations. Below we highlight additional topics from four areas that we believe are critical for studies of morphological evolution. We hope that our fellow evolutionary biologists will join with us to address these interesting and important questions.

1. MORPHOLOGICAL DIVERSITY

Despite the new emphases on developmental and genetic paradigms (e.g., Raff 1996; Atchley and Hall 1991), there is still no coherent theory about the evolution f morphology and the production of morphological diversity. We believe that morphological integration provides a conceptual framework for the study of phenocovariation that will lead to a set of critical hypotheses about the mechanistic production of form. How does integration relate to morphological diversity? Is morphological diversity a result of numerous but separate building blocks, each with a high degree of plasticity? Or is it the result of a highly integrated whole, historically decoupled to form flexible descendants?

We believe that studies of large, monophyletic, morphologically diverse radiations, such as Darwin's finches, *Geospiza*, or Mexican silverside fishes, *Chirostoma*, may provide key information. Of particular interest will be studies that are able to infer ancestral characteristics. Were the ancestors of such radiations flexible with regard to patterns of developmental covariation and expression of traits?

2. MORPHOSPACE

There have been many studies of the morphospaces occupied by organisms through time. Many of these studies have sought a pattern in changes to morphological diversity before and after extinction events, geologic epochs, etc. (Foote 1997). Despite the many advances made in studies of disparity (variance of phenotypes in morphospace; Foote 1997), there have been few if any attempts to link given developmental patterns among populations or species to patterns of disparity. Perhaps this is due to sampling problems. Nevertheless, we see questions relating morphological integration to the position of phenotypes in morphospace as fundamental. Is morphological integration associated with the amount of morphospace occupied by species or by clades? Is morphological integration associated with probabilities of extinction or survival? What are the integrative characteristics associated with speciating (recovering) clades after extinction events? Olson and Miller suggested that overall levels of integration declined after cladogenesis. What is the generality of their conclusion?

Foote (1997) reviews the literature and shows convincingly that morphospace is not occupied uniformly. Whether this is due to generative developmental programs as suggested by Raup (1966) or to mechanical aspects of morphometric strain (Bookstein et al. 1985), the contribution of morphological integration to this issue has been unexplored. If we view integration as the modularization of developmental covariance, then is non-uniform distribution of phenotypes in morphospace to be expected? In morphospace, are saltations and foldings a function of the integrative patterns of development or do we need special theories of explanation (e.g., Stanley 1979)?

3. MORPHOLOGICAL NOVELTY

Morphological novelty arises despite strong patterns of developmental constraint (Müller and Wagner 1991; Raff 1996). Müller and Wagner (1991) describe morphological evolution as the progressive origination, transformation, and loss of homologues. While we agree, we note that integration is lacking from this perspective. Where do novelties arise? Do novelties arise as traits of systems that are more or less coupled? Must traits become decoupled in order to form novelties? Does the evolution of novelty involve the reformulation of an integrative pattern? Are there morphological or developmental predictors of novelty arising as an elaboration of an integrative pattern or as reorganization to form a novel modular structure?

4. EVOLVABILITY

Evolvability refers to the ability of an ancestral developmental system to give rise to diverse future generations. How is evolvability affected by morphological integration? Riedl (1978) introduced the notion of "burden" to relate the flexibility, and hence evolvability, of developing systems to their degree of constraint. If as demonstrated by Olson and Miller there are historical patterns of strengthening integrated systems, then a burden becomes placed on such systems. That is, as morphological integration increases, the flexibility of the systems to evolve is diminished (Wagner 1988a,b, 1989). Thus, does integration present a double-edged sword—initially increasing rates of evolution, but then presenting a narrower and narrower corridor for variation which can only be escaped via dissociation and reorganization of covariation?

We conclude with a consideration of Olson and Miller's closing words. They recognized that the measure of their success would be the extent to which their ideas stimulated new research. Towards this end we judge them successful—Olson and Miller's vision and insights have sparked investigations in numerous biological systems, from the level of the developing embryo to the level of populations and species. Their most difficult task, as they saw it, would be "to cast thinking about morphology into the framework of covariance . . ." (p. 268). These words have proven prophetic; rare is the phenotypic research program which does not consider the covariation of multiple traits. We hope that the reprinting of their book will make their ideas readily available to a new generation of evolutionary biologists who will take to heart their admonishments regarding the importance of trait covariation for the evolution of phenotypes.

ACKNOWLEDGEMENTS

We wish to thank James M. Cheverud, Michael Foote, Michael LaBarbera and K. Rebecca Thomas for their discussions and critical comments on the manuscript. We also wish to express our gratitude to Christina M. Henry, University of Chicago Press, for her enthusiasm and support for this project.

References Cited

AIROLDI, J.-P., and B. K. FLURY. 1988. An application of common principal component analysis to cranial morphometry of *Microtus californicus* and *M. ochrogaster* (Mammalia, Rodentia). J. Zool., Lond. 216:21–36.

ARCHIE, J. W. 1985. Methods for coding variable morphological features for numerical taxonomic analysis. Syst. Zool. 25:563–583.

ATCHLEY, W. R., D. E. COWLEY, C. VOGEL, and T. MCLELLAN. 1992. Evolutionary divergence, shape change, and genetic correlation structure in the rodent mandible. Syst. Biol. 41:196–221.

ATCHLEY, W. R., and B. K. HALL. 1991. A model for development and evolution of complex morphological structures. Biol. Rev. 66:101–157.

ATCHLEY, W. R., S. XU, and C. VOGL. 1994. Developmental quantitative genetic models of evolutionary change. Developmental Genetics 15:92–103.

BELL, E. T. 1934. Exponential numbers. Amer. Math. Monthly 41:411–419.

BERG, R. L. 1960. The ecological significance of correlation pleiades. Evolution 14:171–180.

BIEWENER, A. A. 1990. Biomechanics of mammalian terrestrial locomotion. Science 250:1097–1103.

BOLLEN, K. A. 1989. Structural equations with latent variables. New York: John Wiley and Sons.

BOOKSTEIN, F. L. 1989. Principal warps: Thin-plate splines and the decomposition of deformations. IEEE Trans. Pattern Analysis and Machine Int. 11:567–585.

———. 1991. Morphometric tools for landmark data: Geometry and Biology. New York: Cambridge University Press.

———. 1994. Can biometrical shape be a homologous character? Pp. 197–227 *in* B. K. Hall, ed., The hierarchical basis of comparative biology. New York: Academic Press.

BOOKSTEIN, F. L., B. CHERNOFF, R. ELDER, J. HUMPHRIES, G. SMITH, and R. STRAUSS. 1985. Morphometrics in evolutionary biology. Philadelphia: Acad. Nat. Sci. of Philadelphia, Spec. Publ. No. 15.

BROOKS, D. R., and D. A. MCLENNAN. 1991. Phylogeny, ecology, and behavior. Chicago: University of Chicago Press.

CANE, W. P. 1993. The ontogeny of postcranial integration in the common tern, *Sterna hirundo*. Evolution 47:1138–1151.

CHERNOFF, B., and A. MACHADO-ALLISON. 1999. *Bryconops colaroja* and *B. colanegra*, two new species (Teleostei, Characiformes) from the Cuyuni and Caroni drainages of South America. Ichthyol. Exploration. Freshwaters 10: in press.

CHERNOFF, B., A. MACHADO-ALLISON, and W. G. SAUL. 1991. Redescription of *Leporinus brunneus* Myers (Characiformes: Anostomidae) with a biogeographic analysis of the fish fauna of the Upper Rio Orinoco. Ichthyol. Explor. Freshwaters 2:295–306.

CHEVERUD, J. M. 1982. Phenotypic, genetic, and environmental morphological integration in the cranium. Evolution 36:499–516.

———. 1984. Quantitative genetics and developmental constraints on evolution by selection. Journal of Theoretical Biology 110:155–172.

———. 1988a. A comparison of genetic and phenotypic correlations. Evolution 42:958–968.

———. 1988b. The evolution of genetic correlation and developmental constraints. Pp. 94–101 *in* G. d. Jong, ed., Population genetics and evolution. Berlin: Springer-Verlag.

———. 1989a. A comparative analysis of morphological variation patterns in the Papionines. Evolution 43:1737–1747.

———. 1989b. The evolution of morphological integration. Pp. 196–197 *in* Splecthna and Hilgers, eds., Fortschritte der Zoologie: Trends in vertebrate morphology. New York: Gustav Fischer Verlag.

———. 1995. Morphological integration in the saddle-back tamarin (*Saguinus fuscicollis*) cranium. Am. Nat. 145:63–89.

———. 1996. Developmental integration and the evolution of pleiotropy. American Zoologist 36:44–50.

CHEVERUD, J. M., and E. ROUTMAN. 1993. Quantitative trait loci: Individual gene effects on quantitative characters. Journal of Evolutionary Biology 6:463–480.

CHEVERUD, J. M., E. J. ROUTMAN, and D. J. IRSCHICK. 1997. Pleiotropic effects of individual gene loci on mandibular morphology. Evolution 51:2006–2016.

CHEVERUD, J. M., J. J. RUTLEDGE, and W. R. ATCHLEY. 1983. Quantitative genetics of development: Genetic correlations among age-specific traits values and the evolution of ontogeny. Evolution 37:895–905.

CHEVERUD, J. M., G. P. WAGNER, and M. M. DOW. 1989. Methods for the comparative analysis of variation patterns. Syst. Zool. 38:201–213.

DOW, M. M., and J. M. CHEVERUD. 1985. Comparison of distance matrices in studies of population structure and genetic microdifferentiation: Quadratic assignment. American Journal of Physical Anthropology 68:367–374.

ELDREDGE, N., and J. CRACRAFT. 1980. Phylogenetic patterns and the evolutionary process. New York: Columbia University Press.

FALCONER, D. S. 1989. Introduction to quantitative genetics. London: Longman Press.

FELSENSTEIN, J. 1988. Phylogenies and quantitative characters. Ann. Rev. Ecol. Syst. 19:445–471.

FINK, W. L., and M. L. ZELDITCH. 1996. Historical patterns of developmental integration in piranhas. Amer. Zool. 36:61–69.

FLURY, B. 1987. A hierarchy of relationships between covariance matrices. Pp. 31–43 *in* A. K. Gupta, ed., Advances in multivariate statistical analysis. Dordrecht, Neth.: Reidel.

———. 1988. Common principal components and related multivariate models. New York: Wiley.

FOOTE, M. 1997. The evolution of morphological diversity. Ann. Rev. Ecol. Syst. 28:129–152.

GABRIEL, K. R., and R. R. SOKAL. 1969. A new statistical approach to geographic variation analysis. Syst. Zool. 18:259–278.

GROMKO, M. H. 1995. Unpredictability of correlated response to selection: pleiotropy and sampling interact. Evolution 49:685–693.

HULL, D. L. 1988. Science as a process. Chicago: University of Chicago Press.

HUXLEY, J. S. 1932. Problems of relative growth. New York: The Dial Press.

KLINGENBERG, C. P., and M. ZIMMERMAN. 1992. Static, ontogenetic, and evolutionary allometry: A multivariate comparison in nine species of water striders. Am. Nat. 140:601–620.

KOHN, L. A. P., and W. R. ATCHLEY. 1988. How similar are genetic correlation structures? Data from mice and rats. Evolution 42:467–481.

KOOTS, K. R., and J. P. GIBSON. 1996. Realized sampling variances of estimates of genetic parameters and the difference between genetic and phenotypic correlations. Genetics 143:1409–1416.

LANDE, R. 1979. Quantitative genetic analysis of multivariate evolution, applied to brain-body size allometry. Evolution 33:402–416.

———. 1980. The genetic covariance between characters maintained by pleiotropic mutations. Genetics 94:203–215.

LEAMY, L., and W. R. ATCHLEY. 1984. Morphometric integration in the rat (*Rattus* sp.) scapula. Journal of Zoology 202:43–56.

LOFSVOLD, D. 1986. Quantitative genetics of morphological differentiation in Peromyscus. I. Tests of homogeneity of genetic covariance structure among species and subspecies. Evolution 40:559–573.

MANTEL, N. 1967. The detection of disease clustering and a generalized regression approach. Cancer Res. 27:209–220.

MARCUS, L. F. 1990. Traditional morphometrics. Pp. 77–122 *in* F. J. Rohlf and F. L. Bookstein, eds., Proceedings of the Michigan Morphometrics Workshop. Ann Arbor, MI: University of Michigan Museum of Zoology, Spec. Publ. No. 2.

MARTIN, W. R. 1949. The mechanics of environmental control of body form in fishes. Publ. of Ontario Fish. Res. Lab. 70:1–72.

MARTINS, E. P. 1996. Phylogenies and the comparative method in animal behavior. New York: Oxford University Press.

MAYR, E. 1969. Principles of systematic zoology. New York: McGraw-Hill.

MÜLLER, G. B., and G. P. WAGNER. 1991. Novelty in evolution: restructuring the concept. Ann. Rev. Ecol. Syst. 22:229–256.

NEEDHAM, J. 1933. On the dissociability of the fundamental process in ontogenesis. Biol. Rev. 8:180–233.

NELSON, G. 1994. Homology and systematics. Pp. 101–149 *in* B. K. Hall, ed., Homology: The hierarchical basis of comparative biology. New York: Academic Press.

NIJHOUT, H. F., and D. J. EMLEN. 1998. Competition among body parts in the development and evolution of insect morphology. Proc. Natl. Acad. Sci. 95:3685–3689.

OLSON, E. C., and R. L. MILLER. 1951. A mathematical model applied to a study of the evolution of species. Evolution 5:325–338.

———. 1958. Morphological Integration. Chicago: University of Chicago Press.

PATTERSON, C. 1982. Morphological characters and homology. Pp. 21–74 *in* K. A. Joysey and A. E. Friday, eds., Problems in phylogenetic reconstruction. New York: Academic Press.

PRICE, T., and T. LANGEN. 1992. Evolution of correlated characters. Trends in Ecol. and Evol. 7:307–310.

RAUP, D. M. 1966. Geometric analysis of shell coiling: general problems. J. Paleontol. 40:1178–1190.

RIEPPEL, O. 1994. Homology, topology and typology: The history of modern debates. Pp. 63–100 *in* B. K. Hall, ed., Homology: The hierarchical basis of comparative biology. New York: Academic Press.

RISKA, B. 1986. Some models for development, growth, and morphometric correlation. Evolution 40:1303–1311.

ROFF, D. A. 1995. The estimation of genetic correlations from phenotypic correlations: A test of Cheverud's conjecture. Heredity 74:481–490.

———. 1996. The evolution of genetic correlations: An analysis of patterns. Evolution 50:1392–1403.

ROHLF, F. J. 1993. Relative warp analysis and an example of its application to mosquito wings. Pp. 131–159 *in* L. F. Marcus, E. Bello, and A. Garcia-Valdecasas, eds., Contributions to morphometrics. Madrid: Museo Nacional de Ciencias Naturales.

———. 1998. On applications of geometric morphometrics to studies of ontogeny and phylogeny. Syst. Biol. 47:147–158.

ROTA, G.-C. 1964. The number of partitions of a set. Amer. Math. Monthly 71:498–504.

SMITH, G. R. 1990. Homology in morphometrics and phylogenetics. Pp. 325–338 *in* F. J. Rohlf and F. L. Bookstein, eds., Proceedings of the Michigan Morphometrics Workshop. Ann Arbor, MI: University of Michigan Museum of Zoology, Spec. Publ. No. 2.

SMITH, J. M., R. BURIAN, S. KAUFFMAN, P. ALBERCH, J. CAMPBELL, B. GOODWIN, R. LANDE, D. RAUP, and L. WOLPERT. 1985. Developmental constraints and evolution. Quart. Rev. Bio. 60:265–287.

SNEATH, P. H., and R. R. SNOKAL. 1973. Numerical taxonomy. San Francisco: W. H. Freeman & Co..

STANLEY, S. M. 1979. Macroevolution. San Francisco: Freeman.

STEPPAN, S. J. 1997a. Phylogenetic analysis of phenotypic covariance structure. I. Contrasting results from matrix correlation and common principal component analyses. Evolution 51:571–586.

———. 1997b. Phylogenetic analysis of phenotypic covariance structure. II. Reconstructing matrix evolution. Evolution 51:587–594.

STRAUSS, R. E. 1992. Developmental variability and heterochronic evolution in poeciliid fishes (Cyprinodontiformes). Pp. 492–514 *in* R. L. Mayden, ed., Systematics, historical ecology, and North American freshwater fishes. Stanford, CA: Stanford University Press.

THOMPSON, D. A. 1917. On growth and form. Cambridge, UK: The University Press.

TSUGE, T., H. TSUKAYA, and H. UCHIMIYA. 1996. Two independent and polarized processes of cell elongation regulate leaf blade expansion in *Arabidopsis thaliana* (L.) Heynh. Development 122:1589–1600.

TSUKAYA, H. 1995. Developmental genetics of leaf morphogenesis in dicotyledonous plants. J. Plant. Res. 108:407–416.

TURELLI, M. 1984. Heritable genetic variation via mutation-selection balance: Lerch's zeta meets the abdominal bristle. Theor. Pop. Biol. 25:138–193.

———. 1985. Effects of pleiotropy on predictions concerning mutation-selection balance for polygenic traits. Genetics 111:165–195.

VALEN, L. V. 1982. Homology and causes. Journal of Morphology 173:305–312.

WAGNER, G. P. 1984. On the eigenvalue distribution of genetic and phenotypic dispersion matrices: Evidence for a nonrandom organization of quantitative character variation. Journal of Mathematical Biology 21:77–96.

WAGNER, G. P. 1988a. The influence of variation and of developmental constraints on the rate of multivariate phenotypic evolution. Journal of Evolutionary Biology 1:45–66.

WAGNER, G. P. 1988b. The significance of developmental constraints for phenotypic evolution by natural selection. Pp. 222–229 *in* G. d. Jong, ed. Population genetics and evolution. Berlin: Springer-Verlag.

WAGNER, G. P. 1989. Multivariate mutation-selection balance with constrained pleiotropic effects. Genetics 122:223–234.

WAGNER, G. P. 1990. A comparative-study of morphological integration in *Apis mellifera* (Insecta, Hymenoptera). Zeitschrift fur Zoologische Systematik und Evolutionsforschung 28:48–61.

WAGNER, G. P. 1996. Homologues, natural kinds and the evolution of modularity. American Zoologist 36:36–43.

WAGNER, G. P., and L. ALTENBERG. 1996. Perspective: Complex adaptations and the evolution of evolvability. Evolution 50:967–976.

WALKER, J. 1996. Principal components of body shape variation within an endemic radiation of threespine stickleback. Pp. 321–334 *in* L. Marcus, M. Corti, A. Loy, G. Naylor, and D. Slice, eds., Advances in Morphometrics. New York: Plenum Press. NATO ASI Series, vol. 284.

WAXMAN, D., and J. R. PECK. 1998. Pleiotropy and the preservation of perfection. Science 279:1210–1213.

WESTNEAT, M. W. 1991. Linkage biomechanics and evolution of the unique feeding mechanism of *Epibulus insidiator* (Labridae: Teleostei). Journal of Experimental Biology 159:165–184.

WRIGHT, S. 1932. General, group and special size factors. Genetics 15:603–619.

ZELDITCH, M. L. 1987. Evaluating models of developmental integration in the laboratory rat using confirmatory factor analysis. Syst. Zool. 36:368–380.

———. 1988. Ontogenetic variation in patterns of phenotypic integration in the laboratory rat. Evolution 42:28–41.

ZELDITCH, M. L., F. L. BOOKSTEIN, and B. L. LUNDRIGAN. 1992. Ontogeny of integrated skull growth in the cotton rat *Sigmodon fulviventer*. Evolution 46:1164–1180.

———. 1993. The ontogenetic complexity of developmental constraints. J. Evol. Biol. 6:621–641.

ZELDITCH, M. L., and W. L. FINK. 1995. Allometry and developmental integration of body growth in a piranha, *Pygocentrus nattereri* (Teleostei: Ostariophysi). J. Morphology 223:341–355.

———. 1996. Heterochrony and heterotopy: stability and innovation in the evolution of form. Paleobio. 22:241–254.

ZELDITCH, M. L., W. L. FINK, and D. L. SWIDERSKI. 1995. Morphometrics, homology, and phylogenetics: Quantified characters as synapomorphies. Syst. Biol. 44:179–189.

ZELDITCH, M. L., W. L. FINK, D. L. SWIDERSKI, and B. L. LUNDRIGAN. 1998. On applications of geometric morphometrics to studies of ontogeny and phylogeny: A reply to Rohlf. Syst. Biol. 47:159–167.

Biographical Information

Everett C. Olson (1910–1993). Everett Olson received his bachelor's, master's, and doctoral degrees at the University of Chicago, where his graduate supervisor was Alfred Romer. Following Romer's departure for Harvard in 1934, Olson joined the faculty of the Department of Geology as the resident vertebrate paleontologist. Olson's research focused on studies of late Permian tetrapods and faunas. Olson trained numerous graduate students, and was responsible for the establishment of the Interdivisional Committee on Paleozoology, an interdisciplinary program which was the forerunner of the current Committee on Evolutionary Biology at the University of Chicago. Everett Olson published numerous articles as well as several books, and served as editor for the journals *Evolution* (1953–1958) and *Journal of Geology* (1962–1967). In 1969, Olson left Chicago to become the first chair of UCLA's Department of Biology (formerly two separate departments—zoology and botany), where he continued to pursue his research program as well as train another generation of evolutionary biologists.[1,2]

Robert L. Miller (1920–1976). Robert Miller graduated with his bachelor's degree from the University of Illinois in 1942, and received his Ph.D. from the University of Chicago in 1950. Following completion of his doctoral degree, Miller spent a short stint (1950–1952) as a research associate in the Department of Geology at the University of Chicago, after which he became a member of the faculty, where he served until the time of his death. Much of his work focused on statistical problems in geology as well as the sedimentology of near-shore marine environments. Besides *Morphological Integration*, Robert Miller co-authored a second book, *Statistical Analysis in the Geological Sciences* (with J. S. Kahn), and served as the editor for the *Journal of Geology* (1957–1962).

---

[1] Rainger, R. (1993). Biology, geology, or neither, or both: vertebrate paleontology at the University of Chicago, 1892–1950. Persp. Sci. 1(3): 478–819.

[2] Bell, M. A. (1998). Everett C. Olson 1910–1993. Biographical Memoirs, vol. 75. National Academy Press, Washington, D.C., pp. 1–24.